Molecular Biology For Oncologists

Molecular Biology For Oncologists

John Yarnold
Michael Stratton
Trevor McMillan
Editors

1993
ELSEVIER
AMSTERDAM – LONDON – NEW YORK – TOKYO

ELSEVIER SCIENCE PUBLISHERS B.V.
Sara Burgerhartstraat 25
P.O. Box 211, 1000 AE Amsterdam, The Netherlands

ISBN 0 444 81507 4

This book is printed on acid-free paper
Printed in the Netherlands

Foreword

The pathophysiology, growth kinetics and metastasis of cancers and their responses to the currently employed therapies are increasingly being understood in terms of their genetic control mechanisms. The resultant greatly enhanced understanding is leading to clinically useful predictors of prognosis and the beginnings of new therapeutic strategies against tumours. This includes not only manoeuvres which deploy the currently available treatment methods with greater effect, but also the introduction into the clinic of conceptually different treatment strategies, with the potential of great impact.

Developments in the field of basic genetics of cancer biology have been extraordinarily rapid with exceptional importance to biology. This area has attracted some of the keenest intellects in biology and medicine. To illustrate the vigour and the growth rate of the research activity in this area, note that the numbers of journal articles published with the words gene and cancer (carcinoma, sarcoma, neoplasm, tumour) in the title or abstract for the years 1972, 1982 and 1992 were 16, 588 and 13 186, respectively (Medline search). The entire field of oncology (clinical and laboratory) has, likewise, been growing at a remarkable although slower pace. This is reflected by the number of articles published with the word cancer (carcinoma, sarcoma, neoplasm, tumour) in the title or abstract for the same three years, viz. 10 894, 22 871 and 101 295. There is hardly an issue of a clinical journal in the field of oncology which does not have at least one article devoted in part, as a minimum, to some aspect of the genetic factors of cancer. Further, our newspapers rarely have less than one major story per week on some aspect of molecular biology, largely related to oncology.

This volume provides the clinical oncologist with a clearly needed and comprehensive statement of the current status of research into the genetic basis of cancer. Without knowledge of this subject, so well presented here, there is little prospect for the clinician to understand the developments in diagnostics, therapeutics, family counselling, epidemiology, etc. which are impinging on the clinical practice of oncology. Indeed, the standards of practice are evolving at a readily perceived rate due to these researchers and, with a virtual certainty, the style, substance and efficacy of practice will be changed even more dramatically over the next decade.

Herman Suit
Massachusetts General Hospital
Boston

Preface

Molecular Biology for Oncologists introduces clinicians to the extraordinary power, elegance and importance of molecular biology in the research and treatment of cancer. The volume explains how new techniques involving the manipulation of genetic material in cells and organisms are providing dramatic insights into how cell behaviour is subverted in common human cancers.

The book also shows how techniques of cell and molecular biology are improving the understanding and effectiveness of radiation and drug actions, as well as leading to the development of gene therapy. Ultimately, techniques of molecular biology will help to identify the specific genetic and environmental factors predisposing to cancer, offering the means of cancer prevention.

Molecular Biology for Oncologists is based on an annual three-day teaching course in London held at the end of June which shows us how quickly and keenly clinicians with no previous knowledge of the subject grasp its scope and opportunities. The book does not attempt a comprehensive coverage of its subject, but chooses topics of particular interest and importance. These illustrate the applications of molecular biology to an understanding of conventional therapies as well as to the development of new strategies of prevention and cure.

About half the authors are scientists established in their fields. Other authors are oncologists applying the techniques of molecular biology to their own research. Each has been asked to provide a teaching document rather than a detailed review.

The editors invite readers to comment on the most useful and least useful features of this volume, as well as to make suggestions for improvements in content and style. We shall endeavour to respond to comments readers may have.

John Yarnold
Michael Stratton
April 1993   *Trevor McMillan*

vii

Contents

Section I: Oncogenes

Chapter 1.   What are cancer genes and how do they upset cell
                            behaviour?
                            John Yarnold

Chapter 2.   Mechanisms of activation and inactivation of dominant
                            oncogenes and tumour suppressor genes
                            Michael Stratton

Chapter 3.   Approaches to proto-oncogene and tumour suppressor gene
              identification
              Helen Patterson

Chapter 4.   Transgenic modelling of dominant and recessive oncogene
              function
              Francisco S. Pardo and Emmett V. Schmidt

Chapter 5. The BCL-2 gene in clinical oncology
Michael Brada and Martin Dyer

Chapter 6.   Identification and significance of EGFR and *erb*B2
                   overexpression
                   Donal P. Hollywood and William J. Gullick

Chapter 7.   GTP-binding proteins and cancer
                   Julian Downward

Chapter 11. Human papillomavirus (HPV) and cervical cancer
Rachel C. Davies and Karen H. Vousden

Chapter 12. Cell cycle control and cancer
Antony M. Carr

Chapter 13. Differentiation and cancer
Malcolm D. Mason

Section II: Therapeutics

Chapter 14. Molecular aspects of radiation sensitivity
Catherine Mort and Trevor J. McMillan

Chapter 15. Cells differ in their susceptibility to DNA damage induction
and in their ability to repair it
Stephen J. Whitaker

Chapter 21.   Antibody technology is being transformed
Robert E. Hawkins and Stephen J. Russell

Chapter 22.   Prospects for gene therapy of cancer
Jonathan D. Harris and Karol Sikora

Section III: Gene Structure and Techniques

Chapter 23.   Introduction to gene structure and expression
Richard Wooster

Chapter 26. Techniques of molecular biology
Philippe J. Rocques

SECTION I
ONCOGENES

Molecular Biology for Oncologists
Edited by John Yarnold, Michael Stratton, Trevor McMillan
© *1993, Elsevier Science Publishers B.V. All rights reserved*

CHAPTER 1

What are cancer genes and how do they upset cell behaviour?

John Yarnold

Academic Unit of Radiotherapy, The Royal Marsden Hospital & Institute of Cancer Research, Downs Road, Sutton, Surrey, SM2 5PT, UK

Cell behaviour can be disrupted in cancer by mutations and oncogenic viruses

Advances in cell and molecular biology are increasing our understanding of how normal cells work. These advances are also beginning to explain the changes in cell behaviour that occur in cancer. Many of these changes are caused by mutations in normal cellular genes or by the introduction of viral oncogenes [1]. The term oncogene is used to describe any gene sequence contributing directly to malignant change in a cell. Many oncogenes causing human cancer are mutated versions of normal cellular genes called proto-oncogenes. Proto-oncogenes are normal cellular genes whose cellular functions are enhanced by mutation. Examples include the over-expression of growth factor receptors and the expression of mutant *ras* proteins. These mutations are referred to as activating or gain-in-function mutations because they activate or enhance the cellular functions of the encoded proteins.

Tumour suppressor genes are normal cellular genes whose cellular functions are inactivated by mutation. Examples include inactivation of the retinoblastoma protein and inactivation of p53. These mutations are referred to as inactivating or loss-of-function mutations because they inactivate the cellular functions of the encoded proteins.

The term oncogene is sometimes restricted to proto-oncogenes in their activated, mutated states and to the oncogenic sequences carried by viruses. This excludes

3

Table 1. Examples of oncogene nomenclature.

v-H-*ras*	Viral oncogene identified by Harvey that causes sarcoma in rats
c-H-*ras*	Cellular homologue of the v-H-*ras* gene located on short arm of chromosome 11 in humans
v-*myc*	Viral oncogene causing avian myelocytosis
c-*myc*	Cellular homologue of v-*myc* located on the long arm of chromosome 8 in humans
BCL-2	Second B-cell lymphoma gene identified (cf BCL-1)
Rb	Retinoblastoma gene on the long arm of chromosome 13 in humans
p53	Named after its protein product, mol. wt 53 kDa and located on the short arm of chromosome 17 in humans
NM23	23rd hamster melanoma clone examined that yielded a non-metastasis (metastasis suppressor) gene
DCC	Deleted in some colon cancers

tumour suppressor genes for no good reason. The term oncogene is used here in a general sense to include any cancer gene, including tumour suppressor genes. Oncogene nomenclature is varied and arbitrary; see Table 1.

Despite the remarkable progress of the last decade, there are large gaps in our understanding of how oncogenes cause the biological and clinical manifestations of cancer. Nevertheless, it is possible to explain the biology of some aspects of tumour growth. For example, some oncogenes increase the rate of cell proliferation. Other cancer genes disrupt different growth control mechanisms; see Table 2.

Several genetic events are required for the development of human cancer

Changes in the histological appearances of normal tissue ranging from atypical hyperplasia to carcinoma-in-situ can be observed over a period of years before the appearance of some cancers. Examples include cancers of the cervix, uteri and colon. These changes are consistent with the concept of multistage carcinogenesis,

Table 2. Growth control mechanisms disrupted in cancer.

Increased positive growth signals lead to increased cell proliferation
Decreased negative growth signals remove constraints on cell proliferation
Differentiation blocks maintain cell proliferative capacity
Inhibition of programmed cell death maintains cell viability

whereby independent cellular events lead to cumulative changes in cell behaviour over a period of years, if not decades [2,3]. Most of the cellular events underlying multistage carcinogenesis can be explained by the accumulation of mutations in cancer genes [4]. The mutations are caused by exposure to environmental carcinogens, or arise endogenously, e.g. from spontaneous errors in DNA replication [5].

Mutations accumulate as independent random events in every somatic cell during the course of an individual's lifetime. The majority of somatic mutations are in non-coding DNA and have no effects on the cell. Some disrupt specific genes and have profound effects on cell behaviour, for example growth regulation [6].

Germline mutations are responsible for inherited cancer predispositions

Not all mutations occur in somatic cells. Mutations also occur in ova and spermatocytes. Some of these mutations can be transmitted via the germline to cause an inherited cancer predisposition [7]. There are two ways in which this can happen. An individual may inherit a mutation in one allele of a cancer gene from one or other parent. Every cell in the body will be heterozygous for the same mutation. Every cell will have taken one step along the pathway to malignancy. This leads to a high risk of at least one cell in the body acquiring the remaining mutation(s) needed for the development of cancer, often quite early in life.

The second way in which germline mutations contribute to cancer is via rare inherited defects that make a person more susceptible to mutation. This explains the inherited cancer predisposition caused by some genetic defects in DNA repair mechanisms, e.g. ataxia telangiectasia.

Activated oncogenes are often called dominant oncogenes

Cancer genes that are activated by mutation are also referred to as dominant oncogenes. Borrowing a term from Mendelian genetics, dominant oncogenes exert their cellular effects despite the presence of normal gene product from the homologous allele. For example, mutation of one c-H-*ras* allele on chromosome 11 exerts its effects despite a mixture of mutant and wild-type (normal) *ras* protein in the cell. Activating mutations can also work by stimulating the overproduction of wild-type gene product in the cell. An example is the over-expression of epidermal growth factor receptor (EGFR, also called c-*erb*B1) in some breast cancers.

Activated cellular proto-oncogenes are important examples of dominant oncogenes, but exogenous gene sequences introduced by viruses also act as gain-in-function mutants. Retroviruses are RNA tumour viruses causing cancer in animals.

They carry modified forms of cellular proto-oncogenes captured from their animal hosts in the course of viral evolution. For example, the v-*erb*B oncogene causes erythroleukaemia in chickens and is a mutated version of chicken EGFR.

Oncogenic DNA viruses are important causes of human cancer. Epstein-Barr virus, hepatitis-B virus and papilloma virus contribute to Burkitt's lymphoma, hepatocellular carcinoma and cancer of the cervix uteri, respectively. These viruses introduce cellular genes that so far have few known homologies with human cancers. DNA viral oncoproteins interfere with important growth control pathways in the cell, and are examples of gain-in-function or dominantly acting oncogenes.

Inactivated oncogenes (tumour suppressor genes) are often called recessive oncogenes

Normal cellular genes whose protein products and cellular functions are inactivated by mutation during carcinogenesis are called tumour suppressor genes [8,9]. Like proto-oncogenes, tumour suppressor genes exist in normal, wild-type forms as well as in the mutated forms responsible for their oncogenic properties. Tumour suppressor genes are referred to as recessive oncogenes because they exert their cellular effects only in the absence of normal gene product. This usually means that both alleles must be inactivated by mutation in order to exert a cellular effect. This contrasts with proto-oncogenes where only one allele needs to be activated by mutation in order to exert its cellular effect (see Table 3). The retinoblastoma (Rb) gene was the first tumour suppressor gene to be isolated. Tumour suppression is dramatically demonstrated by the introduction of a normal Rb gene into Rb-defective tumour cells lacking functional Rb protein. This suppresses malignant behaviour in vitro, including the ability of the cells to form tumours in animals.

Table 3. Characteristics of proto-oncogenes and tumour suppressor genes.

	Proto-oncogenes	Tumour suppressor genes
Number of alleles in normal somatic cells	Two	Two
Effect of mutations on cellular function of gene product	Enhanced	Reduced
Adjectives used to describe mutations	Activating Gain-in-function Dominant	Inactivating Loss-of-function Recessive

Mutations in tumour suppressor genes can be transmitted via the germline and result in cancer predisposition (retinoblastoma)

Somatic cells are subject to sporadic mutations in both tumour suppressor genes and proto-oncogenes. However, only mutations in tumour suppressor genes are known to be transmitted in the germline. Mutations in proto-oncogenes are not known to be transmitted via the germline presumably because their dominant mode of action is incompatible with normal embryonic development. Heterozygous transmission of a tumour suppressor gene mutation to a fertilised egg is usually functionally silent, perhaps because there is enough gene product expressed from the wild-type allele for normal embryogenesis.

Germline transmission of a mutation in one allele of the retinoblastoma (Rb) suppressor gene in hereditary retinoblastoma is the paradigm for hereditary pre-disposition [7]. The disease is inherited as a Mendelian autosomal dominant trait which usually presents within the first few months of life. It is important to understand how a disease inherited as a dominant clinical trait is caused by recessive mutations at the cellular level (see Figure 1).

In the hereditary form of retinoblastoma, every cell in the affected foetus inherits an inactivated Rb allele on one of its chromosomes 13. This genetic defect is compatible with normal embryonic development. Through random somatic mutation, one or more cells in the retina lose the normal allele as well. This is a relatively rare event considering there are approximately 2×10^6 retinoblasts, but usually more than one cell sustains the second event, leading to multiple tumours.

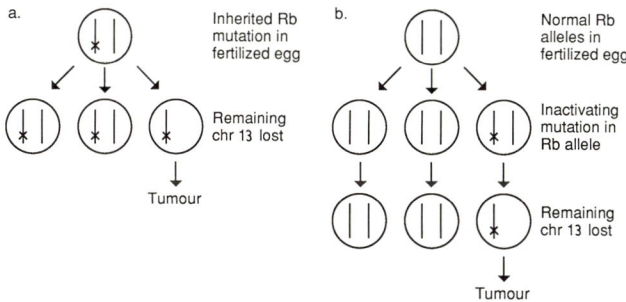

Figure 1. Hereditary and sporadic retinoblastoma. In the hereditary form (a), every developing retinal cell inherits the same inactivating mutation in one of its two Rb alleles. There is a very high chance of at least one retinal cell sustaining a somatic loss of the normal Rb allele, leading to sustained proliferation, i.e. a tumour. Several tumours may arise in the same patient by the same process. In sporadic retinoblastoma (b), both mutations are somatic. The chances of both somatic events occurring independently in more than one retinal cell in the same patient are remote, hence tumours are unilateral and single.

J. Yarnold

The removal of functional Rb protein removes a natural constraint on cell proliferation which contributes to tumour formation.

The germline mutation is recessive at the cellular level because it does not cause the transformed phenotype unless both Rb alleles are inactivated. The predisposition to retinoblastoma is transmitted as an autosomal dominant trait in a family pedigree because of the very high probability of a somatic mutation involving the homologous wild-type locus in at least one retinal cell. Independent inactivations of the homologous Rb allele in several retinal cells result in tumours randomly distributed between the two retinas and explain the bilateral clinical presentation. Sporadic retinoblastoma arises from two somatic mutations inactivating both Rb alleles in the same retinoblast. This occurs very rarely and explains why these tumours are single and unilateral.

It is not clear whether mutations inactivating both alleles of the retinoblastoma locus are sufficient to explain all manifestations of the disease, but it is likely that other oncogenes are also involved. Patients with hereditary retinoblastomas are also susceptible to other cancers, especially osteosarcoma.

The biological functions of many cancer genes are concerned with growth regulation

Many viral oncogenes, activated proto-oncogenes and some inactivated tumour suppressor genes interfere with normal growth control processes. Many activated proto-oncogenes and retroviral oncogenes stimulate cell proliferation by up-regulating growth signal pathways. Some inactivated tumour suppressor genes remove constraints on cell division by lifting cell cycle controls. The onco-proteins of DNA viruses such as papilloma virus also act by interfering with natural constraints on proliferation.

Not all oncogenes are concerned with the regulation of cell proliferation, even though their biological functions are concerned with tumour growth. For example, overexpression of BCL-2 protein in nodular B-cell lymphoma appears to prolong B-cell survival by interfering with programmed cell death (apoptosis) [10]. This causes tumour growth by delaying cell death rather than by increasing proliferation. The control of cell differentiation is another process that can be upset by cancer genes. For example, inappropriate or overexpression of c-*myc* protein appears to block differentiation pathways in some cell types. This sustains cell division in response to growth factors to which differentiated cells lose responsiveness.

Very little is known about the biological processes leading to invasion and metastasis [11]. Some oncogenes [12] have been identified that are probably concerned with these processes. For example, the DCC gene (Deleted in Colon Cancer) is a tumour suppressor gene encoding a cell adhesion molecule, loss of which may enable cells to detach from the surrounding matrix and migrate through tissues [13].

Oncoproteins have a wide variety of biochemical functions

The protein products of cancer genes represent a wide range of molecules in terms of structure and their biochemical functions. They include tyrosine kinase trans-membrane receptors, cytoplasmic messenger molecules, nuclear transcriptional regulators, cell adhesion molecules, etc. Some oncoproteins fit into well character-ised biochemical pathways that can be related to their biological functions. Examples include the tyrosine kinase transmembrane receptor pathways that transfer peptide growth factor signals across the cell membrane, e.g. epidermal growth factor receptor (EGFR or c-*erb*B1) and c-*erb*B2 [14,15]. These connect with cytoplasmic pathways that transmit proliferative signals to the nucleus. In the nucleus, these signals interact with biochemical pathways that control passage through the cell cycle.

Phosphorylation is a recurring biochemical theme in cell metabolism. The transfer of a high energy phosphate group from ATP or GTP to another molecule is

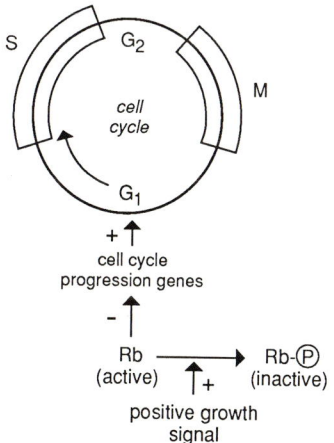

Figure 2. The normal cellular function of retinoblastoma protein (Rb). Rb protein interacts with genes that control cell cycle progression. Rb protein acts as a constraint on cell cycle progression in its unphosphorylated form. Arrival of a positive growth signal inside the nucleus from outside the cell is associated with phosphorylation of Rb (Rb-P), in which form it loses its growth constraining properties.

catalysed by enzymes called kinases. Phosphorylation of proteins on amino acids serine and threonine is widespread, but phosphorylation on tyrosine residues is restricted and very significant. The phosphorylation of tyrosine residues is performed by tyrosine kinases specific for particular substrates, for example specific growth factor receptors. Tyrosine phosphorylation changes the biochemical activity of a molecule, activating or inactivating an enzymic function or modifying the binding of one protein to another. For example, phosphorylation and dephosphorylation modulate the physiological activity of the retinoblastoma protein (see Figure 2).

Different oncogenes function in the same biochemical pathway

Different oncogenes can function in the same biochemical pathway, although not necessarily in the same tumour type. For example, the product of the c-H-*ras* gene on chromosome 11 has a molecular weight of 21 kDa and is called p21*ras* [16,17]. It is a small GTP-binding protein located on the inner surface of the plasma membrane which responds to mostly unknown signals at the cell surface. In many types

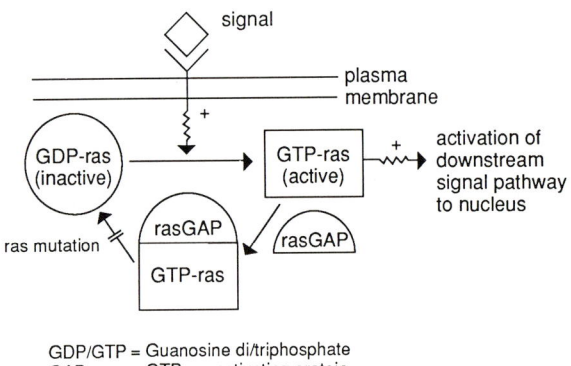

GDP/GTP = Guanosine di/triphosphate
GAP   = GTPase activating protein

Figure 3. Schema of normal ras function and its disruption by mutations in ras and GAP. Ras is a small GDP/GTP binding protein which functions as a molecular switch in diverse biochemical pathways, including those transmitting growth signals. It is located on the inner aspect of the plasma membrane and binds GTP after stimulation, e.g. by a specific growth signal. This activates a downstream signal pathway, as well as activating an intrinsic GTPase activity which catalyses the conversion of active GTP-ras to inactive GDP-ras. A GAP protein family member (GTPase activating protein) is an important cofactor for the activation of ras-GTPase activity. Inactivation of ras GTPase activity by selective point mutations in ras locks it in the active GTP-ras configuration. Inactivation of GAP function by homozygous mutations in both alleles has the same effect. Neurofibromin is a GAP family member and functions as a tumour suppressor gene in von Recklinghausen's disease.

Protein active   Protein inactive
(cell cycle arrest)   (cell cycle progression)

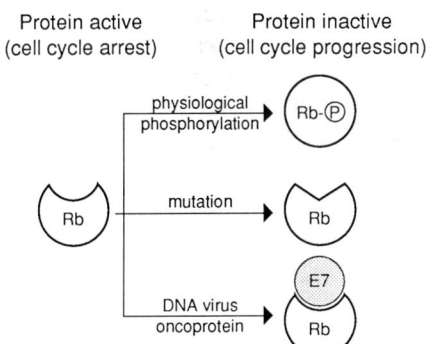

Figure 4. Retinoblastoma protein can be inactivated in at least three ways. In normal cells, it is inactivated physiologically by phosphorylation in response to appropriate growth signals. In retinoblastoma, it is inactivated by mutations in the Rb gene. In cancer of the cervix, DNA tumour virus oncoproteins bind and inactivate Rb protein.

of cancer, p21*ras* is activated by a point mutation that keeps it in the "switched on" position, without a physiological growth signal. In this activated state it transmits a continuous growth signal to the nucleus (see Figure 3).

The tumour suppressor gene responsible for von Recklinghausen's disease codes for a protein called neurofibromin located very close to the p21*ras* on the inner plasma membrane [18–20]. Neurofibromin is a member of a family of proteins called GAP proteins (see Figure 3). Its normal biochemical function is to down-regulate p21*ras* shortly after physiological activation by a growth signal. In the Schwann cells of patients with von Recklinhausen's disease, an inherited deficiency of neurofibromin allows normal p21*ras* to remain in the activated state and transmit a continuous growth signal to the nucleus. This contributes to the development of neurofibromas and other benign and malignant tumours in affected individuals. The inactivation of pRb and p53 releases a physiological constraint on cell cycle progression in infected cervical squamous epithelium (see Figure 4)

Environmental factors influence tumour development and progression

Conditions prevailing in the environment of the tumour cell have a very potent impact on tumour development even though they involve no changes in the structure of DNA. Environmental influences are described as epigenetic factors. Examples include the induction of fibroblast growth factors and angiogenetic factors in breast cancer. The down-regulation of oestrogen receptor protein in some breast cancers is also an epigenetic effect.

References

1   Bishop JM. The Molecular Genetics of Cancer. Science 1987, **235**, 305–311.
2   Foulds L. Tumour progression. Cancer Res 1957, **17**, 355–356.
3   Nowell PC. Mechanisms of tumour progression. Cancer Res 1986, **46**, 2203–2207.
4   Fearon ER, Vogelstein B. A genetic model for colorectal tumorigenesis. Cell 1990, **61**, 759–767.
5   Vogelstein B, Kinzler KW. Carcinogens leave fingerprints. Nature 1992, **355**, 209–210.
6   Tabin CJ, Bradley SM, Bargmann CI et al. Mechanisms of activation of a human oncogene. Nature 1982, **300**, 143–149.
7   Knudson AG. Hereditary cancer, oncogenes, and anti-oncogenes. Cancer Res 1985, **45**, 1437–1443.
8   Ponder B. Gene losses in human tumours. Nature 1988, **335**, 400–402.
9   Weinberg RA. Tumour suppressor genes. Science 1991, **254**, 1138–1146.
10 Hockenbery D, Nunez G, Milliman C, Schrieber RD, Korsmeyer SJ. Bcl-2 is an inner mitochondrial membrane protein that blocks programmed cell death. Nature 1990, **348**, 334–336.
11 Marx JL. How cancer cells spread in the body. Science 1989, **244**, 147–148.
12 Marx JL. New clue to cancer metastasis found. Science 1990, **249**, 482–483.
13 Pullman WE, Bodmer WF. Cloning and characterisation of a gene that regulates cell adhesion. Nature 1992, **356**, 529–532.
14 Druker BJ, Mamon HJ, Roberts TM. Oncogenes, growth factors, and signal transduction. N Engl J Med 1989, **321**, 1383–1391.
15 Ullrick A, Schlessinger J. Signal transduction by receptors with tyrosine kinase activity. Cell 1990, **61**, 203–212.
16 Hoffman M. Getting a handle on *ras* activity. Science 1992, **255**, 159.
17 McCormick F. Gasp: not just another oncogene. Nature 1989, **340**, 678–679.
18 Wigler MH. GAPs in understanding Ras. Nature 1990, **346**, 696.
19 Ponder B. Neurofibromatosis gene cloned. Nature 1990, **346**, 703–704.
20 Bollag G, McCormick F. NF is enough of GAP. Nature 1992, **356**, 663–664.

Molecular Biology for Oncologists
Edited by John Yarnold, Michael Stratton, Trevor McMillan
© 1993, Elsevier Science Publishers B.V. All rights reserved

Mechanisms of activation and inactivation of dominant oncogenes and tumour suppressor genes

Michael Stratton

Section of Chemical Carcinogenesis, Institute of Cancer Research,
15 Cotswold Road, Belmont, Sutton, Surrey, SM2 5NG, UK

Introduction

In the course of a lifetime, a variety of mutations accumulate in cells throughout the body. These arise through exposure to chemicals, radiation, viruses and as a consequence of mistakes made by the DNA replicative machinery. Most mutations make no difference to the functioning of the individual cell in which they occur. A few may alter its viability so that the cell dies. In either case, there are no noticeable ill effects on the well-being of the body as a whole. Occasionally, however, a single cell suffers a series of mutations which, when converted into abnormal cellular machinery, cause the cell and its descendants to proliferate in the unrestrained and uncoordinated fashion characteristic of cancer cells. The subject of this chapter is the nature of the mutations, the patterns in which they occur and some of the functional consequences that they entail.

Categories of mutation involved in human tumours

Types of mutation involved in the activation or inactivation of oncogenes in human tumours include point mutations, translocations (gene rearrangements), gene amplification and deletions.

Point mutation

The term point mutation describes the substitution of one base pair of a DNA sequence by another, for example, substitution of a G:C base pair by an A:T. In the protein, this single base pair substitution may have a number of effects depending upon its position.

The mutation may result in the substitution of one amino acid for another (a missense mutation) resulting in a protein identical to the wild type except for the single amino acid change. The functional consequences of this change will depend on the nature of the amino acid substitution and the biological activity of the altered part of the protein.

If the mutated sequence encodes a signal for the termination of translation (a stop codon) then the protein will be prematurely terminated at this position. As a consequence, a substantial part of the protein may be omitted rendering it nonfunctional.

During the production of mRNA, two exons of coding sequence are joined together and the intron between them is removed. The process is known as splicing and takes place at certain consensus sequences near intron-exon junctions. Should a mutation arise in one of these sequences, the correct splicing events may not take place and a whole exon may be missed out of the mRNA. When translated into protein, the amino acids encoded by this exon will be absent and a frame shift may be introduced, altering substantially the predicted protein sequence downstream of the mutation.

Translocation

In a translocation, part of one chromosome is joined to another. The outcome is a hybrid chromosome that may be detectable by karyotypic (cytogenetic) analysis of tumour cells. At the molecular level, a gene located at or near the breakpoint in one of the two chromosomes is fused to sequences from the other chromosome. This rearrangement of DNA sequences may generate a structurally altered version of the gene and its protein, or may place it under new transcriptional control.

Gene amplification

The diploid genome of each eukaryotic human cell normally carries two copies of each gene, one originating from each parent (genes on X and Y chromosomes are exceptions). Under certain circumstances one copy may be multiplied up to several thousand fold, a phenomenon known as gene amplification. Amplified genes may form two types of microscopically abnormal chromosomal configuration known as double minutes (DM) and homogeneously staining regions (HSR). DMs are tiny, paired, extrachromosomal chromatin bodies that segregate randomly during mitosis and are not linked to a centromere. HSRs are expanded chromosomal regions which do not exhibit the usual banding pattern of normal chromosomes but which, being linked to a centromere, segregate in the normal way. Gene amplification is believed to contribute to oncogenesis by increasing the levels of mRNA that are transcribed from the gene, and as a consequence increasing the levels of protein that it encodes.

Increased or decreased expression of certain proteins without amplification or any other form of mutation of the DNA encoding them is a common feature of many tumours. In some cases, this may have a role in the generation of the neoplastic phenotype. In others, however, changes in expression are more likely to be a consequence of neoplastic change rather than a cause.

Deletion

A wide range of DNA deletions occur in tumour cells. At one extreme, a single base pair may be removed. Larger deletions may encompass part or all of a gene. Finally, a deletion may be large enough to be visible under the microscope by karyotype analysis or may remove a whole chromosome.

Whilst a large deletion will obviously remove from a cell and hence inactivate all the genes that are included within it, smaller, intragenic deletions may have a number of effects. If the DNA fragment that is deleted results in removal from the mRNA of a segment that is a multiple of three base pairs, then only the amino acid sequence encoded by that segment will be absent from the protein. The consequences of this structural abnormality will depend upon the functional properties of this segment in the normal protein.

If, however, the deleted segment in the mRNA is not a multiple of three base pairs, then a frame shift will be introduced. The likely consequence is that the amino acid sequence of the protein downstream from the mutation will bear little resemblance to the normal and will often be prematurely terminated because of the presence of a stop codon. The effects of insertions into DNA are subject to the same considerations as deletions but tend to be less common as mutational events in cancer genes.

Much of this repertoire of genetic events is used in both the activation of dominantly acting oncogenes and in the inactivation of tumour suppressor genes. However, because of the different outcomes as far as gene function is concerned the spectrum and pattern of changes differs markedly between the two classes of gene. These differences are emphasised in the account below.

Activating mutations in dominant oncogenes and their functional consequences

The proteins encoded by genes of the ras family are activated by missense point mutations at only three amino acids

Genes of the *ras* family, H-, K- and N-*ras*, provide the most celebrated examples of oncogene activation by point mutation in human tumours (reviewed in [1]). Ras proteins are composed of 188 or 189 amino acid residues. Normally, they undergo a cycle of activation and inactivation associated with the hydrolysis of GTP to GDP. Mutated *ras* proteins in human tumours, however, are fixed in the activated conformation. The alterations in DNA that result in this change in biochemical activity are almost exclusively point mutations. All are missense and hence allow translation of a full length *ras* protein. Of particular importance, however, is the restriction of the mutations to certain sites within the gene, namely codons 12, 13 and 61 (that encode amino acids 12, 13 and 61 of the *ras* protein). Mutations at other sites are not found in human tumours, nor do most of them result in onco-genic activation in experimental models. Thus it appears that the biological characteristics of *ras* proteins place considerable constraints upon the type and location of activating point mutations. Moreover, these structural and biochemical alterations are apparently sufficient for an activated *ras* protein to contribute to onco-genesis. Changes in the level of expression of the protein are unnecessary (although occasionally present) and other mutational mechanisms such as translocation or gene amplification are rare.

Genes can be activated by translocation

Cytogenetic studies have indicated that chromosomal abnormalities are common in human tumour cells. Closer inspection has revealed that some of these abnormalities are consistently associated with a particular tumour type (reviewed in [2]). For example, the majority of cases of chronic myelogenous leukaemia carry a translocation between particular regions of chromosomes 9 and 22 termed the Philadelphia translocation. Similarly, translocations involving chromosomes 8 and 14 are found in Burkitt's lymphoma and between chromosomes 14 and 18 in other B cell lymphomas. Chromosomal translocations have also been reported in solid tumours, for example, the t(11:22) in Ewing's sarcoma and the t(X:18) in synovial sarcoma. The role of the BCL-2 gene in the t(14:18) of B cell lymphoma is de-scribed in another chapter. Here the t(9:22) of CML and the t(8:14) of Burkitt's lymphoma will be used to illustrate the ways in which chromosomal translocations activate genes in oncogenesis.

*The translocation between chromosomes 9 and 22 in chronic myelogenous
leukaemia (CML) generates a hybrid protein encoded by sequences from both
chromosomes*

Between 90 and 95% of CML carry the translocation between chromosomes 9 and
22 termed the Philadelphia translocation. A clue to the important molecular events
reflected by this translocation was provided by the localisation of the human
homologue of the v-*abl* gene to precisely the region in which the breakpoint was
located on chromosome 9. (v-*abl* is the oncogene carried by the abelson murine
leukaemia virus). It was subsequently demonstrated that in the t(9:22) translocation, one allele of the c-*abl* gene is split and joined to a gene on chromosome 22
known as *bcr* (for breakpoint cluster region). The consequence of the rearrangement is that the *abl* gene loses its most 5' exon and is joined to an allele of the *bcr*
gene which has lost its 3' region (Figure 1). When the region is transcribed, the resulting mRNA is a hybrid, with the 5' region composed of *bcr* and the 3' region of
abl sequences. Similarly the protein is a hybrid of *bcr* amino acid sequence at the
amino terminus and *abl* amino acid sequence at the carboxy terminus. A biochemical feature of the *abl* protein that is believed to be of importance both in its
normal functions and in oncogenesis, is its ability to phosphorylate other proteins
on tyrosine residues. The *bcr-abl* fusion protein has an elevated tyrosine kinase
activity and this may contribute to its oncogenic action.

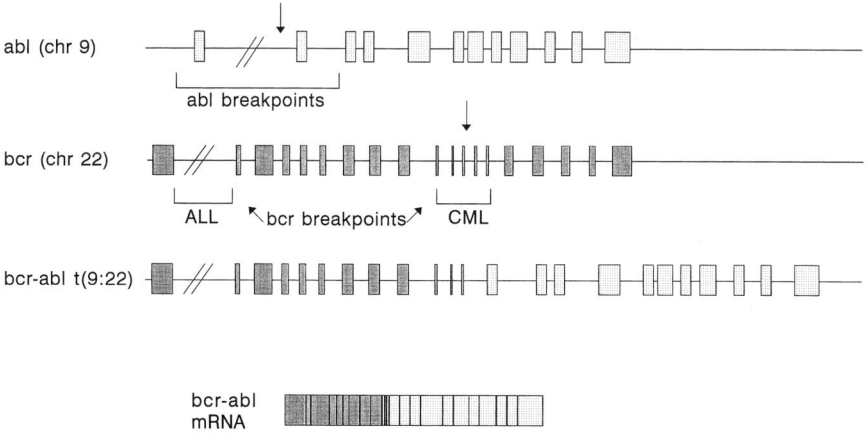

Figure 1. Translocation of the *abl* gene on chromosome 9 to the *bcr* gene on chromosome
22 results in the formation of a fusion gene that encodes a hybrid mRNA and protein. The
abl exons are represented as pale boxes and the *bcr* exons as dark boxes. The usual sites of
the breakpoints in *abl* and in *bcr* (in both CML and ALL) are indicated as are the locations
of the breakpoints in the particular case illustrated (vertical arrows).

The breakpoints underlying the Philadelphia translocation are usually within introns. In the *abl* gene, the breakpoint separates the 3' part of the gene carrying exons 2–11 from the 5' part carrying exons 1a or 1b. However, in the *bcr* gene there is more than one option with the breakpoint located in different introns in different tumours. Consequently several *bcr-abl* fusion mRNAs and proteins have been reported in which the 3' region, composed of *abl* sequences, is constant and the 5' region is constituted by different *bcr* sequences. Notably, one set of breakpoints is associated predominantly with the t(9;22) in CML and generates a 210 kDa protein, while the breakpoints in a more 5' intron are often associated with the t(9;22) in acute lymphoblastic leukaemia (ALL).

The translocation between chromosomes 8 and 14 in Burkitt's lymphoma results in deregulated expression of a structurally normal c-myc gene

In 75% of cases of Burkitt's lymphoma, the distal end of chromosome 8 is translocated to the long arm of chromosome 14q32. In the remaining 25%, a cytogenetically indistinguishable segment of chromosome 8 translocates to the long arms of either chromosome 2 or 22. Thus translocation of the segment 8q24-ter is a feature common to most and probably all Burkitt's lymphomas. The loci to which this segment is transferred are the positions of the immunoglobulin heavy (chromosome 14), lambda (chromosome 22) and kappa (chromosome 2) chains.

The c-*myc* gene is located at the common Burkitt's lymphoma breakpoint at chromosome 8q24. (c-*myc* is the human homologue of an oncogene carried by an acutely transforming retrovirus known as avian myelocytomatosis virus). Although some of the Burkitt's lymphoma breakpoints interrupt the c-*myc* gene itself, this is usually in untranslated regions and rarely in the coding sequence. Indeed many of the breakpoints are not within the c-*myc* gene at all and some are many thousands of kilobases away from it. Irrespective of these differences, the regulation of expression of the translocated c-*myc* allele is abnormal and is influenced by factors controlling the expression of the immunoglobulin gene to which it has been joined. Thus in contrast to the situation in chronic myelogenous leukaemia with *bcr-abl*, a fusion protein is not produced by the Burkitt's lymphoma translocations, but instead the deregulation of a structurally normal c-*myc* protein occurs.

Gene amplification can lead to overexpression of the oncoprotein

Gene amplification contributes to the transformed phenotype by increasing the level of mRNA and hence protein that is expressed from a proto-oncogene. However, expression of the amplified allele is sometimes increased out of proportion to the increase in gene copy number, indicating that abnormalities of transcriptional regulation may also be operative in amplified genes. Moreover, the amplicon, the segment of DNA that is amplified, is often large extending over several hundred kilobases or even megabases. The amplified fragment therefore usually stretches well beyond the boundaries of the proto-oncogene and encompasses several other genes.

Many types of oncogene are activated by gene amplification. These include the genes for growth factor receptors (for example epidermal growth factor receptor (EGFR) in malignant gliomas and *erb*-B-2 in breast cancer), and nuclear oncogenes such as N-*myc* in neuroblastoma. Gene amplification is usually a late and hence progressive step in oncogenesis rather than an initiating or early event. Moreover, amplification can sometimes be correlated with poor outcome and late stage, for example, amplification of the N-*myc* gene in neuroblastoma is now recognised as an independent clinical indicator of outcome.

Intragenic deletions can activate oncogenes

Whilst there can be no transcript or protein from a gene that has been completely deleted, small intragenic deletions may result in activation of a proto-oncogene protein. In principle this could occur through removal of a domain of the protein that normally mediates inhibition of the protein's activity, or by inducing conformational changes that mimic those normally brought about by stimulatory signals.

The clearest example of this phenomenon in human tumours that has been reported is in the EGFR gene in brain tumours. Amplification of EGFR is a common late event in the development of glioblastoma multiforme. It transpires, however, in a proportion of cases, that not only is the gene amplified, but an in-frame deletion is present within the amplified copy. The deletion results in the removal of exons 2–7 which encode part of the extracellular domain. As a result, a truncated EGFR protein is produced [3]. Studies of the functional consequences of this change are in progress, but it is presumed that it contributes further to the constitutive activation of the protein.

Inactivating mutations in tumour suppressor genes (recessive oncogenes) and their functional consequences

The mutations described above that result in activation of dominant oncogenes are usually found in only one of the two alleles of the gene that are present in diploid human cells. By contrast, inactivation of a tumour suppressor protein usually requires mutation of both alleles of the gene that encodes it. Therefore two independent mutations or "hits" are required. There are differences in the patterns of mutation between dominantly acting genes and recessive (tumour suppressor) genes. In addition, there are differences between the first and the second mutations that result in the inactivation of tumour suppressor genes. At present there do not, however, appear to be clear differences in the nature or pattern of the first hit in a tumour suppressor gene depending upon whether it is present in the germline or has been acquired in a tumour as a somatic mutation. Nevertheless one would predict that mutations which are incompatible with development of the embryo will be selected out from the spectrum of germline mutations but perhaps still have a role to play as somatic events within tumours.

The first inactivating event in a tumour suppressor gene is usually a "small mutation"

The mutation or hit that inactivates the first allele of a tumour suppressor gene is usually confined to the gene itself or involves the gene and chromosomal DNA that is immediately adjacent to it. In contrast to the dominantly acting genes which are mutated in a consistent manner either by point mutation (*ras*), by translocation (*abl*) or by gene amplification (N-*myc*), the types of alterations that inactivate a particular tumour suppressor such as the retinoblastoma gene are diverse and include point mutations, deletions and rearrangements. Moreover, in contrast to the point mutations, deletions and translocations of dominantly acting oncogenes, the mutations in tumour suppressor genes are much less constrained both in their type and position within the gene. For example, the point mutations that inactivate the retinoblastoma or p53 genes are located at a large number of positions within the gene in contrast to the three codons that are point mutated in activated *ras* genes. Moreover, although they may be missense, they can also be nonsense (encoding translational stop codons) or affect splice sites. Similarly, deletions may affect part or all of the gene and may be in or out of frame. Translocations may occur at virtually any site that will interrupt the production of an intact mRNA.

The reasons for this difference in pattern are not difficult to appreciate but may be elucidated by considering the parallel of a car travelling along a road. "Activation" of the car (as in activation of a dominant oncogene) so that it fails to respond when instructed to stop, requires tampering with one particular part of the mechanism, the brake system. On the other hand, "inactivation" of the car (as in inactivation of a tumour suppressor gene) so that it is stopped dead can be the result of a variety of malfunctions from subtle electrical defects, to glue in the ignition, a puncture, or an axe being driven into the engine.

The pattern of inactivating mutations differs in different tumour suppressor genes

Whilst the overall pattern described above is applicable to most tumour suppressor genes, there are differences between tumour suppressor genes in the predominance of types of mutation. For example, the adenomatous polyposis coli (APC) gene, which is the gene responsible for the syndrome of familial polyposis coli and is mutated in a large proportion of colonic tumours, is subject to a preponderance of nonsense mutations such as stop codons, or small deletions and insertions which introduce frameshifts [4]. Translation of the APC protein is usually prematurely terminated and hence the protein inactivated.

By contrast, in the p53 gene there is a much higher proportion of missense mutations that result in a full length protein with just a single amino acid substitution [5]. The reasons for this contrasting pattern are probably complicated. It seems likely, however, that some p53 proteins with missense mutations can interact with and functionally inactivate normal p53 protein in the cell without the need for a mutation at the DNA level in the remaining allele (see Chapter 8). In order to do this, however, the carboxy terminus of the protein needs to remain intact, which will not be the case if there is a stop codon upstream. Conversely, most missense mutations in APC may not be sufficient to inactivate the function of the protein encoded by the mutated allele and hence there is selection for the major disruptions of protein structure introduced by stop codons and frameshifts.

The second inactivating event in a tumour suppressor gene is usually a "large" mutation

The mutation that inactivates the second allele of a tumour suppressor gene is in only a minority of cases similar in character to the first mutation, i.e. confined to the gene and neighbouring chromosomal DNA. More commonly it involves loss of a large part and even all of the chromosome upon which the second allele of the tumour suppressor gene is situated. The reason for this difference is probably that the second hit often arises due to errors of mitosis. Such errors, which include non-disjunction (incorrect separation of chromatids during mitosis) or mitotic recombination (exchange of segments of DNA by homologous chromosomes) usually involve large stretches of DNA up to and including a whole chromosome.

This phenomenon has been of considerable importance in the detection and localisation of tumour suppressor genes. By using probes to parts of the genome that are polymorphic (see Chapter 23), it is possible to visualise on Southern blots both parental copies of a particular region independently. If one copy has been lost during oncogenesis, this will manifest itself as loss of heterozygosity (reduction from two bands to one) in the tumour compared to the germline DNA. Because the second hit in a tumour suppressor gene is large, the probe can be located several millions of base pairs away from the gene itself and still detect its presence by loss of heterozygosity. By performing studies in which a set of tumours are examined using at least one polymorphic probe on each chromosomal arm (termed an allelo-type), it is possible to obtain an approximate picture of the number and genetic locations of the tumour suppressor genes involved in the development of a tumour class.

Type of mutation

☐ Deletions ▨ Transitions ■ Transversions

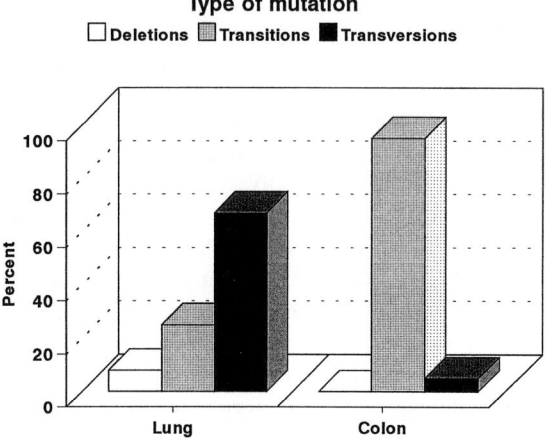

Mutations at indicated base

☐ meC ▨ G or C ■ A or T

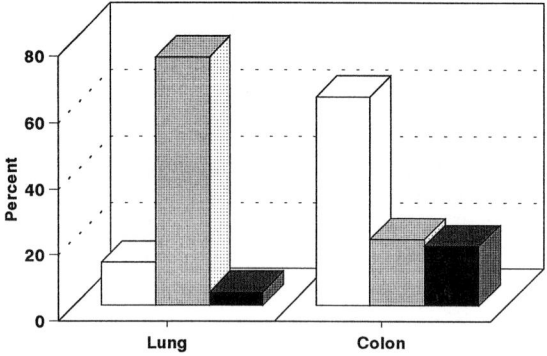

Figure 2. Exposure to certain types of carcinogen can affect the the type and location of a point mutation. In the upper panel, the types of mutation in lung and colon cancer are compared. In colon cancer most of the changes are transition mutations whilst, by contrast, in lung cancer most are transversions. In the lower panel, the sites of the mutations in these tumours are compared. In colon cancer, most mutations are at CpG dinucleotides at which the C may become methylated whilst in lung cancer most are not. The most likely interpretation of these differences is that the mutations in colon cancer are endogenous and due to errors made during DNA replication due to deamination of methyl cytosine. In contrast, the mutations in lung cancer may be due to the particular pattern of DNA binding by mutagenic carcinogens in tobacco smoke.

Mutations in oncogenes may also reflect the mutagen that induced them

The emphasis so far has been on the way constraints introduced by the modes of function of oncogenic proteins determine the type and pattern of mutations that result in oncogenic change. However, superimposed upon these influences are the effects of involvement of certain types of mutagen. Evidence for this notion in human tumours has emerged recently through study of mutations in the p53 gene in many types of cancer. Together these investigations suggest that in many tumours (e.g. colon cancer), the predominant mutation type is attributable to errors introduced by the DNA replicative machinery rather than to exogenous carcinogenic influences. This type of mutation is a G:C to A:T transition mutation which characteristically targets CpG dinucleotides (where a C is just 5' (or upstream) to a G). The proposed reason for this change is as follows. The C in CpG dinucleotides can become methylated. If it is subsequently deaminated, DNA polymerases cannot distinguish between the deaminated methylated C and a T residue. The polymerase will then insert an A opposite the C rather than the correct G resulting in a G:C → A:T transition mutation. (A transition mutation is one in which a purine is replaced by a purine or a pyrimidine by a pyrimidine; a transversion mutation is one in which a purine is replaced by a pyrimidine or a pyrimidine by a purine.)

By contrast, a number of tumour types are found to have a different spectrum of mutations that in some cases can be related to a known mutagen. For example, a proportion of squamous carcinomas of the skin show a particularly characteristic CpC → TpT double transition mutation which has not yet been reported in any type of visceral neoplasm. This is highly reminiscent of the type of mutation induced by UV light, a known risk factor for skin cancer [6]. Similarly, in smoking-induced lung cancers, the predominant mutation type is a G:C → T:A transversion, consistent with the type of mutation induced by mutagens in tobacco smoke (see Figure 2) [5]. Moreover, in the case of hepatoma, it is believed that exposure to aflatoxin results not just in a particular type of base change, but also in clustering at a particular codon of p53 (codon 249) around which the DNA sequence may preferentially encourage carcinogen binding.

References

1   Bos JL. The *ras* gene family and human carcinogenesis. Mutat Res 1988, **195**, 255–271.
2   Korsmeyer S. Chromosomal translocations in lymphoid malignancies reveal novel protooncogenes. Annu Rev Immunol 1992, **10**, 785–807.
3   Sugawa N, Ekstrand AJ, James CD, Collins VP. Identical splicing of aberrant epidermal growth factor receptor transcripts from amplified rearranged genes in human glioblastomas. Proc Natl Acad Sci USA 1990, **87**, 8602–8606.
4   Powell S, Zilz N, Beazer-Barclay Y, Bryan T, Hamilton S, Thibodeau SN, Vogelstein B,

M. Stratton

Kinzler K. APC mutations occur early during colorectal tumorigenesis. Nature 1992, **359**, 235–237.
5 Hollstein M, Sidransky D, Vogelstein B, Harris CC. p53 mutations in human cancers. Science 1991, **253**, 49–53.
6 Brash DE, Rudolph JA, Simon JA et al. A role for sunlight in skin cancer: UV-induced p53 mutations in squamous cell carcinoma. Proc Natl Acad Sci USA 1991, **88**, 10124–10128.

Molecular Biology for Oncologists
Edited by John Yarnold, Michael Stratton, Trevor McMillan
© *1993, Elsevier Science Publishers B.V. All rights reserved*

CHAPTER 3

Approaches to proto-oncogene and tumour suppressor gene identification

Helen Patterson

Section of Molecular Carcinogenesis and Section of Cell Biology and Experimental Pathology, Institute of Cancer Research, 15 Cotswold Road, Belmont, Sutton, Surrey, SM2 5NG, UK

Introduction

Cancer genes have been identified and isolated by several different approaches. The demonstration in tumour cells of mutations in a gene sequence is often taken as evidence that this sequence is acting as an oncogene or tumour suppressor gene. This is complemented by the demonstration of transforming (neoplastic) effects following expression of the cloned gene into the appropriate non-transformed cells. Conversely, proof that a gene has suppressor function can only be obtained from experiments in which tumour cells are reverted to a normal phenotype following introduction and expression of the normal cloned gene.

Oncogenes originally demonstrated in retroviral bioassays were identified as transduced and mutated copies of cellular genes. Following these initial observations, a limited number of approaches have been used to clone new oncogenes, notably the NIH3T3 transfection assay, the analysis of structural alterations in primary tumours and the search for novel members of cancer gene families. These methods have proved extremely successful and over 60 oncogenes have now been characterized. Malignant suppression by tumour suppressor genes was first demonstrated in hybrid cells created from the fusion of malignant and normal parent cells. However, the difficulty in devising conditions that positively select for the suppressed phenotype has undermined the cloning of suppressor genes by this route. As a result, highly individual approaches, including linkage analysis in cancer families and the analysis of genetic loss in primary tumours, followed by extensive physical mapping and cloning projects have been employed. A handful of genes, all of which have been implicated in human malignancy, have now been characterized and the strategies used in their isolation are considered.

Bioassays

Bioassays involve the introduction of foreign DNA into recipient cells and analysing the DNA of cells showing a change in biological behaviour as a means of tracing oncogenes and tumour supressor genes.

Tumour-inducing retroviruses carry mutated copies of cellular genes

The study of the transforming genes of tumour-inducing retroviruses has given us a rich harvest of oncogenes and our first glimpse of the role of gene activation by mutation in tumorigenesis. Retroviruses possess a diploid genome of single-stranded RNA which replicates via a double-stranded DNA intermediate integrated into the host cell genome, the provirus [1]. The essential features of free viral genomic RNA and of the provirus are shown in Figure 1. Retroviruses are implicated in the genesis of naturally occurring leukaemias and tumours in many avian and mammalian species including humans (HTLV1). However, the acutely transforming retroviruses, so-called because they rapidly induce tumours in experimental animals and efficiently transform cells in culture, are essentially laboratory constructs. These are derived from weakly or non-oncogenic retroviruses by the repeated passage of increasingly malignant tumour cells in susceptible birds or mammals.

The first tumour-inducing retrovirus was isolated by Peyton Rous in 1911. He demonstrated that a filtrable agent, derived from a serially passaged spontaneous sarcoma, could induce sarcomas in young chickens and thereby implicated a viral aetiology. Several decades later, dissection of the Rous sarcoma virus (RSV) genome ascribed its transforming ability to a distinct portion of the viral genome not involved in viral replication. A close homologue of this sequence appeared to be present as a single copy in the haploid genome of all vertebrate species studied, establishing a cellular rather than a viral origin for the transforming sequence of RSV, called v-*src*.

Subsequent analysis of a variety of acutely transforming retroviruses derived by the serial passage of avian and mammalian leukaemias and sarcomas has led to the characterization of over 20 distinct oncogenic sequences. The paradigm of RSV has held true. In each case, the viral oncogene (v-*onc*) was derived by transduction (Figure 2) from a cellular gene, which is referred to as the cellular proto-

A. RETROVIRAL GENOMIC RNA.

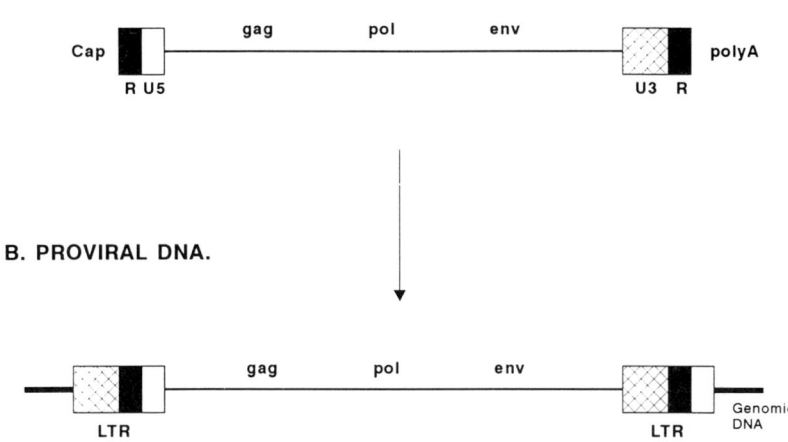

B. PROVIRAL DNA.

Figure 1. Retroviral genomic RNA has a capped 5' end and a polyadenylic acid tail as do eukaryotic messenger RNAs. U5 and U3, unique sequences at the 5' and 3' ends of viral RNA, are duplicated at each end during proviral DNA synthesis to form the long terminal repeats (LTRs) which are important for proviral transcriptional regulation. R, repeated sequences at both ends of viral RNA. The viral genome carries three open reading frames: the *gag* frame encodes viral core proteins; the *pol* frame encodes viral enzymes such as reverse transcriptase and integrase; the *env* frame encodes viral envelope glycoproteins.

oncogene (c-*onc*). Sequences originally identified as viral oncogenes include v-*myc*, v-H-*ras*, v-K-*ras*, v-*abl*, v-*src*, v-*erb*A and v-*erb*B [2] (see Table 1). The v-*onc* and corresponding c-*onc* usually differ in their level of expression as v-*onc* sequences are under viral transcriptional control. In addition, retroviral transduction appears able to efficiently mutate cellular proto-oncogenes and as a result, viral oncogenes are frequently truncated or contain point mutations when compared to their cellular counterparts.

Retroviruses induce tumours by insertional mutagenesis

Chronically transforming retroviruses, so-called because they induce tumours in animals with long latency, lack oncogenes but act via proviral integration in the host genome to disrupt cellular proto-oncogene sequences and their transcriptional control (Figure 3). Two lines of enquiry have identified cellular genes which appear to be activated by proviral insertion. The first assumed that the cellular counterparts of acute viral oncogenes might be targets for proviral insertion. For example, Southern analysis of DNA from tumour cells of avian leukosis virus (ALV)-

H. Patterson

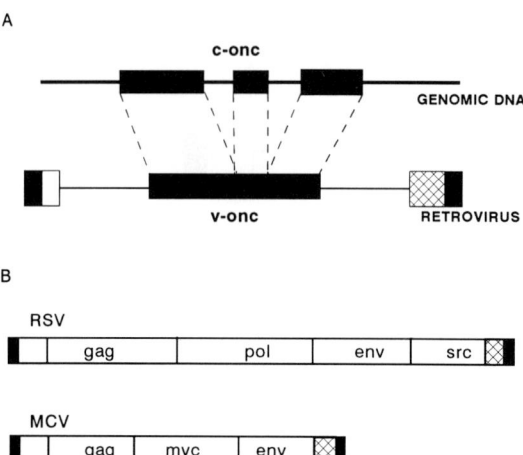

Figure 2. (A) The diagram indicates how the v-*onc* of an acutely transforming retrovirus is derived from the exons of a cellular gene by transduction. (B) Schematic representation of the genomes of the Rous sarcoma virus (RSV) and the avian myelocytomatosis virus (MCV). Transduction of the *src* gene in RSV does not disrupt the viral genome leaving the retrovirus competent for replication. In MCV, which is typical of other acutely transforming retroviruses, transduction of the *myc* gene results in disruption of the retroviral genome rendering the virus defective. These defective viruses require co-infection with a competent retrovirus to produce mature acutely transforming viral particles.

induced B-cell lymphomas demonstrated that the c-*myc* locus was consistently rearranged by proviral sequences [3]. Similar analyses have identified the c-*erb*B locus as the common integration site in ALV-induced erythroblastosis. A second type of analysis was used to clone novel sequences at the sites of proviral integration. In these studies, viral sequences were used as probes to clone the junction between exogenous proviral and cellular DNA. Oncogenes cloned by this route in clude *int*1 [4] and *int*2, the common integration sites in mouse mammary tumour

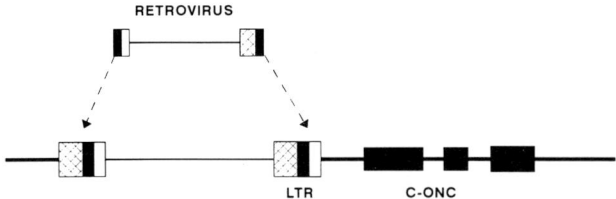

Figure 3. Schematic representation of oncogene activation by proviral insertion. The retrovirus has been inserted immediately upstream of a cellular proto-oncogene c-*onc*, and transcription of the proto-oncogene becomes driven by the viral LTR.

Table 1. Human proto-oncogenes.

Proto-oncogene	Involvement in human malignancy	Function
Proto-oncogenes first identified as viral oncogenes of acutely transforming retroviruses		
c-*src*	–	Tyrosine kinase
c-*abl*	Rearranged by translocation in 95% CGL and 10% ALL	Tyrosine kinase
c-*myc*	Amplified: 20% small cell lung cancer; 10% breast cancer; rearranged in 100% Burkitt's lymphomas	Transcription factor
c-H-*ras*	Point mutations in codons 12, 13, 61 in several	GTP-binding protein
c-K-*ras*	classes of human tumour, e.g. 70% pancreatic cancer, 40% colorectal cancer, 20% leukaemias (AML and ALL), also lung, melanoma and thyroid tumours	GTP-binding protein
c-*erb*B1	Amplified: 40% glial tumours, 20% squamous cell lung cancer	Growth factor receptor
c-*erb*A	–	Thyroid hormone receptor
c-*fos*	–	Transcription factor
c-*jun*	–	Transcription factor
c-*fms*	Point mutations in 20% chronic myelo-monocytic leukaemia and 20% M4 AML	GMCSF
c-*raf*	–	Serine/threonine kinase
Proto-oncogenes detected by analysis of chronically transforming retrovirus integration sites		
int-1	–	Growth factor
int-2	–	Growth factor
pim-1	–	–
Proto-oncogenes first identified by the NIH3T3 DNA transfection-transformation assay		
N-*ras*	Point mutations in several classes of human tumour	GTP-binding protein
met	Amplified in some gastric carcinoma cell lines	Tyrosine kinase
trk	Papillary carcinoma of the thyroid	Tyrosine kinase
dbl	Rearranged during transfection	–
ret	Papillary carcinoma of the thyroid	–
hst	Rearranged during transfection	Growth factor
mas	Rearranged during transfection	Receptor-like

(continued on next page)

Table 1, continued

Proto-oncogene	Involvement in human malignancy	Function
Proto-oncogenes cloned by analysis of structural alterations		
BCL-1	Rearranged in 80% diffuse centrocytic B-cell lymphomas and 30% SLVL	–
BCL-2	Rearranged in 80% follicular lymphomas	–
TCL-1	Rearranged in T-cell CLL	–
gli	Amplified in human glioma cell line	–
Proto-oncogenes identified through sequence homology		
c-*erb*B2	Amplified: 25% breast carcinomas	Receptor-like
N-*myc*	Amplified: 50% late stage neuroblastomas, 20% small cell lung cancer and 20% retinoblastoma	Transcription factor
L-*myc*	Amplified: 15% small cell lung cancer	Transcription factor

[a]Percentages are given only when a large number of tumours have been examined.

virus-induced breast tumours, and *pim*1 the integration site in murine leukaemia virus-induced T-cell lymphomas.

The DNA transfection-transformation assay detects activated oncogenes in tumour DNA

By providing proof of the existence of proto-oncogenes, the retroviral model led to the search for activated oncogenes in cells transformed by non-viral agents. DNA-mediated gene transfer or "transfection" depends upon the application of DNA to non-tumorigenic cells, usually NIH3T3 cells, as a co-precipitate with calcium phosphate. This allows the uptake and stable incorporation of exogenous DNA into the recipient cell genome. Using this type of analysis, DNA from cell lines and primary human tumours has been shown to induce cellular transformation, identified by the appearance of foci of morphologically transformed cells on a background of non-transformed cells (Figure 4) and the ability of cells from these foci to produce tumours in nude mice. Transformation by transfection does not represent one-step tumorigenesis. NIH3T3 cells are not primary cultures, they are immortal aneuploid cells which appear to have undergone many of the changes required for tumorigenesis. They nevertheless retain contact inhibition in culture and do not produce tumours in nude mice. Primary cultures, e.g. human fibroblasts, are not transformed by DNA transfection experiments.

These experiments demonstrated that many types of tumour contain activated cellular genes that can confer the transformed phenotype on the recipient NIH3T3

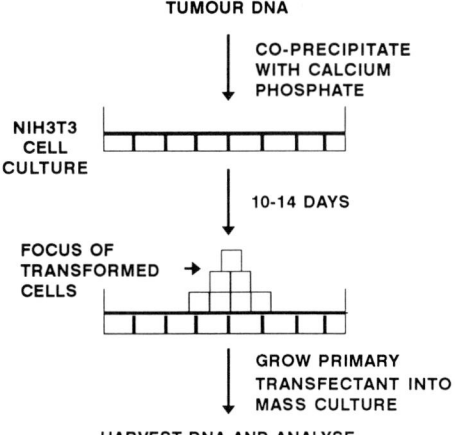

Figure 4. The figure illustrates the essential features of DNA-mediated gene transfer. Tumour DNA, as a fine precipitate with calcium phosphate, is endocytosed by NIH3T3 cells in culture. Foci of morphologically transformed cells can then be identified within the culture after 10–14 days. The stable uptake and incorporation of human DNA sequences into the genome of these focus forming cells can be confirmed, without prior knowledge of the specific sequences transferred, by the use of probes homologous to human repeat sequences that are interspersed throughout the genome. If human sequences are confirmed, these same probes provide a starting point for cloning the gene.

cells [5]. In most cases, the genes detected by this method are activated *ras* genes; K-*ras* and H-*ras* previously detected as viral oncogenes, and N-*ras* [6] first identified in this assay. Several other genes including *neu*, *met*, *ret*, *raf*, *mas*, *hst*, *trk* and *dbl* (Table 1) have been identified by this route. Several of these genes carry abnormalities which can be traced back to the original tumour, but others appear to be activated by the transfection procedure. The spectrum of genes cloned by transfection may only reflect the category of genes to which NIH3T3 cells are sensitive. In addition, genes spanning large regions of genomic DNA are unlikely to be transfected intact and for this reason will not be active in this assay.

Malignant suppression was first demonstrated in somatic cell hybrids

The earliest indication that malignant transformation might involve loss of normal gene function was provided by evidence of malignant suppression in somatic cell hybrids. The basic methodology is to fuse pairs of cells, one normal and one tumorigenic. Hybrid clones are selected by screening for genetic markers from both parents and examined for their ability to form tumours in immunologically appropriate hosts, usually nude mice. Early cultures of hybrid clones continued to grow in culture but were frequently non-tumorigenic when injected into nude mice, indicating that tumorigenicity had been suppressed. However, on continued growth in culture, suppressed clones had a tendency to re-express tumorigenicity. Subsequent cytogenetic analysis of non-tumorigenic hybrids and tumorigenic revertants showed that reversion correlated with loss of specific chromosomes which could be traced back to the normal fibroblast parent [7]. The technique has been further refined through microcell transfer experiments [8] in which single intact chromosomes, presumably bearing wild-type tumour-suppressor genes, are transferred to tumorigenic cells as interphase micronuclei and are retained in subsequent generations. Using this method, tumorigenicity has been suppressed in HeLa cells by the introduction of human chromosome 11, in a renal cell carcinoma cell line by chromosome 3p, and in melanoma cell lines by chromosome 6. The main problem with these studies is how to proceed from chromosomal location to molecular cloning. Although some genes cloned by other approaches may be active in this assay, none have been directly isolated through this route.

The analysis of structural alterations in tumour and germline DNA

Chromosomal translocations are the cytogenetic hallmark of gene rearrangement

Cytogenetic analyses have identified numerous specific chromosomal translocations in several classes of human malignancy and many of these translocations have been shown to result in oncogene activation by rearrangement. The genes involved in these translocations have been identified using two types of approach.

The first reciprocal translocations were characterized at the molecular level by assessing the evidence for the involvement of candidate proto-oncogenes cloned by other routes. For example, the search for the genes involved in the translocation t(9;22) in chronic granulocytic leukaemia focussed on the c-*abl* proto-oncogene previously mapped to chromosome 9. The c-*abl* gene was used as a probe to clone DNA at the c-*abl* locus in tumour DNA. Analysis of these clones confirmed that the c-*abl* gene was disrupted by sequences from the *bcr* gene on chromosome 22 [9]. The translocation results in the production of a chimeric *bcr-abl* fusion protein which has elevated tyrosine kinase activity. In Burkitt's lymphomas, translocations involve breakpoints at 8q24 which become transposed to one of the immunoglobulin gene loci at 14q32, 2p13 or 22q11. Analysis of tumour DNA using a candidate chromosome 8 proto-oncogene c-*myc* as a probe confirmed its role in these translocations [10].

When neither gene involved in a reciprocal translocation is known, the efforts required to clone the breakpoint are considerably greater. For example during the analysis of the translocations t(15;17) in promyelocytic leukaemia and t(11,22) in

Ewing's sarcoma, the analysis of candidate genes was unhelpful. In such cases, the successful approach to the characterization of the breakpoint typically involves the construction of long range physical maps of the breakpoint region, the cloning of extensive segments of DNA in the vicinity of the breakpoint and the analysis of this cloned DNA for the presence of a gene disrupted by the translocation.

Gene amplification can activate proto-oncogenes

Genes whose copy number is amplified in human tumours usually over-express the same gene. Hence, amplification provides a mechanism by which oncogenes may be activated. Southern analysis of tumour DNA using oncogene probes and in situ hybridization have been used to show amplification of the c-*myc* gene in a number of tumour types, the c-*erb*B gene in glioblastomas and a *myc*-like gene N-*myc* in neuroblastomas. However, one putative proto-oncogene *gli* was first cloned by manipulation of amplified DNA from a human glioma cell line [11].

Genetic loss is the hallmark of tumour suppressor genes

The first hit in tumour suppressor gene inactivation is usually a small deletion or point mutation. However, the loss of the second copy of the gene in tumour cells often involves loss of a large segment or all of the remaining normal chromosome [12]. Such losses can be revealed by cytogenetic analysis, but also as loss of heterozygosity (LOH) for adjacent polymorphic alleles in blot-hybridization experiments of tumour DNA when compared to normal DNA (Figure 5). Because the second hit can involve the loss of quite extensive regions of DNA from the remaining normal chromosome, polymorphic markers need not be closely linked to the tumour suppressor gene to provide valuable information in LOH analyses.

Only a handful of tumour suppressor genes have been cloned but the existence of many other suppressor genes has been inferred by the consistent finding of LOH at particular loci in many different tumour types. Using this approach, early studies mapped the inherited cancer syndrome multiple endocrine neoplasia 1 (MEN1) to chromosome 11q, the Von-Hippel Lindau syndrome to 3p and neurofibromatosis type 2 (NF-2) to chromosome 22. In addition, using one or more highly poly-morphic probes for every non-acrocentric chromosome arm, allelotypes in which the incidence of loss of heterozygosity for each chromosome arm is recorded, have been developed for many types of neoplasm. It has become apparent from these analyses that for most tumour types, several loci exhibit LOH, implicating the loss of several tumour suppressor genes in the development of these tumours. The cloning of the genes alluded to by these studies will have a considerable impact on our understanding of molecular tumorigenesis over the next decade.

A

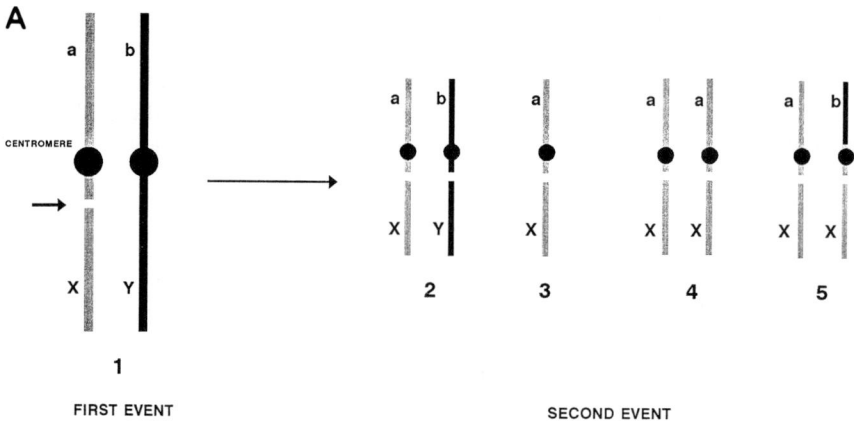

FIRST EVENT   SECOND EVENT

B

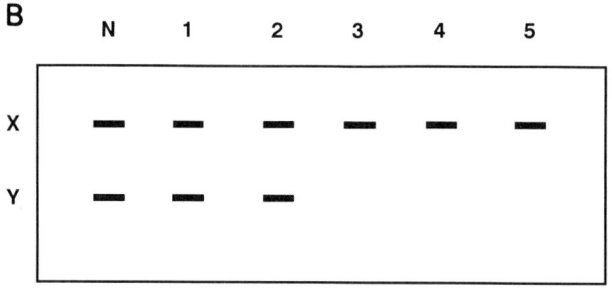

Figure 5. (A) Diagram (1) represents a pair of homologous chromosomes. The first mutation (arrowed) is usually a very small deletion or point mutation in one allele of the gene. In inherited malignancies, this first mutation will be carried in the germline, while in sporadic malignancies it will be an acquired somatic mutation. The second mutation can arise in a variety of ways. Occasionally, as a result of a second point mutation or small deletion in the remaining normal allele of the gene (2). More commonly the second copy of the gene is lost by deletion of the whole chromosome as a result of non-disjunction during mitosis. In example (3) the chromosome has been lost in this way without reduplication of the mutated chromosome and the abnormality is visible both cytogenetically and by RFLP analysis. In (4) the mutated chromosome is reduplicated and although this event is invisible cytogenetically, the tumour will show LOH for informative polymorphic alleles along the entire length of the chromosome, e.g. a/b → a/a and X/Y → X/X. Alternatively the second copy of the gene may be lost as a result of mitotic recombination between mutant and wild type chromosomes with subsequent segregation. This results in a cell which maintains heterozygosity proximal to the point of recombination a/b → a/b, but exhibits LOH distal to this point X/Y → X/X (5). (B) Schematic representation of Southern blot analysis of LOH using the polymorphic probe with alleles X and Y. The expected band pattern for normal DNA (N) and in each of the examples 1, 2, 3, 4 and 5 from (A) are demonstrated.

Linkage analysis in cancer families

Linkage analysis of affected families [13] can be used to pinpoint genetic loci associated with familial malignancy. Briefly, linkage analysis exploits the fact that genetic loci which lie closely together on the same chromosome are rarely segregated by homologous recombination at meiosis. In large pedigrees, the co-inheritance of the same allele of a specific polymorphic locus with a genetic trait provides evidence that the polymorphic locus and the disease locus are linked. In this way, neurofibromatosis was mapped to chromosome 17q11.2 [14], MEN II to chromosome 10 [15] and familial adenomatous polyposis (FAP) to chromosome 5q21-q22 [16]. The genes have subsequently been cloned.

The majority of common malignancies arise as a result of spontaneous or environmentally induced somatic mutations. However, a key to identifying the targets for these mutations may lie in the study of families at very high risk of developing a common cancer, e.g. familial breast cancer and non-FAP colon cancer. Familial cancer although epidemiologically distinct is histologically and pathologically identical to cancer in the general population and therefore the identification of genetic alterations which predispose to these familial tumours may have very important implications for related but much more common sporadic tumours. For example, extensive linkage analysis has recently mapped early onset familial breast cancer to chromosome 17q21 [17].

Other approaches

Structural homology to known proto-oncogenes has been used to clone novel oncogenes

Under stringent conditions, single-stranded DNA (or RNA) probes will only hybridize to other copies of single-stranded DNA to which they are exactly complementary. In this way, unique sequences are identified by Southern and Northern analysis and in DNA libraries. However, if the stringency of hybridization is reduced (by reducing the temperature or increasing the salt concentration in which hybridization takes place), then DNA or RNA probes will hybridize to DNA species to which they have only partial homology. In this way genes identified as oncogenes by other methods have been used as probes to clone structurally related genes with sequence homology from DNA libraries. These genes are then examined for transforming activity or alteration in human tumours. *Myc* gene probes have been used to isolate three related genes, N-*myc* amplified in neuroblastomas, L-*myc* amplified in small cell lung cancer, and a third gene, R-*myc* [18]. c-*erb*B1 was used to isolate c-*erb*B2, which is amplified and over-expressed in 30% of breast cancers. *Ras* related genes such as *rho*, *ral*, and R-*ras* have been cloned, but these have not yet been implicated in human malignancy.

A functional approach has identified novel oncogenes

Functionally, proto-oncogenes appear to act in a limited number of ways: protein phosphorylation by tyrosine, serine or threonine kinases, signal transduction by GTP-binding proteins, transcriptional activation by DNA binding proteins and trophic stimulation by growth factors and their receptors. Theoretically, any gene has the potential to become an oncogene providing its product contributes to the biochemical pathways which mediate cellular proliferation. For example, many trophic hormones mediate their effects via the cAMP pathway. Attempts to implicate cAMP as a mediator of oncogene action have identified point mutations in a new putative proto-oncogene *gsp* in pituitary tumours [19]. These point mutations, like point mutations in *ras* genes, destroy the intrinsic GTPase activity of Gs, the stimulatory regulator of adenylyl cyclase. This results in elevated levels of cAMP and, it is proposed, promotes pituitary cell proliferation in these tumours.

Acknowledgements

I would like to thank Christine Bell for typing this manuscript and Dr Colin Cooper for his helpful advice and comments. HP is supported by grants from the Cancer Research Campaign and the Medical Research Council.

References

1　Varmus H. Retroviruses. Science 1988, **240**, 1427–1435.
2　Bishop JM, Varmus HE. Functions and origins of retroviral transforming genes. In: Weiss R, Teich N, Varmus, H, Coffin J, eds. Molecular Biology of Tumour Viruses. RNA Tumour Viruses. Cold Spring Harbour Laboratory, 1984, Vol. 1, 999–1108.
3　Hayward WS, Neel BG, Astrin SM. Activation of a cellular onc gene by promoter insertion in ALV-induced lymphoid leukosis. Nature 1981, **290**, 475–480.
4　Nusse R, Varmus HE. Many tumours induced by the mouse mammary tumour virus contain a provirus integrated in the same region of the host genome. Cell 1982, **31**, 99–109.
5　Shih C, Padhy LC, Murray M, Weinberg RA. Transforming genes of carcinomas and neuroblastomas introduced into mouse fibroblasts. Nature 1981, **290**, 261–264.
6　Hall A, Marshall CJ, Spurr NK, Weiss RA. Identification of transforming gene in two human sarcoma cell lines as a new member of the *ras* gene family located on chromosome 1. Nature 1983, **303**, 396–400.
7　Stanbridge EJ. Genetic analysis of human malignancy using somatic cell hybrids and monochromosome transfer. Cancer Surv 1988, **9**, 317–324.

8   Fournier REK, Ruddle FH. Microcell-mediated transfer of murine chromosomes into mouse, Chinese hamster, and human somatic cells. Proc Natl Acad Sci USA 1977, **74**, 319–323.

9   Heisterkamp N, Stephenson JR, Groffen J et al. Localization of the c-*abl* oncogene adjacent to a translocation breakpoint in chronic myelocytic leukaemia. Nature 1983, **306**, 239–242.

10 Marshall C. Human oncogenes. In: Weiss R, Teich N, Varmus H, Coffin J, eds. Molecular Biology of Tumour Viruses. RNA Tumour Viruses. Cold Spring Harbour Laboratory, 1985, Vol. 2, 487–558.

11 Kinzler KW, Bigner SH, Bigner DD et al. Identification of an amplified, highly expressed gene in a human glioma. Science 1987, **236**, 70–73.

12 Cavanee WK, Dryja TP, Phillips RA et al. Expression of recessive alleles by chromosomal mechanisms in retinoblastoma. Nature 1983, **305**, 779–784.

13 King M-C. Genetic analysis of cancer in families. Cancer Surv 1990, **9**, 417–435.

14 Goldgar DE, Green P, Parry DM, Mulvihil JJ. Multipoint linkage analysis in neurofibromatosis type 1: an international collaboration. Am J Hum Genet 1989, **44**, 6–12.

15 Mathew CGP, Chin KS, Easton, DF et al. A linked genetic marker for multiple endocrine neoplasia type 2A on chromosome 10. Nature 1987, **328**, 527–528.

16 Bodmer WF, Bailey CJ, Bodmer J et al. Localization of the gene for familial adenomatous polyposis on chromosome 5. Nature 1987, **328**, 614–616.

17 Hall JM, Lee MK, Newman B et al. Linkage of early-onset familial breast cancer to chromosome 17q21. Science 1990, **250**, 1684–1689.

18 DePinho R, Mitzock L, Hatton K et al. *myc* family of cellular oncogenes. J Cell Biochem 1987, **33**, 257–266.

19 Landis CA, Masters SB, Spada A, Pace AM, Bourne HR, Vallar L. GTPase inhibiting mutations activate the c chain of Gs and stimulate adenylyl cyclase in human pituitary tumours. Nature 1989, **340**, 692–696.

Molecular Biology for Oncologists
Edited by John Yarnold, Michael Stratton, Trevor McMillan
© *1993, Elsevier Science Publishers B.V. All rights reserved*

CHAPTER 4

Transgenic modelling of dominant and recessive oncogene function

Francisco S. Pardo and Emmett V. Schmidt

Departments of Radiation Oncology and The MGH Cancer Center,
Massachusetts General Hospital/Harvard Medical School, Boston, MA 02116, USA

Introduction

Transgenic mice provide a unique opportunity for studying the functions of proto-oncogenes and tumour suppressor genes in vivo. Genetic information can be introduced into somatic tissues by microinjection into the germline, by microinjection of DNA into the pronucleus of the fertilized egg, or by infecting embryos with retroviral vectors. The ability to alter the germline in such fashion allows one to investigate the phenotypic effects of altered gene expression.

The effects of gain in function mutations are modelled by introducing an oncogene sequence attached to a promoter which determines the developmental stage and tissue specificities of expression. Alternatively, loss of gene function can be modelled by interrupting coding sequences of particular genes (these are often called knockout mice). Such approaches provide new experimental strategies, not only for studying basic questions in areas of mammalian biology, but also allowing the production of animal models of human disease. It is the latter that is emphasized in this chapter, in particular modelling the effects of particular dominant and recessive oncogenes.

How are transgenic mice made?

Transgenic mice are commonly created by microinjection of DNA directly into the pronuclei of fertilized mouse eggs (Figure 1A). Following the harvesting of a zygote, the DNA of choice is microinjected into the pronucleus. After identification of viable eggs with visible pronuclei, the tip of the injection pipet is filled with the DNA of interest (in this case a purified oncogene or anti-oncogene construct). A holding pipet "sucks" the egg onto its end and an injection pipet is gently

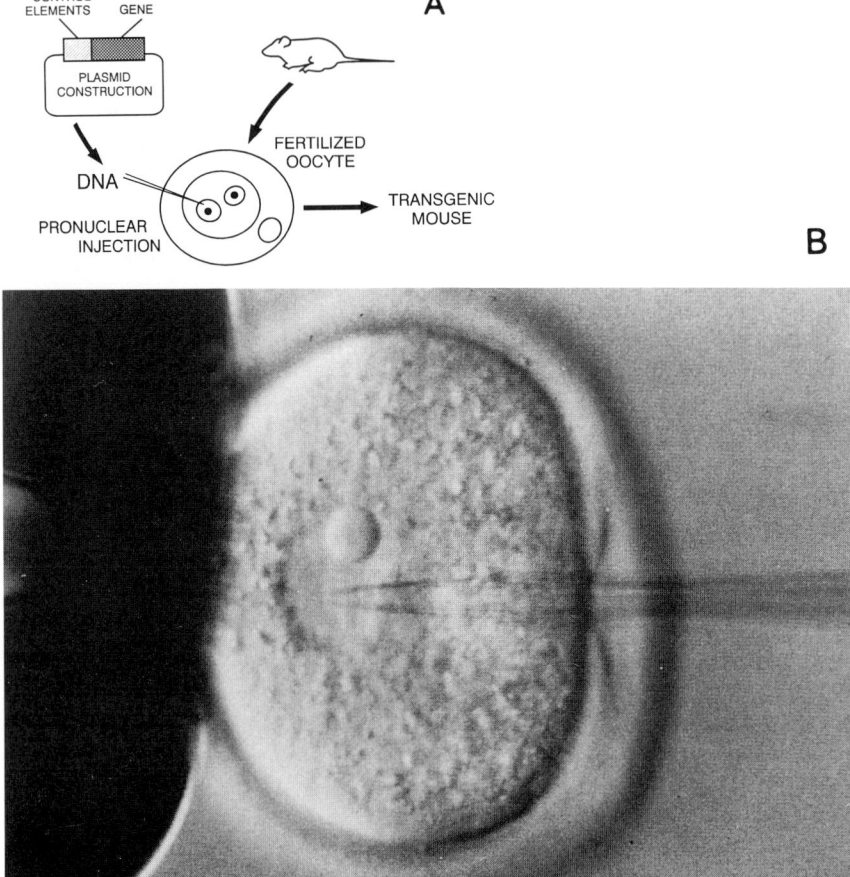

Figure 1. (A) The overall basic experimental paradigm. An injection fragment consisting of control or promoter elements linked to a particular oncogene or anti-oncogene construct is microinjected into the pronucleus. The resultant fertilized egg develops in the pseudo-pregnant female. Subsequent offspring are screened for the presence of the transgene of interest. (B) The process of microinjection into the pronucleus is illustrated. Once viable eggs with visible pronuclei are identified, the tip of the injection pipet is filled with the DNA of interest. A holding pipet sucks the egg onto its end and the injection pipet is gently pushed into the zona pellucida towards the target nucleus, whereupon the DNA solution is injected.

pushed into the zona pellucida towards and into the target nucleus (Figure 1B). In general, this approach leads to stable integration of the injected DNA in 10–40% of the injected embryos. In the majority of animals, such integration takes place at the one cell stage, thus leading to the integration of foreign DNA into every cell of the transgenic animal, including the primordial germ cells. The foreign DNA can integrate almost anywhere. Integration is usually at a single site but multiple copies of the transgene can be incorporated. Theoretically, virtually any DNA molecule should be able to be efficiently inserted into the mouse genome by microinjection into the zygote. The length of DNA for microinjection is limited only by DNA cloning technology. The quality of the DNA being microinjected can affect the rate of success. Linearization of injection fragments, purification of DNA, and optimization of both DNA and microinjection buffer concentrations are all critical to success.

Mice that develop from injected eggs are often termed "founder mice." It is essential to confirm that the founder mouse carries the transgene. This is usually established by isolating DNA from a small piece of its tail and testing for the presence of the gene of interest. In accord with basic Mendelian genetics, most transgenic founders will transmit the foreign gene to approximately 50% of their offspring. Mice that are heterozygous for the transgene of interest can be mated to each other to create homozygous transgenic mice. Approximately 5–15% of the DNA integration events in transgenic mice produce lethal mutations in homozygous transgenic offspring [1]. Consequently, many of the transgenic lines are preferentially maintained in the heterozygous state, in which the transgene will be passed on to approximately 50% of resultant offspring following mating to a wild type animal.

Much of the evidence that somatically altered oncogenes play a major role in the initiation and progression of human malignancy comes from transgenic technology, which directly tests the efficacy of proto-oncogenes or tumour suppressor genes. The incorporation of dominant or recessive oncogenes into particular tissues often depends upon the design of expression constructs used for microinjection. Such constructs often have a proto-oncogene (or tumour suppressor gene) under the transcriptional control of a promoter, which is relatively tissue-specific for a particular target population of cells. In general, the promoter region of DNA constructs used for microinjection renders tissue specificity for targeting of transgene expression. Thus, the promoter region of the mouse mammary tumour virus (MMTV) upstream of the coding sequences for c-*myc* or H-*ras*, preferentially targets *ras* expression to those tissues in which MMTV is expressed in greatest quantities, i.e. the breast, salivary epithelium and other lymphoid tissues. Similarly, one could use the promoter of the albumin gene to confine transgene expression to liver. Targeting allows the differential expression of the gene of interest, usually an oncogene previously implicated in the pathogenesis of human malignancy, in the

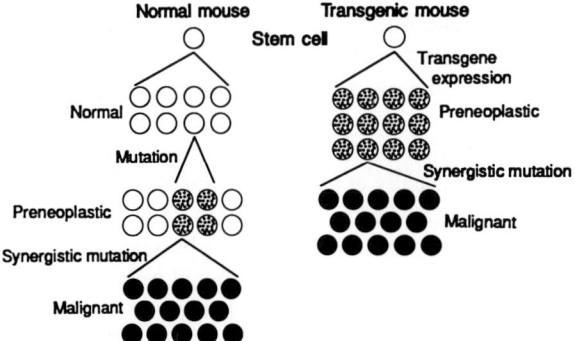

Figure 2. Illustrated here is a representation of the "preneoplastic state". Stem cells from a transgenic mouse carrying the transgene require a synergistic mutation acquired during the development of the animal for full transformation to the malignant phenotype. The preneoplastic state typically reveals polyclonal proliferations of cells which can evolve into clonal tumorigenic cellular populations over the developmental lifespan of the animal. From [2]. Reproduced with permission.

cellular population from which malignancy arises. Thus transgenic mice preferentially targeting B-lymphoid populations bear the c-*myc* proto-oncogene with the En promoter (the lymphoid-specific promoter). Models testing the role of the *ras* or the *neu* oncogene with respect to its role in the pathogenesis of breast cancer are often subjugated to the control of an MMTV promoter.

A preneoplastic state exists prior to tumour formation

Regulation of transgene expression starts during embryogenesis and continues throughout the adult lifespan of the animal. Although promoter control elements help to restrict expression of transgenes to selected cell populations, transgenes are often more widely expressed during embryogenesis and adult life. Expression of the transgene in a particular tissue does not necessarily lead to malignancy, or may do so only after a delay. There may also be a preneoplastic state before malignant progression (Figure 2) [2]. Several transgenic systems develop tumours in a stochastic fashion with time. Many of the cellular proliferations are polyclonal. Only later in the animal's lifespan do monoclonal populations of tumour cells emerge. As these transgenic animals develop, additional factors, including environmental and metabolic factors, modulate transgene expression. In addition, alterations in expression of other dominant or recessive oncogenes interplay with the host genetic background in the formation of the malignant phenotype and subsequent tumour initiation and progression.

Gain-in-function or dominant oncogenes

Murine breast cancer is a model for studying oncogenes in vivo

c-myc is not sufficient for tumorigenesis

The c-*myc* oncogene is not sufficient to transform cells in culture. Deregulation of the c-*myc* oncogene can, however, be modelled in transgenic experiments involving activated c-*myc* genes under the control of the tissue-specific mouse mammary tumour virus (MMTV) promoter [3]. In these animals, solitary adenocarcinomas of the breast arise from morphologically normal mammary epithelium. In addition, testicular tumours, tumours of mast cells and lymphocytes occur at lower frequency. Not all tissues that express the transgene develop neoplasms. Genetic events in addition to activation or over-expression of c-*myc* are required to induce malignancy in the living organism. This suggests a multi-step model of tumour initiation and progression.

Mutated v-H-ras is synergistic with c-myc but the combination is still insufficient for complete malignant transformation

Transgenic mice carrying the v-H-*ras* oncogene driven by the mouse mammary tumour virus promoter (MMTV), begin to develop both mammary and salivary tumours by the fifth week of life [4]. High levels of *ras* expression are consistently observed in both the mammary and salivary glands. In addition, a large percentage of animals have Harderian gland hyperplasia, resulting in the phenotype of bilateral exophthalmos. These hyperplasias, however, are polyclonal in origin and do not invade the orbit. In addition, they are not transplantable in nude mice, and cannot be permanently cultured.

Heterozygous carriers of either the MMTV/v-H-*ras* or the MMTV/c-*myc* transgene have been mated to each other to develop a stable transgenic line heterozygous for both transgenes [3]. While the spectrum of tumours does not differ substantially from those produced by either of the respective transgenic lines, the kinetics of tumour appearance differ dramatically in mice carrying both genes. Tumours occur relatively late in the lifespan of c-*myc* mice, appearing at approximately 113 days of age [4]. One-half of female mice carrying only the v-H-*ras* transgene develop malignancies by 168 days of age [4]. Approximately 50% of all female animals carrying both transgenes develop tumours by 46 days of age, and all female *ras/myc* "dual carriers" develop malignancies by day 163 (Figure 3). Most of these malignancies are mammary adenocarcinomas, although malignant lymphoma and salivary gland adenocarcinomas are also seen. This transgenic model system, using the MMTV/c-*myc* or v-H-*ras* oncogenes, demonstrates the synergistic effects of both oncogenes in tissues expressing the transgenes. It is evident that the effect of combining both oncogenes is highly carcinogenic in target tissue.

F.S. Pardo and E.V. Schmidt

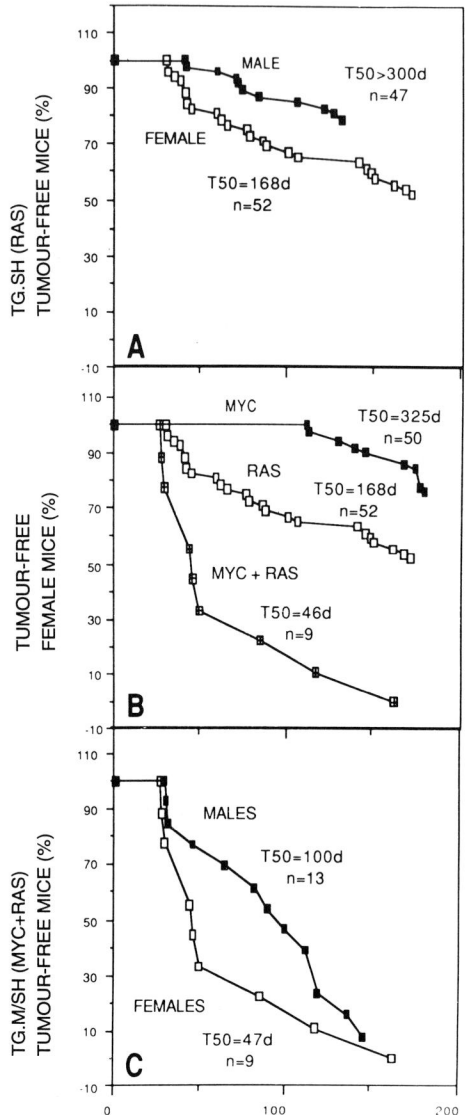

Figure 3. Depicted are the synergistic effects of the c-*myc* and c-Ha-*ras* oncogenes on the overall survival of transgenic animals. The middle figure emphasizes the fact that the combination of c-*myc* and c-Ha-*ras* oncogenes has a cooperative effect which leads to decreased survival. In addition, the *ras* or c-H-*ras* transgenics demonstrate a decreased overall survival and tumour free incidence compared to the c-*myc*-alone transgenics. From [4]. Reproduced with permission.

However, next to the dysplastic foci of frankly malignant cells can be found morphologically normal, non-malignant mammary epithelial cells. Thus, additional events, either genetic or epigenetic, must also come into play in directing particular groups of cells towards the expression of the fully malignant phenotype.

Expression of a mutated c-neu oncogene can be sufficient for tumour formation in transgenic mice

Recent data show that amplification of c-*erb*B2, the normal human homologue of the rat c-*neu* oncogene, correlates with human breast cancer progression [5]. A transgenic line has been created carrying an activated (mutated) version of the c-*neu* oncogene under the transcriptional control of the mouse mammary tumour virus (MMTV) promoter to model the effects of c-*erb*B2 overexpression in human cancer [6]. This promoter sequence is in order to stimulate transgene expression at high levels in mammary epithelium. In contrast to the synergy required in the c-*myc* and v-H-*ras* experiments mentioned above, mice engineered using the MMTV promoter and the coding sequence of the mutated c-*neu* oncogene develop mammary adenocarcinomas that involve the entire mammary epithellium. By 56 days, multiple hyperplastic and dysplastic mammary nodules can be noted in virgin female mammary tissue. By 95 days of age, all multiparous female carriers develop multiple mammary tumours. Interestingly, the kinetics of tumour formation are slightly decreased in males, with only half developing similar mammary tumours by day 114 (Figure 4). The tumours acquired are polyclonal in origin, indicating that mutated c-*neu* expression in breast tissue may be nearly sufficient to induce malignant transformation in a single step. Although the transgene is also expressed in the parotid gland and in the epididymis, only epithelial hyperplasia or benign hypertrophy is noted on long-term follow-up. Genetic and environmental factors in these tissues must play a major role in regulating malignant progression.

Transgenic mice carrying a normal c-*erb*B2 transgene have also been created. When expression is driven by a MMTV promoter, the transgene is expressed in several tissues, including the mammary gland. No mammary carcinomas have been reported, although adenocarcinomas of the lung and lymphomas are seen. The lymphomas develop after considerable latency, implying multi-stage carcinogenesis, whereas the lung cancers develop swiftly and might represent single step carcinogenesis.

Chromosome-specific abnormalities can be modelled in vivo

BCL-2 prolongs lymphocyte survival and results in lymphomagenesis

BCL-2 is an inner mitochondrial membrane protein, believed to extend the survival of certain cell types by blocking programmed cell death or apoptosis. The BCL-2 gene was discovered by its translocation to the IgH locus in B-cell follicular lymphoma. This is the most common of the lymphomas, and typically demon-strates a t(14;18) chromosomal translocation on karyotype analyses. The t(14;18) translocation involves the BCL-2 gene, creating a deregulated BCL-2-immuno-globulin fusion gene. Follicular lymphoma typically exhibits a rather indolent

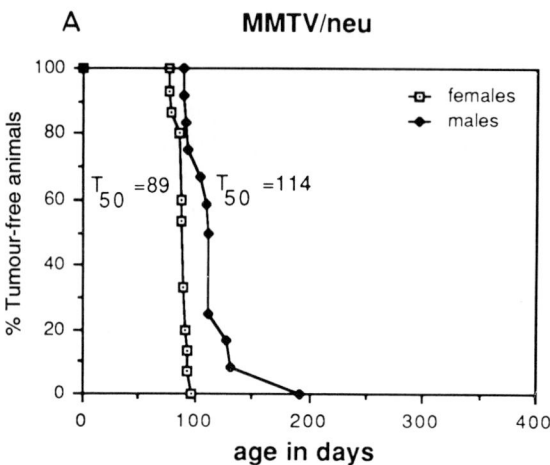

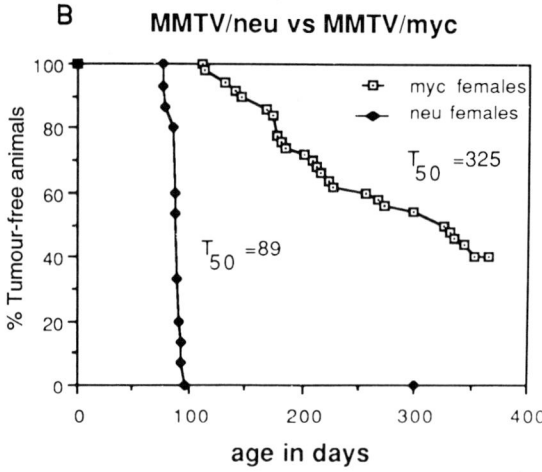

Figure 4. Transmission of the c-*neu* oncogene is sufficient for the induction of murine breast carcinoma. When compared to the MTV/*myc* mice, MMTV/*neu* mice have markedly decreased overall survival times and tumour-free incidences. From [6]. Reproduced with permission.

clinical course, but clinicopathologic progression into an aggressive lymphoma is common.

The translocation has been modelled using a BCL-2 immunoglobulin minigene that structurally mimics the t(14;18) translocation. Transgenic mice harbouring this transgenic allele develop an indolent follicular hyperplasia which, over time, progresses to histopathologically documented malignant large-cell lymphoma [7]. Cells of B-lymphoid lineage in transgenic mice demonstrate enhanced in vitro survival, resulting in an unusually high number of non-cycling B-cells. Few BCL-2 transgenic animals develop tumours within the first year of life, suggesting that BCL-2 might preserve lymphocyte lifespan until other genetic and/or environmental changes transform the particular B-cell clone into a neoplastic proliferation. Subsequent mutations at other genetic loci are responsible for progression to frank neoplasia. Included amongst these are the rearrangements of c-*myc* detected in the lymphomas of BCL-2 transgenic mice [9]. Approximately 50% of the high grade lymphomas studied at a molecular level show rearrangement of the c-*myc* gene.

Mice harboring a c-myc germline transgene demonstrate a preneoplastic state leading to high grade non-Hodgkin's lymphoma
In a large proportion of human Burkitt's lymphomas, the c-*myc* gene is activated by a translocation to an immunoglobulin heavy chain locus. This translocation can be modelled in transgenic mice containing a normal c-*myc* transgene coupled to the immunoglobulin heavy chain enhancer (En). These mice develop fatal non-Hodgkin's lymphoma at a mean of 90–100 days, although during the first several weeks of life only a relatively small percentage of animals develop clinically overt disease [10]. Eventually, virtually all animals die with gross mediastinal, axillary, and abdominal adenopathy (Figure 5A,B). Cytofluorometric analyses of lymphoid tissues, bone marrow, spleen and thymus from these animals reveals abnormalities in the number of both immmature and mature B-cells. Interestingly, when an additional transgene encoding the membrane-bound form of human IgM (rather than the secreted form) is bred into the susceptible strain, the incidence of lymphomas in c-*myc* bearing transgenic mice decreases [11] (Figure 6).

Tumour suppressor genes

p53 has dominant-negative effects

Transgenics created with p53 missense mutations and two normal p53 alleles develop normally, but acquire tumours postnatally
Alterations in the p53 gene are one of the most common genetic alterations found in human cancer. The p53 gene, although classified under the group of recessive

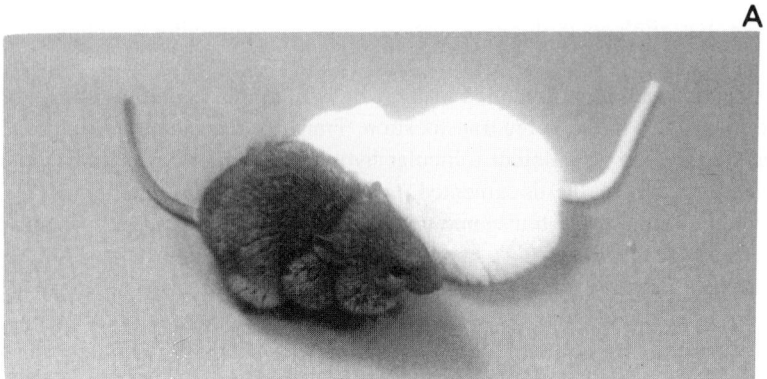

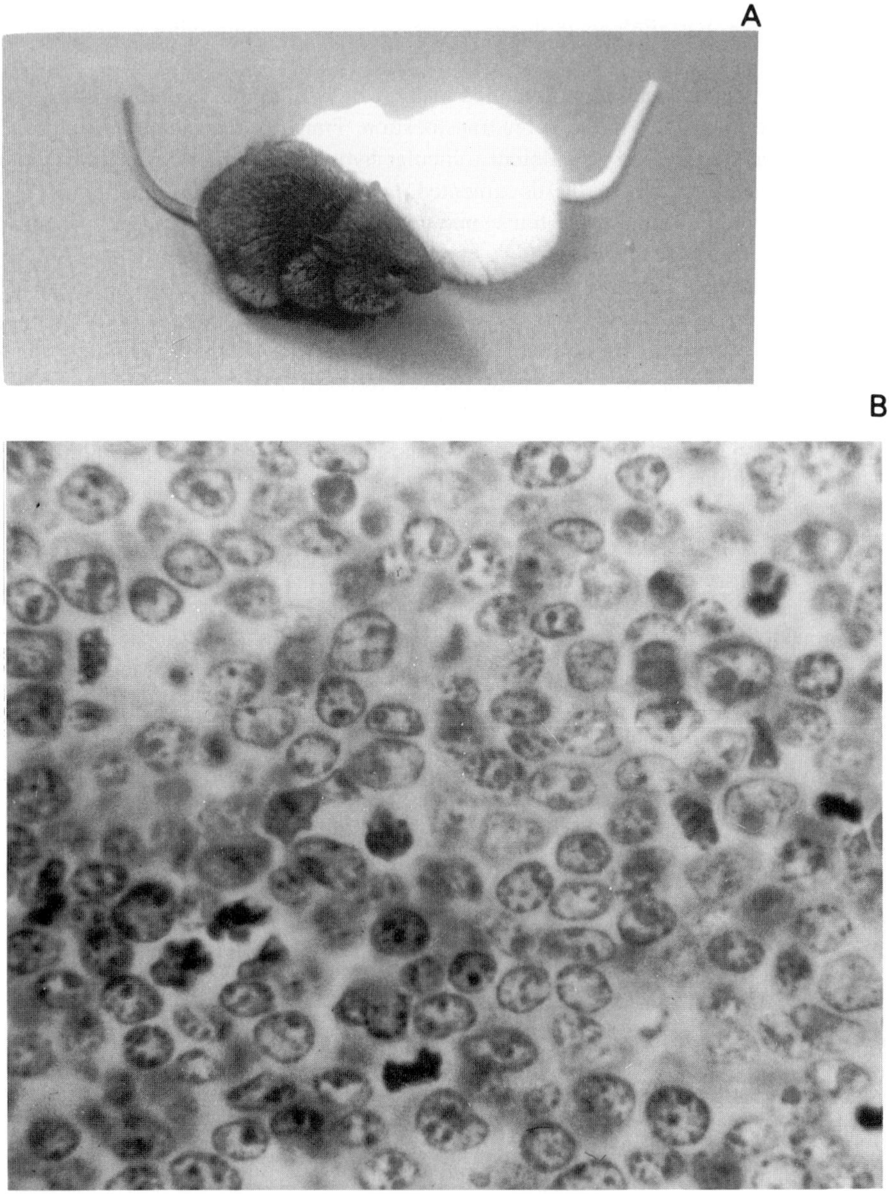

Figure 5. (A) The phenotype of Eµ *myc* transgenic mice is illustrated, typically revealing massive mediastinal, axillary, and abdominal adenopathy. (B) The histologic subpattern of these predominantly pre-B-cell lymphomatous proliferations, revealing hypercellularity, nuclear atypia and mitoses.

oncogenes, can also act in dominant fashion. The "dominant negative" effects of certain p53 mutations are caused by inactivation of the wild type protein by a mutant protein which complexes to wild type p53 and inactivates it. Two transgenic systems have been devised for modelling p53 function. P53 mice derived from the introduction of a missense mutation into the germline still possess two normal alleles. Lung adenocarcinomas, osteosarcomas (Figure 7) and lymphomas develop in approximately 20% of animals with latencies varying between 4 and 18 months

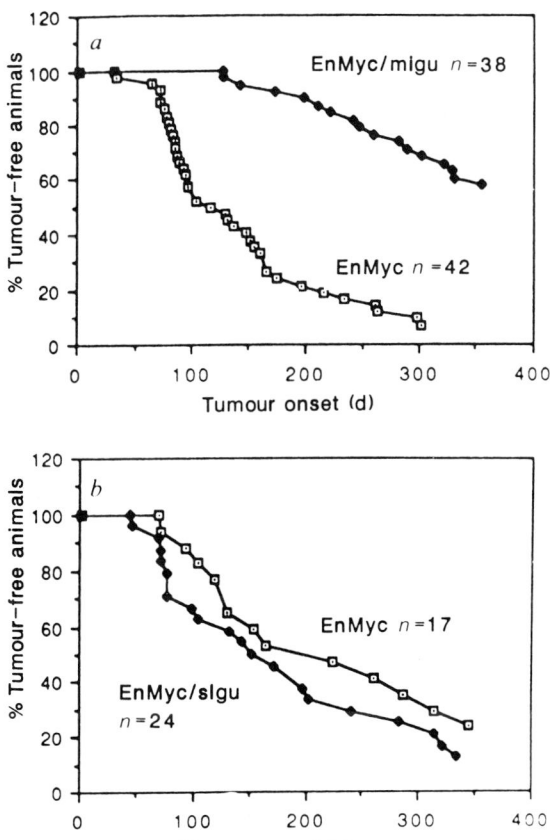

Figure 6. Portrayed here is the overall survival of Eμ *myc*/sIgμ and Eμ *myc*/mIgμ mice demonstrating that introduction of the membrane-bound (m), but not the secreted (s) form of the immunoglobulin enhancer can partially reverse the rapid onset of lymphomagenesis in EμMyc/transgenic mice. The upper graph represents the introduction of the membrane-bound form of the immunoglobulin enhancer which actually confers a protective effect on the onset of tumorigenesis. The bottom curve shows that introduction of the secreted form does not delay tumour onset. From [11]. Reproduced with permission.

F.S. Pardo and E.V. Schmidt

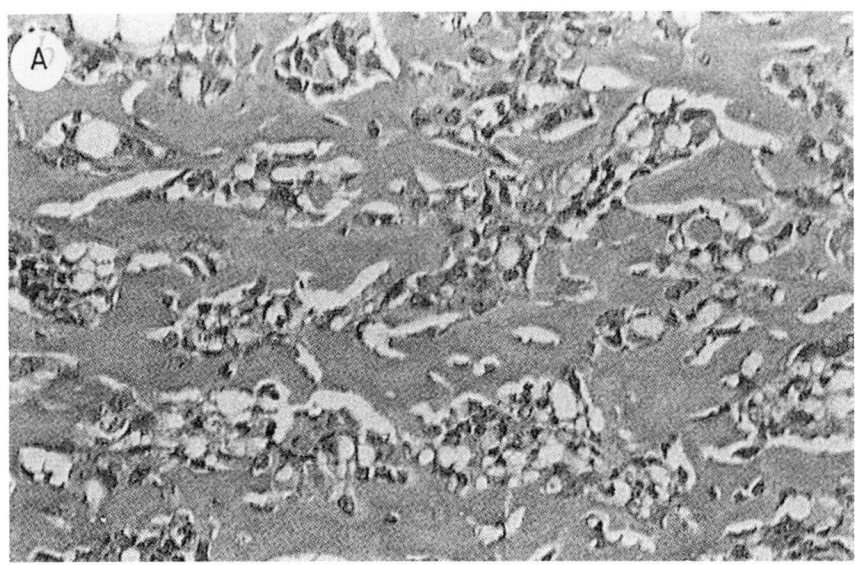

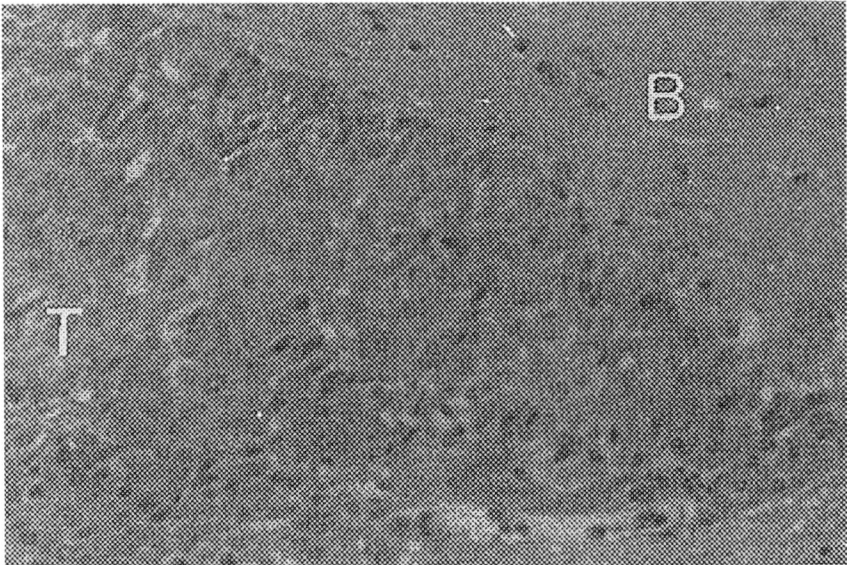

Figure 7. (A) A well-differentiated sacral osteosarcoma with numerous pink island osteoids, obtained from a p53 "knockout" mouse. In addition to osteosarcoma, hemangio-sarcoma, gonadoblastoma and various germ cell tumours also occur in these mice. From [13]. (B) Histologic representation of a pituitary adenocarcinoma from a Rb knockout mouse. (T) illustrates the region of tumour abutment onto the normal brain (B). From [15]. Reproduced with permission.

[12]. Although these animals possess endogenous murine p53, epitope mapping studies have proven that the mutant p53 transgene is expresssed in tumours developing in these animals [12].

Knockout experiments suggest that absent wild type p53 expression is consistent with normal development but is associated with tumorigenesis
A second animal model involves the creation of "knockout" mice that lack p53 production during embryogenesis and subsequent development. Such mice are created by the insertion of the neomycin gene, or neomycin cassette, into the critical p53 coding region, thus ablating expression of the p53 protein product. p53 knockout mice do not express p53 during development, yet they reach term without sequelae. This suggests that p53 function is not essential for normal murine development. The mice develop a spectrum of tumours similar to those mentioned above in the case of mice with endogenous p53 function. Lymphomas, osteosarcomas, adenocarcinomas and germ cell-type tumours generally develop by approximately 6 months of age [13]. Thus, total ablation of p53 function results in the tumorigenic phenotype with relatively short latency.

SV40 T antigen (Tag) transgenics express T antigen (protein) in the retina and develop ocular tumours with ultrastructural features similar to those found in retinoblastoma
The retinoblastoma (Rb) gene is probably the best characterized of the tumour suppressor genes or anti-oncogenes. Early attempts at disruption of Rb function involved expression of an SV40-T antigen transgene in mice. The protein expressed by the T antigen of the oncogenic DNA virus SV40 is known to complex with and inactivate wild type Rb protein in infected monkey cells. This interaction between the viral oncogene product and Rb protein in the host cell is thought to be crucial in the development of virally induced tumours in monkeys. In mice with an SV40 transgene, ocular tumours develop with histopathologic and ultrastructural characteristics of retinoblastoma [14].

Retinoblastoma knockout heterozygotes develop normally, but acquire pituitary adenocarcinomas
Ocular tumours are not found. More recent attempts at disruption of Rb gene function have concentrated on knockout experiments, similar to those engineered for the development of p53 knockout mice. Heterozygous animals have only one inactivated Rb allele. They do not develop retinoblastomas but approximately 25% of the animals develop pituitary adenocarcinomas [15] (Figure 7B). Embryos homozygous for the Rb mutation develop normally for the first 12 days of gestation, dying of heart failure at about day 14 [16]. Homozygous animals that die in utero show significant defects in hematopoeisis, presumably a defect related to absent Rb

expression. Neuronal cell death is common in the central nervous system, although the neural retina is unaffected. Why SV40 transgenics develop retinoblastoma, while none of the Rb gene knockout mice do so is unknown [17].

References

1 Hogan B, Costantini F, Lacy E. Manipulating the Mouse Embryo – A Laboratory Manual. Cold Spring Harbor: Cold Spring Harbor Laboratory, 1986.
2 Adams JM, Cory S. Transgenic models of tumor development. Science 1991, **254**, 1161–1167.
3 Stewart TA, Pattengale PK, Leder P. Spontaneous mammary adenocarcinomas in transgenic mice that carry and express MTV/*myc* fusion genes. Cell 1984, **38**, 627–637.
4 Sinn E, Muller W, Pattengale P, Tepler I, Wallace R, Leder P. Coexpression of MMTV/V-Ha-*ras* and MMTV/c-*myc* genes in transgenic mice: synergistic action of oncogenes in vivo. Cell 1987, **49**, 465–475.
5 Kolata G. Oncogenes give breast cancer prognosis. Science 1987, **235**, 160–161.
6 Muller WJ, Sinne E, Pattengale PK, Wallace R, Leder P. Single-step induction of mammary adenocarcinoma in transgenic mice bearing the activated c-*neu* oncogene. Cell 1988, **54**, 105–115.
7 McDonnell TJ, Korsmeyer SJ. Progression from lymphoid hyperplasia to high-grade malignant lymphoma in mice transgenic for the t(14;18). Nature 1991, **349**, 254–256.
8 Vaux DL, Cory S, Adams JM. BCL-2 gene promotes haemopoietic cell survival and cooperates with c-*myc* to immortalize pre-B cells. Nature 1988, **335**, 440–442.
9 Fanidi A, Harrington EA, Evan GI. Cooperative interaction between c-*myc* and BCL-2 proto-oncogenes. Nature 1992, **359**, 554–556.
10 Schmidt EV, Pattengale PK, Weir L, Leder P. Transgenic mice bearing the human c-*myc* gene activated by an immunoglobulin enhancer: a pre-B-cell lymphoma model. Proc Natl Acad Sci USA 1988, **85**, 6047–6051.
11 Nussenzweig MC, Schmidt EV, Shaw AC et al. Human immunoglobulin gene reduces the incidence of lymphomas as in c-*myc*-bearing transgenic mice. Nature 1988, **336**, 446–450.
12 Lavigueur A, Maltby V, Mock D, Rossant J, Pawson T, Bernstein A. High incidence of lung, bone and lymphoid tumors in transgenic mice overexpressing mutant alleles of the p53 oncogene. Mol Cell Biol 1989, **9**, 3982–3991.
13 Donehower LA, Harvey M, Slagle BL, et al. Mice deficient for p53 are developmentally normal but susceptible to spontaneous tumours. Nature 1992, **356**, 215–221.
14 Windle JJ, Albert DM, O'Brien JM et al. Retinoblastoma in transgenic mice. Nature 1990, **343**, 665–668.
15 Jacks T, Fazeli A, Schmitt EM, Bronson RT, Goodell MA, Weinberg RA. Effects of an Rb mutation in the mouse. Nature 1992, **359**, 295–300.
16 Lee EY-HP, Chang C-Y, Hu N, et al. Mice deficient for Rb are nonviable and show defects in neurogenesis and haematopoiesis. Nature 1992, **359**, 288–294.
17 Harlow E. For our eyes only. Nature 1992, **359**, 270–271.

Molecular Biology for Oncologists
Edited by John Yarnold, Michael Stratton, Trevor McMillan
© *1993, Elsevier Science Publishers B.V. All rights reserved*

CHAPTER 5

The BCL-2 gene in clinical oncology

Michael Brada[a] and Martin Dyer[b]

[a]Academic Unit of Radiotherapy and Oncology, [b]Academic Unit of Haematology and Cytogenetics, Institute of Cancer Research and The Royal Marsden Hospital, Downs Road, Sutton, Surrey, SM2 5PT, UK

BCL-2 gene was discovered at a translocation breakpoint in malignant lymphoma cells

Functional immunoglobulin genes and T-cell receptor (TCR) genes are produced by bringing together dispersed DNA fragments of the variable (V), diversity (D) and joining (J) regions of the gene. This process involving somatic recombination occurs during lymphocyte differentiation [1], and is prone to errors which may result in chromosomal translocations [2]. The immunoglobulin (Ig) loci are targets for translocation in B-cell malignancy and TCR loci are similarly targeted in T-cell neoplasms [3].

Molecular identification of translocations using cloned Ig and TCR probes to define the breakpoints allows the isolation of genes that are important, not only in the pathogenesis of malignancy, but also in normal lymphocyte development and differentiation. As a consequence of the translocation, other genes come under the control of powerful enhancers within the Ig and TCR loci resulting in deregulated (over)expression of the gene. Alternatively, the gene may be physically disrupted or mutated with loss of normal function.

BCL-2 gene, whose name derives from its association with B-cell lymphoma, has been identified in this manner. It was found close to a translocation site on chromosome 18q21 in t(14;18) translocation, which is present in the majority of nodular low grade lymphomas and a proportion of high grade lymphomas. The BCL-2 gene is placed adjacent to the immunoglobulin (Ig) heavy chain gene on

53

chromosome 14q32. The BCL-2 gene encodes an unusual mitochondrial membrane protein of unknown function which is widely expressed in normal and malignant human tissues. As a consequence of translocation, the gene expression is deregulated, with increase in BCL-2 protein which may play an important role in the pathogenesis of lymphoma and some carcinomas.

The BCL-2 gene is composed of three exons [4], and the coding portions of the gene lie within exons II and III (Figure 1). These are separated by a large 370 kilobase intron. The BCL-2 gene lacks a TATA box, the classical transcription initiation site, and transcripts show considerable heterogeneity in the precise transcription initiation sites. The region immediately 5' of exon I is extremely G-C rich and contains several DNA sequence motifs that may bind transcription factors.

BCL-2 protein has no structural similarities to other known proteins

The BCL-2 gene encodes a 26 kDa non-glycosylated protein [5]. At the time of initial sequencing, the function of the BCL-2 protein was completely unknown. The only clue to its possible role in oncogenesis was limited to some weak sequence homology with an EB virus protein.

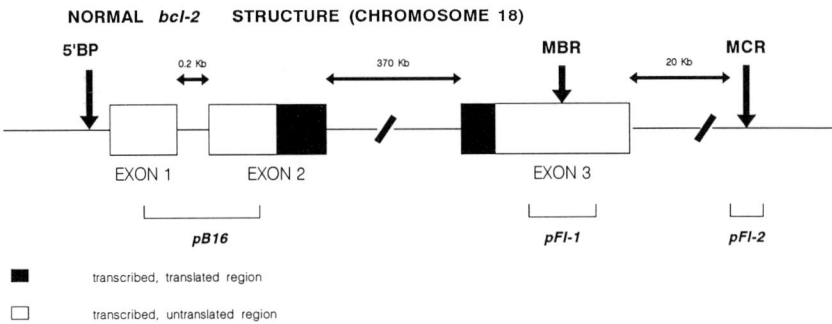

Figure 1. Structure of the BCL-2 gene and the site of translocation breakpoints. The BCL-2 gene has three exons; regions which are translated into protein and are shown by the shaded boxes. Exon 1 and 2 are separated by a short 200 basepair intron; exons 2 and 3 are separated by a 370 kilobase intron. The predominant transcripts initiate from the 5' site upstream of exon 1, but transcripts also initiate from the first intron. The translocation breakpoints are shown by arrows; MBR denotes major breakpoint cluster occurring within the 3' untranslated region of the gene: MCR denotes minor cluster region occurring 25 kilobases 3' to the MBR. pB16, pFL-1 and pFL-2 are probes commonly used to detect rearrangements of both 5' and 3' regions of the BCL-2 gene.

Antibody localisation, cell fractionation experiments and the hydrophobic nature of the carboxy terminal suggest association of the BCL-2 protein with intracellular membrane. The BCL-2 protein lacks any recognisable DNA binding motifs and its role in neoplastic transformation appears to differ from most other potential oncogenes.

BCL-2 protein is localised to the outer mitochondrial membrane

The hydrophobic nature of the carboxy terminus of the BCL-2 protein indicated that it might be associated with a membrane and this has been confirmed. Immunofluorescent studies fail to detect surface expression of antigen indicating that the protein is expressed within the cytoplasm. Laser confocal microscopy of cell lines with a t(14;18) give a punctate cytoplasmic distribution suggesting that the protein is associated with an organelle and probably with mitochondria. Subcellular fractionation experiments have shown that BCL-2 immunoreactivity co-migrates with the mitochondrial membrane fraction and specifically with the inner mitochondrial membrane [6].

However, immunoelectron microscopic studies localise BCL-2 to the outer, not the inner, mitochondrial membrane (Figure 2) with only a small percentage of staining localised in the mitochondrial matrix. The reasons for the discrepancy between the sets of data are difficult to determine but may be due to accessibility of the various BCL-2 epitopes. The precise localisation of BCL-2 protein in mitchondria has important implications for its function and still needs to be fully defined.

BCL-2 is expressed in a wide range of normal and malignant cells

Initial reports using polyclonal rabbit antisera suggested that BCL-2 expression is limited to neoplastic B-cells which exhibit the t(14;18) translocation. Development of specific monoclonal antibodies (MAbs) revealed that BCL-2 is expressed widely among normal haemopoietic and non-haemopoietic human tissues [7]. BCL-2 staining has been observed in immature myeloid cells, immature erythroid cells, medullary but not cortical thymocytes and plasma cells. BCL-2 expression has also been detected in the epithelial cells of breast, thyroid, prostate and the crypt cells, but not more mature cells within the gastrointestinal epithelia. BCL-2 was also demonstrable in some neurons within the brain.

The amount of BCL-2 present in normal lymphocytes fluctuates with stages of differentiation and proliferative activity. Transcription of BCL-2 transiently increases when the pre-B-cell is developing into mature B-cells and then is down-

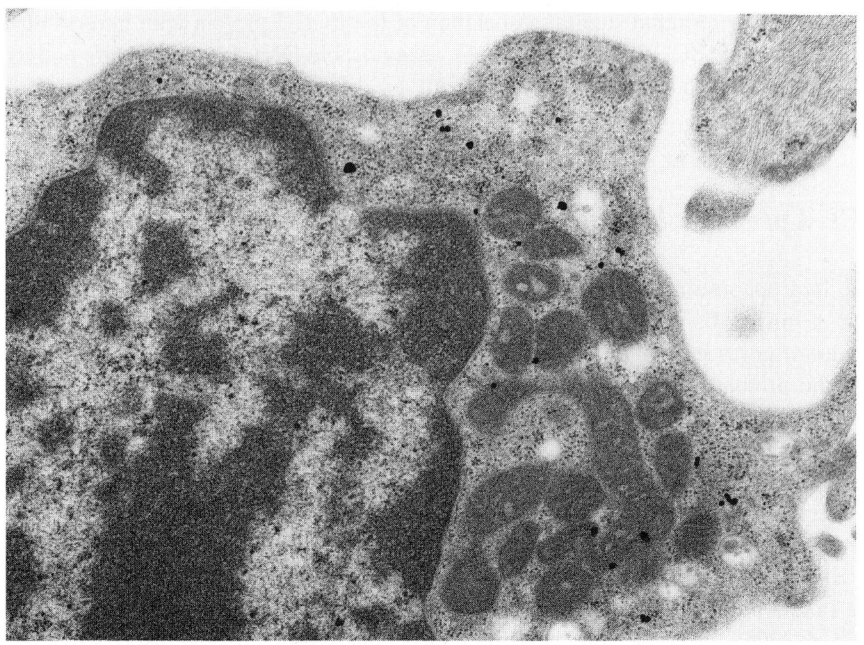

Figure 2. Electron micrograph of a lymphoma cell line labelled with BCL-2 MAb. BCL-2 is localised in three areas: the peripheral region of the mitochondria, the nuclear membrane and to a lesser degree in membrane structures in the cytoplasm. There was no staining of the mitochondrial matrix.

regulated. BCL-2 mRNA levels also increase in B- and T-cells in response to pro-liferation signals, although the protein remains undetectable by immunohisto-chemistry in normal active germinal centres. BCL-2 is therefore a mitochondrial membrane protein which appears to be involved in growth regulation in B- and T-cells.

Among neoplastic cells, overexpression of BCL-2 has been found initially in the malignancies of mature B-cells with the t(14;18). However, BCL-2 expression is observed in virtually all haemopoietic malignancies of all lineages and stages of maturation including most cases of acute myeloid leukaemia, chronic myeloid leukaemia, acute lymphoblastic leukaemia, myeloma and chronic lymphocytic leu-kaemia as well as in B-cell NHL [8]. In B-cell NHL, expression of BCL-2 does not correlate with the presence of the BCL-2 translocations. Interestingly, in trans-formed cases of B-cell NHL, BCL-2 expression is often down-regulated.

BCL-2 protein appears to promote cell survival rather than proliferation

Activated BCL-2 gene appears to prevent apoptosis, particularly in B-cells of the germinal centre. In vivo evidence of involvement of BCL-2 in neoplastic transformation centres around gene transfer experiments and studies in transgenic mice with deregulated BCL-2 gene.

Transfection of BCL-2 into NIH 3T3 fibroblasts failed to induce transformation in vitro although some transfectants may cause tumours in vivo when injected into mice [9]. BCL-2 transfected into growth factor-dependent haemopoietic cells fails to transform directly but prolongs cell survival in the absence of the necessary growth factors [10]. Transfection of BCL-2 in combination with c-*myc* has a synergistic effect on cell transformation [11].

The primary effect of BCL-2 transgene expression under the control of the Ig enhancer in mice is a prolongation of lifespan of B-cells. This produces an increase in their total number, with most animals having between 5 and 10 times the normal number of polyclonal non-cycling B-cells in the blood, marrow and spleen, as well as follicular hyperplasia in lymph nodes [12]. These B-cells also produce greatly amplified and prolonged immune responses. In some transgenes, the increase in polyclonal immunoglobulin is associated with a variety of autoimmune diseases and many animals die of immune complex glomerulonephritis.

B-cell tumours do occur in BCL-2 transgenic mice but only at low frequency and only after a long latency suggesting that BCL-2 per se is insufficient for transformation. These tumours are frequently associated with deregulated c-*myc* expression [13]. In contrast, BCL-2 transgenic mice do not have an increased incidence of T-cell tumours, and it appears that overexpression of BCL-2 in T-cells may not inhibit lymphocyte cell death. Crossing of BCL-2 and c-*myc* transgenic mice to give mice expressing both transgenes results in the rapid onset of tumours which have the features of "primitive" haemopoietic stem cells.

In conclusion BCL-2 appears to extend B-cell survival by blocking programmed cell death. The extended lifespan of the B-cells may increase the probability of acquisition of more directly transforming oncogenic events such as c-*myc* translocations. This hypothesis is compatible with the clinical experience of the behaviour of low grade lymphoma. The long indolent phase of the disease characterised by t(14;18) with BCL-2 deregulation may be due to a prolonged survival of B lymphocytes which will be susceptible to further genetic alteration. The progression to a more malignant phase is associated with new genetic events such as c-*myc* activation. The co-existence of BCL-2 and c-*myc* translocation has been demonstrated in some high grade NHL, and the appearance of c-*myc* translocation has been detected on transformation of a low grade lymphoma bearing t(14;18) [14].

BCL-2 activation by chromosomal translocation in lymphoid malignancies appears confined to B-cell NHL

Despite the broad expression of BCL-2 in both normal and malignant haemopoietic cells, translocations involving the BCL-2 gene are found almost exclusively among malignancies of mature B-cells and principally B-cell NHL. The reason for the restricted distribution of BCL-2 translocations is presently unclear.

The DNA sequence of the t(14;18) translocations has led to the suggestion that the translocations arise as a direct consequence of errors in the process of somatic recombination of the Ig heavy chain genes [3]. Translocation occurs primarily within two short regions of the BCL-2 described as the major breakpoint region (mbr) and the minor cluster region (mcr).

A number of possible mechanisms for translocation have been suggested. The favoured mechanism is the involvement of a common DNA recombinase enzyme, which is responsible for normal Ig and TCR gene rearrangements. In this model, the translocation occurs by recognition of heptamer-spacer-nonamer nucleotide sequences on chromosome 18 which are the signals for the joining of Ig genes. Translocation is therefore considered as a mistake of normal Ig gene rearrangement. If this theory is correct, it has implications for the pathogenesis of B-cell NHL. It suggests that tumour stem cells derive from the B-cell precursor population at a stage when the recombinase enzymes are normally expressed, rather than from more mature B-cells.

There are arguments against this mechanism fully discussed by Tycko and Sklar [3]. In addition, the sequencing of derivative chromosome 18 shows the proximity of D (diversity) sequences [15] which suggest translocation in a more mature B-cell where Ig gene rearrangement has already occurred. Alternative models of translocation suggest an illegitimate pairing of staggered double-strand DNA breaks.

BCL-2 activation in non-lymphoid malignancy is currently being investigated

Overexpression of BCL-2 has been observed in 80% of primary breast carcinomas. Studies on the breast carcinoma line MCF-7 using quantitative immunofluorescence, immunoblotting and RNA blotting experiments show levels of normal sized BCL-2 mRNA and protein comparable to those seen in the B-cell lymphomas (Gusterson, Dyer, O'Hare, Jadayel: unpublished observations).

The wide tissue distribution of BCL-2 indicates that BCL-2 overexpression may be of importance in the primary transformation, not only of B-lymphocytes but also of a wide range of other cells. Overexpression in neoplastic cells is not necessarily linked with translocation, which appears limited to subgroups of malignancies of mature B-cells. The molecular basis for the deregulation of BCL-2 outside of translocation remains unclear. MCF-7 for example retains normal configuration of BCL-2 and there is no evidence for amplification of the BCL-2 gene (Jadayel and Dyer: unpublished observations). Whether the presence or absence of BCL-2 expression is associated with differences in the biological behaviour of the various tumours is not yet known.

Methods of detecting of BCL-2 gene translocation

Translocations involving the BCL-2 gene can be detected by karyotype analysis, DNA blotting with BCL-2 probes and using the polymerase chain reaction (PCR). They have been found in the majority of follicular low grade non-Hodgkin's lymphomas (NHL), in 20–30% of high grade NHL [16], in rare cases of chronic lymphatic leukaemia [17], and 20–30% of cases of Hodgkin's disease [18]. The lack of detection in some follicular NHL with commonly used probes to mbr and mcr does not exclude the involvement of BCL-2, as the BCL-2 product may be detected by antibodies in tissue sections or probes outside the two common breakpoint regions [19]. BCL-2 translocation is not detected in small cleaved follicular centre cell lymphomas (centrocytic lymphoma on Kiel classification). It is also undetected in MALT lymphoma, although the BCL-2 protein can be seen in these tumours by immunostaining. These findings suggest that the increase of BCL-2 protein is a common event in low grade NHL which in the majority of cases is due to t(14;18) and rarely due to other mechanisms.

Transformation of follicular lymphoma to diffuse B-cell NHL with t(14;18) may occur in a proportion of patients. Such transformation has been associated with the acquisition of further cytogenetic abnormalities, such as deregulation of the c-*myc* gene which has been detected on transformation of low grade B-cell lymphoma with t(14;18) [14]. Concurrent deregulation of BCL-2 and c-*myc* is associated with aggressive disease, often of leukaemic phenotype. These data are consistent with the synergistic effects of c-*myc* and BCL-2 observed in gene transfer and mouse transgene experiments.

The clustering of breakpoints within the 3' region of BCL-2 gene has allowed the use of PCR to amplify the t(14;18) breakpoint junction. Gene amplification using the polymerase chain reaction with conventional mbr, mcr and immunoglobulin J_H primers will only detect the subset of t(14;18) breakpoints within the range of primers used. The technique can detect one tumour cell in 10^5–10^6 normal cells [20] (reviewed in [21]), and permits direct DNA sequencing of the amplified material.

The usefulness of BCL-2 in the management of patients with B-cell NHL

Diagnosis

There has been a considerable spin-off from the molecular understanding of BCL-2 translocation. The translocations are specific for each cell clone and can be used as clonal markers. Molecular studies with Southern blotting or PCR can distinguish clonal proliferation of NHL from benign proliferation. However, the absence of translocation is not helpful as it does not exclude the diagnosis of lymphoma. The presence of translocation in 20–30% of high grade tumours means that it is not possible to distinguish low from high grade NHL on molecular probing or immunohistochemistry. BCL-2 can be used as a clonal marker which confirms the persistence of an original clone of cells during histological transformation from low to high grade NHL.

PCR has revealed the possible presence of the t(14;18) in two somewhat unexpected clinical situations. BCL-2-J_H fragments have been detected in normal B-cells in tonsils, reactive lymph nodes and peripheral blood of normal blood donors at a frequency of about 1 cell in 100 000 [22]. These data appear to confirm that the t(14;18) alone is not sufficient to transform individual B-cells and that this event occurs at a relatively high frequency in normal B-cell differentiation.

Positive BCL-2-J_H products have been seen in subgroups of Hodgkin's disease which raises the possibility that B-cell NHL and Hodgkin's disease may share a common pathogenic mechanism [18].

Detection of minimal disease and prognosis

With PCR it is possible to detect up to $1:10^5$–10^6 cells bearing chromosomal translocation. This has been applied to the detection of tumour DNA from lymph nodes, bone marrow and peripheral blood of patients who are apparently disease free. For a detailed review of this application, see [21]. It appears that the presence of circulating lymphoma cells is common in patients with low grade NHL and the detection of small numbers of cells by PCR is not of clear prognostic significance.

The principal value of the PCR method of detecting residual disease is in patients undergoing high dose chemotherapy/autologous bone marrow transplantation to assess the quality of the remission and the harvested bone marrow. Contamination of the harvested bone marrow by low levels of tumour stem cells may contribute to lymphomatous relapse, especially if the lymphoma stem cell resides in the bone marrow. Studies in B-cell precursor acute lymphoblastic leukaemia using immunoglobulin PCR have shown the persistence of low levels of the leukaemic clone for many months after intensive chemotherapy. A similar situation may pertain to B-cell lymphoma [23]. B-cells may be ablated either by in vitro purging of harvested bone marrow or by alternative forms of in vivo therapy administered after attainment of conventional remission. The value of this approach and the contribution of residual lymphoma in harvested marrow to eventual relapse has been shown in a series of patients with B-cell NHL who received autologous bone marrow transplants: in-vitro purging of B-cells was performed using a cocktail of monoclonal antibodies and the re-infused cells were analysed by PCR for the presence of the t(14;18). Persistence of the t(14;18) correlated strongly with eventual relapse [23].

Although the presence of t(14;18) is of no clear prognostic significance in nodular lymphomas, prognostic information can be obtained from the detection of other chromosomal changes that may confer more aggressive biological behaviour. The BCL-2 translocation in high grade lymphoma may predict a relapsing course of disease. Detection of t(14;18) in high grade NHL is associated with worse progression-free survival, although it has no influence on overall survival [24]. In already relapsed high grade lymphoma, the detection of BCL-2 rearrangement seems to confer worse prognosis. These findings, although not entirely consistent, suggest a persistence of BCL-2 bearing clones with indolent behaviour causing clinical recurrences.

Therapy

The excess of BCL-2 protein seems critical in extending and maintaining survival of B-cells. The therapeutic aim should be to reverse the BCL-2 deregulation. At present, this is difficult as the control mechanisms that govern the deregulation are not fully understood. However, a possibility remains to block the BCL-2 production at the translation or transcription level. Early in vitro experiments show successful inhibition of BCL-2 expression by specific antisense oligonucleotides in leukaemic cell culture leading to inhibition of cell growth and survival [25].

Conclusion

The BCL-2 gene has been isolated through its involvement in the t(14;18) translocation observed in B-cell NHL. It now appears that deregulated expression of BCL-2 may be central to the pathogenesis of a much wider range of malignancies. In a subgroup of B-cell NHL, deregulated expression results from chromosomal translocation, but this is not the case in other haemopoietic malignancies nor in some cases of adenocarcinoma. In malignancies of the B-cell lineage, overexpression of BCL-2 in some way prevents programmed cell death. Whether the same is true of other malignancies of other lineages remains to be determined.

The functions of the BCL-2 gene, the mechanism by which it is deregulated in the absence of translocation and the mechanisms of translocation itself remain to be determined. Clinically, the prospect of targeting BCL-2 gene overexpression for therapy remains an interesting idea. At present, molecular knowledge has been used for diagnosis, detection of minimal disease and as a prognostic indicator. The expression of the BCL-2 gene in a wide range of tissues may pose problems. It may be that BCL-2 overexpression leads to the presence of specific BCL-2 peptides on the cell surface expressed in conjunction with major histocompatibility antigens. These may prove to be more suitable targets than either the gene or its mRNA.

References

1   Alt FW, Blackwell TK, Yancopoulos GD. Development of the primary antibody repertoire. Science 1987, **237**, 1079–1987.
2   Tycko B, Sklar J. Chromosomal translocations in lymphoid neoplasia: a reappraisal of the recombinase model. Cancer Cells 1990, **2**, 1–8.
3   Boehm TLJ, Rabbitts TH. The human T-cell receptor genes are targets for chromosomal abnormalities in T-cell tumours. FASEB J 1989, **3**, 2344–2359.
4   Seto M, Jaeger U, Hockett RD et al. Alternative promoters and exons, somatic mutation and transcriptional deregulation of the BCL-2 Ig fusion gene in lymphoma. EMBO J 1988, **7**, 123–131.
5   Chen-Levy Z, Nourse M, Cleary ML. The BCL-2 candidate proto-oncogene product is a 24-kilodalton integral-membrane protein highly expressed in lymphoid cell lines and lymphomas carrying the t(14;18) translocation. Mol Cell Biol 1989, **9**, 701–710.
6   Hockenbery D, Nunez G, Milliman C, Schreiber RD, Korsmeyer SJ. BCL-2 is an inner mitochondrial membrane protein that blocks programmed cell death. Nature 1990, **348**, 334–336.
7   Hockenbery DM, Zutter M, Hickey W, Nahm M, Korsmeyer SJ. BCL-2 protein is topographically restricted in tissues characterized by apoptotic cell death. Proc Natl Acad Sci USA 1991, **88**, 6961–6965.
8   Zutter M, Hockenbery DM, Silverman GA, Korsmeyer SJ. Immunolocalisation of the BCL-2 protein within haemopoietic neoplasms. Blood 1991, **78**, 1062–1068.
9   Reed JC, Cuddy M, Slabiak T, Croce CM, Nowell PC. Oncogenic potential of BCL-2 demonstrated by gene transfer. Nature 1988, **336**, 259–261.
10 Nunez G, London L, Hockenbery D, Alexander M, McKearn JP, Korsmeyer SJ. Deregulated BCL-2 gene expression selectively prolongs survival of growth factor-deprived hemopoietic cell lines. J Immunol 1990, **144**, 3602–3610.
11 Vaux DL, Cory S, Adams JM. BCL-2 gene promotes haemopoietic cell survival and cooperates with c-*myc* to immortalize pre-B cells. Nature 1988, **335**, 440–442.
12 McDonnell TJ, Deane N, Platt FM, et al. BCL-2 immunoglobulin transgenic mice demonstrate extended B cell survival and follicular lymphoproliferation. Cell 1989, **57**, 80–88.
13 McDonnell TJ, Korsmeyer SJ. Progression from lymphoid hyperplasia to high-grade malignant lymphoma in mice transgenic for the t(14;18). Nature 1991, **349**, 254–256.

14 de Jong D, Voetdijk BMH, Beverstock GC. Activation of the c-*myc* oncogene in a precursor B-cell blast crisis of follicular lymphoma, presenting as a composite lymphoma. N Engl J Med 1988, **318**, 1373–1378.

15 Cotter FE, Price C, Zucca E, Young BD. Direct sequence analysis of 14q+ and 18 q– chromosome junctions in follicular lymphoma. Blood 1990, **76**, 131–135.

16 Yunis JJ, Oken MM, Kaplan ME, Ensrud KM, Howe RB, Theologides A. Distinctive chromosomal abnormalities in histologic subtypes of non-Hodgkin's lymphoma. N Engl J Med 1982, **307**, 1231–1236.

17 Adachi M, Tefferi A, Greipp PR, Kipps TJ, Tsujimoto Y. Preferential linkage of BCL-2 to immunoglobulin light chain gene in chronic lymphocytic leukemia. J Exp Med 1990, **171**, 559–564.

18 Stetler-Stevenson M, Crush-Stanton S, Cossman J. Involvement of the BCL-2 gene in Hodgkin's disease. J Natl Cancer Inst 1990, **82**, 855–858.

19 Weiss LM, Warnke RA, Sklar J, Cleary ML. Molecular analysis of the t(14;18) chromosomal translocation in malignant lymphomas. N Engl J Med 1987, **317**, 1185–1189.

20 Lee MS, Chang KS, Cabanillas F, Freireich EJ, Trujillo JM, Stass SA. Detection of minimal residual cells carrying the t(14;18) by DNA sequence amplification. Science 1988, **237**, 175–178.

21 Cotter FE, Price C, Young BD, Lister TA. Minimal residual disease in leukaemia and lymphoma. Ann Oncol 1990, **1**, 167–170.

22 Limpens J, de Jong D, van Krieken JHJM, et al. BCL-2/JH rearrangements in benign lymphoid tissues with follicular hyperplasia. Oncogene 1991, **6**, 2271–2276.

23 Gribben JG, Freedman AS, Neuberg A. Immunologic purging of marrow assessed by PCR before autologous bone marrow transplantation for B-cell lymphoma. N Engl J Med 1991, **325**, 1525–1533.

24 Offit K, Koduru PRK, Hollis R, et al. 18q21 rearrangement in diffuse large cell lymphoma: incidence and clinical significance. Br J Haematol 1989, **72**, 178–183.

25 Reed JC, Stein C, Subasinghe C, et al. Antisense-mediated inhibition of BCL2 protooncogene expression and leukemic cell growth and survival: comparisons of phosphodiester and phosphorothioate oligodeoxynucleotides. Cancer Res 1990, **50**, 6565–6570.

Molecular Biology for Oncologists
Edited by John Yarnold, Michael Stratton, Trevor McMillan
© *1993, Elsevier Science Publishers B.V. All rights reserved*

CHAPTER 6

Identification and significance of EGFR and erbB2 overexpression

Donal P. Hollywood and William J. Gullick

ICRF Oncology Group, Hammersmith Hospital, London, UK

Growth factors and growth factor receptors act as signalling molecules

Control of cell proliferation is dependent on a series of interlinked signalling mechanisms that mediate intracellular and intercellular communication. Collectively these processes are known as mitogenic signal transduction pathways. Many components are now recognised including growth factors, growth factor receptors, membrane-associated and cytoplasmic signal transduction molecules and complex arrays of nuclear proteins including transcription factors and tumour suppressor gene products. The specific role of growth factor receptors is the transmission of growth regulatory "information" in the form of extracellular growth factors to the next signalling component in the intracellular environment.

A diverse family of growth factors and growth factor receptors control cell growth

Growth factors exist in several forms including polypeptides (EGF, PDGF, TGFβ), oligopeptides (bradykinin, bombesin) and steroid molecules (oestrogens, progestagens, androgens) [1,2]. In general, each factor recognises a single type of receptor. During growth stimulation, two classes of growth factor action are considered to operate [3]. "Competence" factors such as epidermal growth factor (EGF), platelet-derived growth factor (PDGF) and fibroblast growth factors (FGF) initiate the critical move from G_0 to G_1. Intermediate phases of G_1 then require the combined action of "competence" and "progression" factors such as insulin and insulin-like growth factor (IGF-1), whereas late G_1 with a commitment to DNA synthesis and passage through G_1/S is dependent on the action of the "progression" factors alone.

Several types of growth factor receptor are also recognised including trans-membrane protein tyrosine kinases, haemopoietic growth factor receptors, seven-pass transmembrane glycoproteins and steroid hormone receptors [3]. The trans-membrane protein tyrosine kinases are the principal receptor type implicated in neoplastic transformation and several "types" or "classes" are now recognised. The members of the type 1 subgroup (EGFR, c-erbB-2 and c-erbB-3) are the subject of this review. Haemopoietic growth factor receptors are also transmembrane glycoproteins and different members of this family bind the interleukins (IL-2, IL-3, IL-4, IL-6, IL-7), granulocyte-macrophage colony stimulating factor (GM-CSF), G-CSF and erythropoietin. They do not possess intrinsic tyrosine kinase activity; however, tyrosine-phosphorylated proteins do appear after ligand-receptor binding, for example p56^{lck} following IL-2 binding. The seven-pass transmembrane recep-tors bind a variety of neurotransmitters (bombesin, bradykinin) that may also act as mitogens. The steroid receptor superfamily differs from the other types of growth factor receptor. Firstly, the receptors exist in the cytoplasm and nucleus as a complex with other intracellular proteins (the heat shock proteins (hsp), hsp90, hsp70 and hsp56). Secondly, the ligands are lipophilic and are capable of travers-

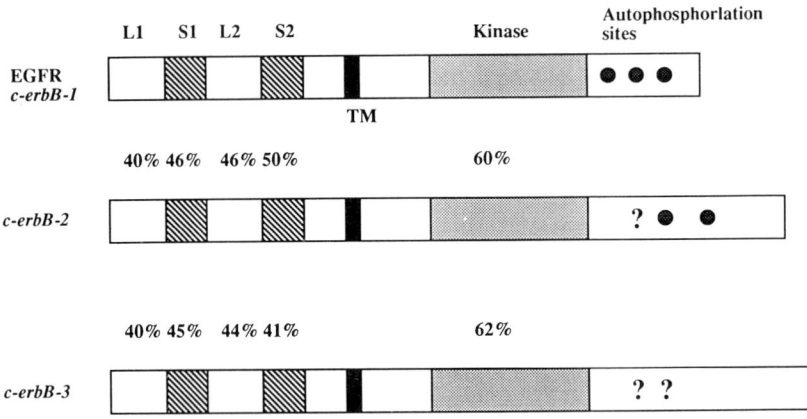

Figure 1. Schematic representation of amino acid sequence similarities and functional domains shared by type 1 growth factor receptors. The percentage homology represents the comparison of c-*erb*B3 with the other two members of the class.

ing the extracellular membrane independent of receptor. Thirdly, the resultant ligand–receptor complex is a transcription factor that binds to specific DNA sequences involved in the regulation of steroid-dependent gene expression.

Three transmembrane protein kinases constitute the type 1 growth factor receptor family

The type 1 growth factor receptor family consists of three genes: c-*erb*B1 (EGFR), c-*erb*B2 and c-*erb*B3. Each gene encodes a distinct protein that functions as a transmembrane protein tyrosine kinase. Considerable homology exists between specific functional regions present in each of the receptor molecules, in particular the kinase domain (Figure 1), suggesting that all are derived from a common ancestral precursor [4].

Different receptor nomenclature exists for each of the type 1 growth factor receptors

Alternative names for *erb*B2 include HER-2, *neu* and NGL. *neu* more correctly describes the rat homologue of human c-*erb*B2; however, in the USA it is occasionally used to describe the human gene and protein product. *erb*B1 describes the gene encoding the epidermal growth factor receptor (EGFR), and HER-3 is an alternative form of nomenclature for *erb*B3.

Abnormalities of EGFR and *erb*B2 overexpression can be detected at the DNA, RNA and protein levels

Abnormalities of the type 1 growth factor receptors can be detected at three levels in human tumours: protein, mRNA and DNA. Assessment of protein levels has been most frequently performed using the techniques of radioimmunoassay, Western blotting and immunocytochemistry. Each approach has advantages and disadvantages so that frequently they are used in combination. Radioimmunoassay, for example, offers a quantitative measure of protein expression but is dependent on the proportion of normal cells in the sample biopsy. In contrast, immunocytochemistry is a semi-quantitative measure of protein expression but does allow the cellular distribution of receptors in a specific tissue to be recognised. This is of particular importance in certain tumours where protein expression is not uniform, for example the expression of erbB2 in stomach adenocarcinoma is frequently patchy. Immunocytochemistry can also highlight the subcellular distribution of protein. For example, the type 1 growth factor receptors classically demonstrate membrane immunoreactivity; however, in some tissues and malignancies, a variable degree of cytoplasmic staining is observed. This is thought to represent intracellular trafficking of newly synthesized receptor and/or the internalisation of ligand-bound receptor.

Although mRNA levels are more difficult to assess due to the inherent instability of mRNA, several approaches have been used to explore overexpression at this level. Northern blotting, RNase protection assays, in situ hybridization and more recently semi-quantitative PCR have all confirmed a spectrum of EGFR and *erb*B2 mRNA levels in many different malignancies. Analysis of the rate of gene expression has required more sophisticated in vitro approaches including nuclear run-on experiments, mRNA half life measurement and the direct analysis of gene regulatory regions using a combination of promoter deletion studies, DNase 1 footprinting and gel retardation assays.

Examination of growth factor receptor genomic DNA is directed at the recognition of gene copy number (single copy or amplification) and/or gene rearrangement. Southern blotting, DNA slot blotting and PCR approaches (using mutant specific oligonucleotides or designed restriction length polymorphisms (RFLP) to detect point mutations) have all been used in this strategy.

Several ligands bind to the type 1 growth factor receptors

EGFR has four structurally related ligands: epidermal growth factor (EGF), transforming growth factor α (TGFα), heparin binding EGF and amphiregulin, all of which are specific to EGFR (Table 1) [5,6]. Several ligands for *erb*B2 have recently been described including heregulin (HRG) of which four forms are recognised (proHRG-α, proHRG-β1, proHRG-β2, proHRG-β3), gp30, p75, neu differentiation factor (NDF) and neu protein-specific activating factor (NAF) [7–11]. All are specific for the erbB2 receptor apart from gp30 which has been reported to interact with both erbB2 and EGFR [12]. The ligands for the remaining type 1 receptor, *erb*B3, remain uncharacterised at present.

A model for type 1 growth factor receptor structure and activation

Mechanistic models of the type 1 receptors suggest the presence of three major regions (Figure 2):

(i)   the extracellular domain responsible for ligand binding;
(ii) the hydrophobic transmembrane region which appears to function in both receptor anchorage in the cell membrane and the dimerisation process; and

(iii) the intracellular domain with its catalytic kinase function and specific amino acid residues that act as recruitment sites for second messengers and juxta-membrane "signal transfer particle" formation.

The glycosylated extracellular region is folded into four domains: L1, L2, S1 and S2 (Figure 1). L1 and L2 are believed to mediate ligand binding whereas S1 and S2 are important in the structure of the extracellular region. Although detailed conformational features of type 1 GFRs await crystallographic analysis, a four domain model for the organisation of the extracellular portion of the EGFR has recently been proposed on the basis of functional analysis of the ligand binding

Table 1. Ligands for EGFR.

Growth factor	Mature protein	mRNA (kb)	Additional information	Normal tissue distribution
EGF	53 amino acids (following proteolytic cleavage from a 1217 amino acid precursor)	4.75	Precursor structure resembles a transmembrane protein	Submandibular salivary gland, gastric epithelium, duodenal Brunner's glands, sweat glands, breast
TGFα	50 amino acids (following proteolytic cleavage of a 160 amino acid cell surface precursor)	4.8	Larger TGFα intermediates exist	Widespread distribution including skin keratinocyte, bronchus, intestine, renal tubule
Amphiregulin	78 or 84 amino acids (following proteolytic cleavage of a 252 amino acid precursor)	1.4	Keratinocyte autocrine factor (KAF) is a closely related or identical peptide	Placenta, testis, ovary, pancreas, colon, breast
Heparin binding EGF-like growth factor	Mature protein is >75 amino acids (208 amino acid primary translation product)	2.5	Diphtheria toxin receptor is probably identical	Expression pattern is not yet known; present in macrophages

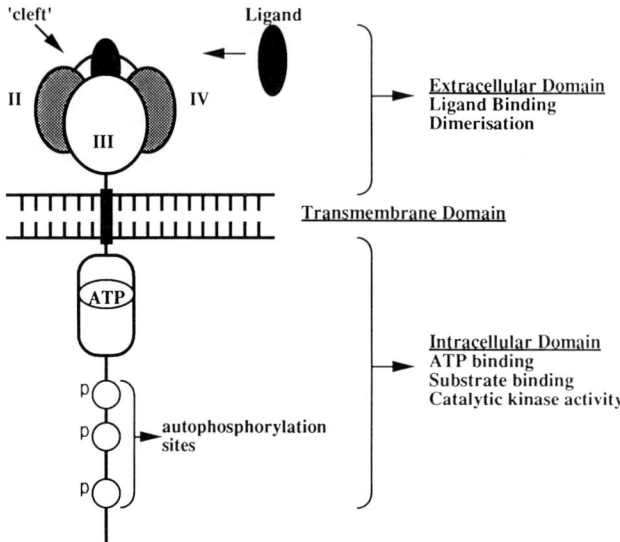

Figure 2. Schematic representation of the structure of EGFR based on a model proposed by Ullrich and Schlessinger. The extracellular region consists of a four domain structure creating a "cleft" for ligand binding.

domain using [^{125}I]EGF affinity labelling experiments, chimaeric EGF receptors and preliminary electron microscopic characterisation of the extracellular domain [13]. The folded extracellular domain is thought to generate a "cleft" for ligand binding (Figure 2).

The transmembrane region consists of a short run of hydrophobic amino acids forming an α-helical bridge across the extracellular membrane lipid bilayer. The relative importance of this structure is a point of debate. In the model proposed by Ullrich and Schleshinger, the transmembrane region plays a passive role in kinase activation and signal transduction, the main function being to anchor the receptor in the cell membrane in the appropriate orientation [13]. In support of this hypothesis, the authors cite the ability to switch EGFR, erbB2, PDGF, insulin and IGF-1 transmembrane regions without altering the capacity for kinase activation. In addition, the possibility of conformational change in the transmembrane region following ligand binding has not been excluded. A more dynamic role for the transmembrane region is proposed in the alternative model proposed by Sternberg and Gullick [14]. Recognising the critical importance of the single activating mutation in the transmembrane region of the oncogenic neu protein (resulting in a valine to glutamic acid substitution), the second model proposes that increased inter-molecular hydrogen bonding exists between adjacent transmembrane regions in a mutant *neu* dimeric complex. The net effect of the enhanced intermolecular bonding in mutant *neu* is oligerimisation and constitutive receptor activation in the absence of ligand. Similar mutations also activate the human erbB2 receptor, the insulin receptor and the *Drosophila* EGF receptor, suggesting that transmembrane interactions also occur in these systems.

The cytoplasmic juxtamembrane region also appears to be important in receptor function. In particular, it contains amino acid residues that function in the down-regulation of EGFR/erbB2 activity following the activation of other receptor types in a process known as receptor "transmodulation" (Figure 3). For example, activation of PDGF receptor or bombesin receptors or alternatively the direct stimulation of protein kinase C (PKC) results in a reduction in the number of high affin-

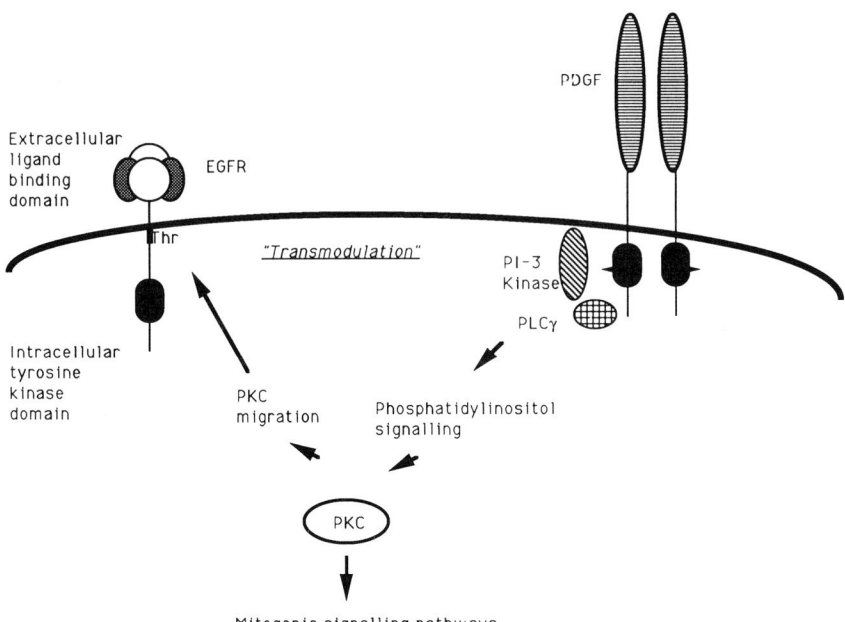

Figure 3. Activation of other transmembrane tyrosine kinase receptors (e.g. PDGF) leads to the activation of specific intracellular pathways including the stimulation and membrane localisation of protein kinase C (PKC, a serine threonine kinase). Transmodulation and down-regulation of a non-ligand bound receptor (e.g. EGFR, *erb*B2) by PKC may follow the phosphorylation of specific threonine residues in the cytoplasmic juxtamembrane region (e.g. Thr 654 of EGFR).

ity EGF binding sites. In this case down-regulation of EGFR is probably dependent on a negative feedback mechanism effected by PKC-induced phosphorylation of the threonine residue (position 654) in the juxtamembrane region [15].

The kinase domain is the most highly conserved region between all receptor tyrosine kinases (Figure 1). It is absolutely essential for signal transduction and the cellular responses that lead to mitogenesis and cellular transformation. Removal of kinase activity from EGFR results in an inability to stimulate Na^+/H^+ exchange, calcium influx, phosphatidylinositol signalling, the induction of the immediate early genes c-*myc* and c-*fos*, S6 ribosomal subunit phosphorylation and DNA synthesis. Kinase activation leads to the recruitment of several intracellular signalling molecules to the intracellular domain thereby altering their subcellular distribution and activation status. In the "signal transfer particle" model proposed

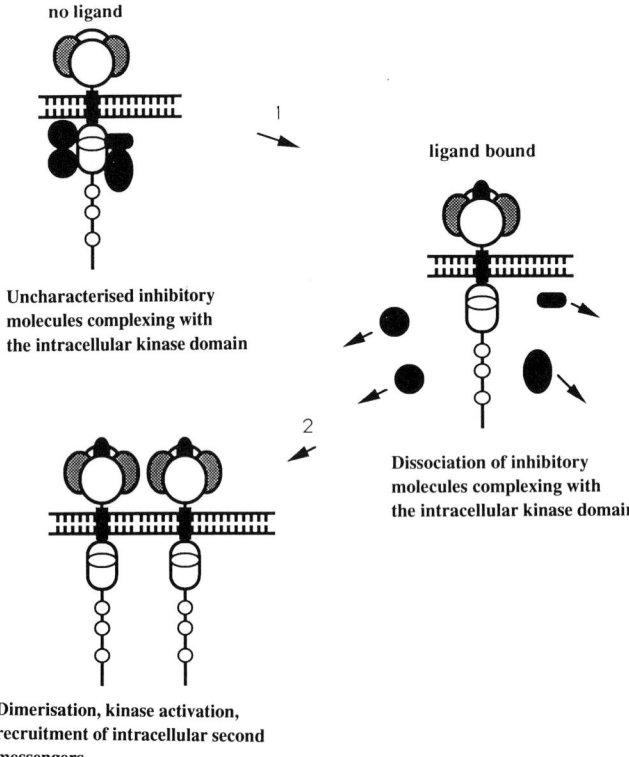

Figure 4. Schematic representation of transition from a non-ligand bound receptor to a ligand-bound dimeric complex based on a model proposed by Ullrich and Schlessinger.

by Ullrich and Schlessinger, the creation of these recruitment sites may follow the dissociation of other uncharacterised molecules from the activated kinase domain (Figure 4) [13]. Interestingly, kinase activity is also essential for the intracellular targeting of EGFR to lysosomes and subsequent intracellular degradation; however, it is unclear if this reflects the presence of lysosomal targets that are kinase-dependent or whether the intracellular degradative pathway only recognises the autophosphorylated dimeric form of the receptor.

The carboxy terminal tails of EGRF and erbB-2 contain several autophosphorylation sites which may limit the interaction with non-receptor targets (signal transfer particle formation) by offering an alternative site for receptor kinase action. In support of this, the alteration of autophosphorylation sites or truncation of the C-terminal tail both lead to a reduction in the EGF concentration required to initiate a mitogenic response. This "competitive model" has to be reconciled with the aforementioned "signal transfer particle" model where the phosphorylated residues act as recruitment sites for several intracellular second messengers. To date this paradox remains unresolved.

EGFR and erbB-2 activate a series of intracellular signal pathways

Growth factor receptors are the origin of a series of interlinked intracellular signal transduction pathways that collectively enable the passage of information from the extracellular milieu to intracellular second messenger molecules [16].

The unoccupied type 1 tyrosine kinase receptor exists as a monomer and is essentially inactive. Binding of ligand to the cleft-like region in the extracellular domain results in conformational changes which in turn facilitate the dimerisation of ligand-bound receptors. Juxtaposition of the intracellular kinase domains leads to kinase activation and the auto- or transphosphorylation of specific carboxy-terminal tyrosine residues. A subset of these phosphorylated amino acids then acts as recruitment sites for the next tier of intracellular second messengers (Figure 5). These include PI-3 kinase and PLCγ (phosphatidylinositol signalling), ras-GAP (guanine nucleotide-based signalling) and others such as c-raf (an intracellular serine-threonine kinase) which may interact with EGFR through uncharacterised intermediary molecules [16,17] . It is highly probable that some of these molecules (c-raf, PI-3 kinase and ras-GAP) are phosphorylated by the EGFR or erbB2 kinase resulting at first in the stimulation of key cytoplasmic signal transduction pathways but eventually in the expression of a large number of genes that constitute the "immediate early gene response" (e.g: c-*fos*, c-*jun*, c-*myc* and ornithine decarboxylase) [18]. Collectively these genes initiate a multitude of cellular processes that together result in cell cycle progression and cell division [19].

Different tyrosine kinase growth factor receptors possess varying abilities to activate the various second messengers so that slight differences in second messenger generation are evident at this level of signalling. In addition to receptor homodimerisation, it is also clear that EGFR and erbB2 can heterodimerise so that different cellular consequences may also follow this alternative form of receptor oligomerisation. In general, however, the signalling consequences share considerable similarities creating an apparent redundancy in the signalling networks (i.e. different signalling influences appear to achieve the same cellular response). Additional unrecognised membrane events may coexist with "signal transfer particle formation" and allow discrimination between different ligand-receptor interactions. Alternatively, specificity may have a greater dependency on downstream signalling components such as the mitogen-activated (MAP) kinases (also known as extracellular signal-regulated kinases or ERKs) and ribosomal S6 kinases (RSKs) [20].

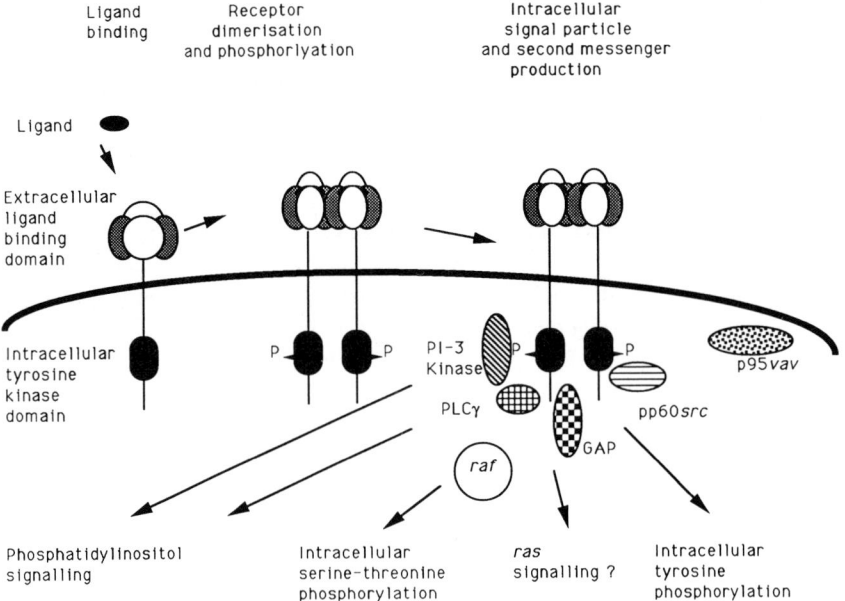

Figure 5. Activation of type 1 transmembrane tyrosine kinase receptors. Ligand binding results in receptor dimerisation, cytoplasmic domain kinase activation, and auto/ transphosphorylation of receptors in the dimeric complex. Secondary messengers are recruited to the intracellular phosphorylated residues leading to their phosphorylation on tyrosine and the activation of specific intracellular pathways.

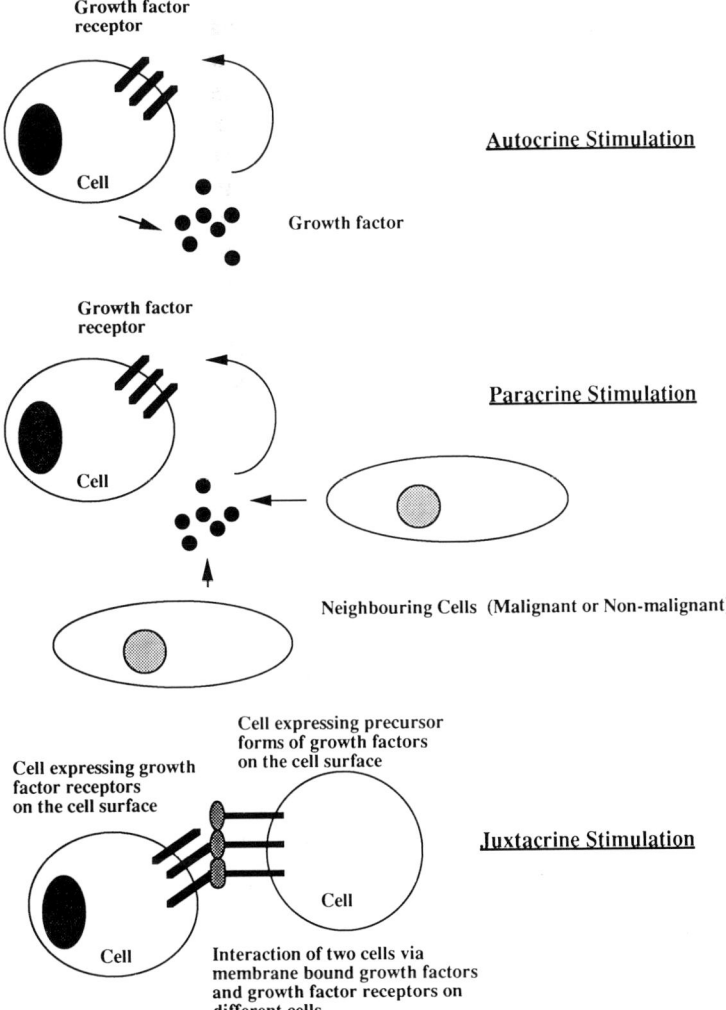

Figure 6. Autocrine, paracrine and juxtacrine models of growth factor production.

Interactions between growth factors and their receptors: autocrine, paracrine and juxtacrine models

Autocrine, paracrine and juxtacrine models have been proposed to explain the in-vivo expression and action of growth factor receptors and their ligands (Figure 6).

In the autocrine system, receptors and their ligands are co-expressed by the same cell creating an autostimulatory or auto-inhibitory loop and a resultant stimulation or inhibition of growth [21]. Alternatively, in the paracrine model, receptor and ligand are expressed by different cells with ligand–receptor interaction resulting from the close proximity of each cell type. The juxtacrine model proposes an interaction between membrane-bound receptors and membrane-bound growth factors on adjacent cells (EGF, TGFα and amphiregulin are initially synthesised in precursor forms that permit membrane localisation) [22]. Each of the systems is believed to exist in normal and malignant tissues. Breast carcinoma for example has been found to express both EGFR and TGFα suggesting a possible autocrine pathway [23].

EGFR and erbB2 are expressed in a wide range of normal tissues

The normal tissue distribution of each of the type 1 GFRs has been described in both adult and foetal tissue [24]. EGFR expression is seen in cells derived from all three germ layers but at particularly high levels in epithelial tissues. erbB2 expression is also observed in a wide variety of tissues; however, expression is particularly associated with secretory epithelium. Interestingly the sites of erbB3 expression also include tissue sites where EGFR or erbB2 are not usually detected (adrenal cortex, pancreatic islets, testis, ovary, central nervous system and spinal basal ganglia) [24].

TGFα/EGFR interactions have been implicated in wound healing, cell migration, angiogenesis and bone resorption. EGF/EGFR interactions have been studied in a number of animal systems and recognised consequences include eyelid opening and tooth eruption in mice, skin development in foetal lambs and lung maturation in foetal rabbits. In man, an increase in EGF protein expression has been observed in regenerating gastroduodenal epithelial surfaces following peptic ulceration suggesting a role in epithelial repair and regeneration [25]. The physiological roles of the other EGFR ligands (amphiregulin and heparin-binding EGF), and the recently characterised erbB2 ligands are not yet clear. Although the normal tissue distribution of erbB3 has recently been described, full characterisation of function awaits the recognition of specific ligands.

Type 1 growth factor receptors: the evidence for a role in cancer

A biological role for EGFR and erbB2 in human neoplasia is based on several observations:

(i)   Overexpression of EGFR and *erb*B2 mRNA/protein is observed in a range of human epithelial tumours as a result of gene amplification, transcriptional up-regulation or a combination of both processes [26]. In contrast protein overexpression of the same order is not documented in normal tissue.

(ii) Transfection of EGFR into immortalised NIH3T3 fibroblasts in the presence of an activating ligand results in a transformed phenotype [27]. Similarly transfection of erbB-2 into NIH3T3 fibroblasts is also transforming.

(iii) Transgenic animals with the mutated rat *erb*B2 gene (*neu*) under the control of the MMTV promoter develop bilateral mammary tumours with a very high frequency [28].

(iv) Antibodies directed against the erbB2 protein limit the growth of malignant cells in vitro and in vivo [29].

Different mechanisms result in EGFR and *erb*B2 overexpression in human cancers

Overexpression of EGFR and *erb*B2 has been documented in a wide variety of human tumours (Tables 1 and 2) [26]; however, the mechanisms underlying this phenomenon appear to vary since increases in mRNA and protein may co-exist with either a normal single copy gene or alternatively an amplified gene (Figure 7) [30]. Recent experimental evidence has highlighted some of the processes that lead to gene duplication. Analysis of fibroblasts from patients with Li–Fraumeni syndrome has suggested that one of the normal functions of p53 (a tumour suppressor gene) is the limitation of cellular events that can result in gene amplification [31,32]. In particular, wild-type p53 expression is associated with a "normal" G1 arrest during cell cycle progression. In contrast, mutant p53 may allow inappropriate S phase entry with an enhanced probability of abnormal chromosomal breakage, segregation and reassembly. In support of this hypothesis, the introduction of wild-type p53 into cells "at risk" of amplification prevented the development of amplification. It is clear, however, that p53 is not the only participant in this process since other cell types with wild type p53 are also capable of developing amplification. Beyond the events that initiate amplification, the regulatory mechanisms that operate to allow selective overexpression of specific genes within amplicons are not yet clear. Amplification of EGFR, for example, frequently involves large regions of DNA of about 1000 kb encompassing several genes [33], but only a subset of these are overexpressed.

It is clear that other mechanisms must also operate to explain the increased growth factor receptor expression evident in a subset of tumours with a single copy of the EGFR or *erb*B2 genes. Several possibilities exist including transcriptional up-regulation, enhanced post-transcriptional stability of mRNA or changes in post-translational processing leading to enhanced protein half-life (Figure 7). Recent work analysing the mechanisms leading to *erb*B2 overexpression in human breast adenocarcinoma cell lines has suggested that an increase in gene transcrip-

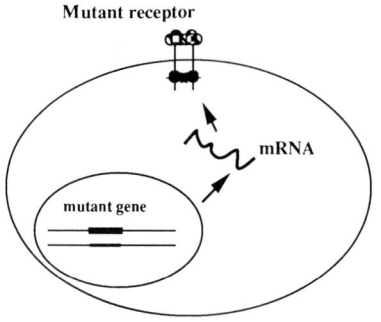

Formation of Mutant Receptors

eg deletion of part of the extracellular domain of EGFR in human glial tumours

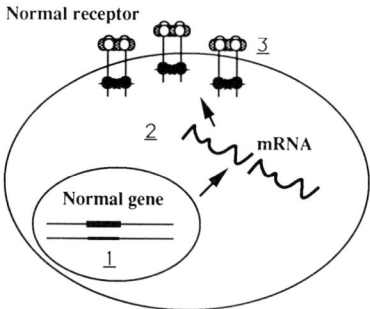

Overexpression of Normal Receptor with a Single Copy Normal Gene

eg: erbB2 overexpression in 10-15% breast carcinoma

1. Transcriptional deregulation

2. Post-transcriptional
   increased mRNA half-life ?
   altered mRNA transport ?

3. Post-translational
   altered protein half-life ?

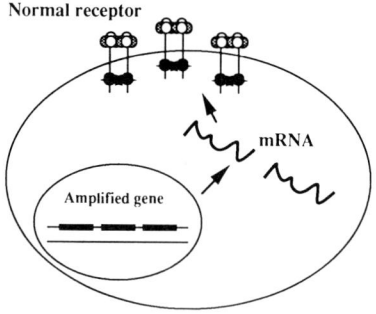

Overexpression of Normal or Mutant Receptors as a result of Gene Amplification

1. Amplicon may be very large eg:1Mb

2. Gene may also be mutated (eg EGFR amplification in human glial tumours) or normal (eg *erbB-2* amplification in human breast carcinoma)

Figure 7.

tion is the principal phenomenon underlying overexpression from a single gene copy. In a subset of patients, for example 10–15% of breast adenocarcinomas that do not demonstrate gene amplification, this form of overexpression may also exist, with the gene "switched on" at a higher rate. This mechanism may also contribute to *erb*B2 overexpression in other non-mammary tumours and where EGFR overexpression exists in the absence of amplification.

Amplification of the type 1 growth factor receptor genes is observed in many human cancers

Amplification of EGFR is frequently observed in glioblastoma multiforme (40%), but is uncommon in bronchogenic tumours (10%), squamous cell carcinomas of the head and neck (10%) and oesophageal carcinoma (8–14%), and quite rare in breast adenocarcinoma (2%), bladder transitional cell carcinoma (<5%) and gastric adenocarcinoma (<5%) [26]. Amplification of *erb*B2 is observed in a more restricted range of human tumours, principally those derived from secretory epithelium including breast adenocarcinoma, gastric adenocarcinoma, ovarian adenocarcinoma and bladder transitional cell carcinoma, where the prevalence of amplification is about 20% in each tumour type. In addition, *erb*B2 amplification is occasionally observed in bronchogenic adenocarcinoma, and rarely in renal adenocarcinoma (<5%), colorectal adenocarcinoma (<5%), pancreatic carcinoma (<5%) and salivary gland adenocarcinoma (<5%). To date, only a small series of tumours and cell lines has been examined for *erb*B3 gene status with no evidence of gene amplification; however, mRNA expression was noted in cultured keratinocytes, melanocytes, breast adenocarcinoma and sarcoma cells [34,35]. It should be noted that with both EGFR and *erb*B2 the relative frequency of amplification as the mechanism underlying overexpression varies considerably between different tumours (Tables 2 and 3).

Clinical consequences of EGFR/*erb*B2 overexpression

Experimental data strongly supports the theory that overexpression of EGFR or *erb*B2 is of profound biological importance in the genesis and progression of certain human tumours. The hypothesis that EGFR and erbB2 subserve a critical role is further supported by the correlation of receptor overexpression to other adverse tumour characteristics and to poorer patient survival (Table 2). For example with regard to other cytopathological characteristics, EGFR overexpression in transitional cell carcinoma of the bladder correlates with degree of invasion and undifferentiated histology. In breast carcinoma, EGFR expression shows an inverse relationship with oestrogen receptor expression. The relationship between erbB2 protein expression and other cytopathological tumour characteristics has been most intensively investigated in breast adenocarcinoma (for a recent review see [36]).

Table 2. Overexpression of EGFR in human malignancies.

Tumour type	Incidence of EGFR overexpression (%)	Clinical consequence of overexpression
Glioblastoma multiforme (Grade 4 astrocytoma)	40 (principally gene amplification); frequent gene rearrangement leading to expression of a truncated EGFR)	
Lung		
Bronchogenic squamous cell carcinoma (SCC)	80	Overexpression associated with poor prognosis in SCC
Adenocarcinoma	55	
Large cell carcinoma	70	
Bladder transitional cell carcinoma		Overexpression associated with poor prognosis?
pT1 - pT3	30–88	
Breast adenocarcinoma	30 (amplification 2%)	Overexpression correlates with reduced relapse-free survival and overall survival; inverse relationship between oestrogen receptor and EGFR expression
Gastric adenocarcinoma	35 (amplification 3%)	Overexpression more common in advanced disease
Oesophageal carcinoma	70 (amplification 8–14%)	Overexpression associated with poor survival ?
Colonic adenocarcinoma	Rare	Overexpression possibly associated with higher grade tumours
Head and neck (squamous cell carcinoma)	20–30 (amplification 10–20%)	
Ovarian adenocarcinoma	35–40	
Endometrial adenocarcinoma	30	
Cervix, vulva, vaginal squamous cell carcinoma	50–65	
Sarcomas	51	
Schwannoma, ependymoma, medulloblastoma, meningioma, pituitary adenoma	Rare	

Several large studies have suggested a consensus where erbB2 overexpression is correlated with higher tumour grade and nuclear grade, and inversely with oestrogen receptor expression [37,38]. In contrast, no specific association is evident with nodal status or primary tumour size, and only statistically non-significant trends are generally recognised when erbB2 is compared with various indices of cell proliferation (ploidy, S-phase fraction, Ki 67 staining) [39].

In a number of tumours including breast adenocarcinoma, bronchogenic squamous cell carcinoma, gastric adenocarcinoma and bladder transitional cell carcinoma, overexpression of EGFR correlates with reduced relapse-free survival and overall survival [26]. Overexpression of erbB2 in node-positive breast carcinoma and ovarian adenocarcinoma is also predictive of reduced relapse-free intervals and overall survival [37]. In node-negative breast carcinoma and most of the other tumour types where erbB2 overexpression has been documented, trends towards poorer survival have also been noted; however, formal testing of this hypothesis within larger trials remains to be established. The prognostic stratification permitted by these studies suggests that high risk groups may be more precisely selected for specific therapies that might afford greater disease control and improved survival. For example, overexpression of either EGFR (>10 fmol/mg membrane protein) or erbB2 appears to predict a poorer therapeutic outcome, following standard endocrine or chemotherapeutic approaches, in both ER+ and ER– primary breast adenocarcinoma [40,41]. Ultimately it is the elucidation of their biological role that may enable the derivation of therapeutic strategies that directly address the growth factor receptor abnormality.

New therapies may interrupt abnormal EGFR/erbB2 signalling

The proposed role of EGFR and *erb*B2 in human malignancies has lead to the intensive study of a number of therapeutic strategies that aim to abrogate the mitogenic signalling pathway activated following ligand-receptor interaction [42,43]. These include:

  (i)   monoclonal antibodies (e.g. MAb 4D5);
 (ii)   growth factor antagonists (e.g. pentosan polysulphate);
(iii)   receptor dimerisation inhibitors;
 (iv)   protein tyrosine kinase inhibitors (Bryostatin, tryphostins);
  (v)   antisense oligonucleotides; and
 (vi)   transcriptional inhibitors.

A large number of monoclonal antibodies against the erbB2 protein have been selected and examined for anti-tumour cytotoxicity [29,44]. Several show in

Table 3. Overexpression of *erb*B2 in human malignancies.

Tumour type	Incidence of *erb*B2 overexpression (%)	Clinical feature of overexpression
Breast adenocarcinoma Invasive ductal	20–30 (amplification 20%, no evidence of gene mutations rearrangement extremely rare <1%)	Overexpression more common in high grade, and inflammatory tumours; inverse correlation with oestrogen and progesterone receptors associated with poorer relapse-free survival and overall survival, and a lesser response to endocrine and/or cytotoxic chemotherapy
Invasive lobular	Rare	
Comedo/large cell DCIS	95–100	
Ovarian adenocarcinoma	20–30 (amplification 10–20%)	Overexpression in serous, mucinous and endometrioid but rare or not present in clear cell variant Overexpression associated with early relapse and reduced survival
Gastric adenocarcinoma	20 (principally gene amplification, occasionally gene rearrangement)	Overexpression more common in advance disease Immunohistochemical overexpression may be patchy suggesting that it may be a late phenomenon Overexpression associated with a poor prognosis
Pancreatic adeno-carcinoma	20 (amplification 2%)	Uniform immunohistochemical staining
Bladder transitional cell carcinoma	30	Overexpression more common in poorly differentiated tumours
Endometrial adeno-carcinoma	10	
Non-small cell lung carcinoma, colorectal adenocarcinoma, renal adenocarcinoma, salivary gland adenocarcinomas, CNS/glial tumours, sarcomas	Rare	Apart from bronchogenic non-small cell carcinoma, overexpression of *erb*B2 has not been reported in tumours of non-secretory epithelium
Head and neck tumours, bronchogenic tumours	Overexpression very rare or not present	
Cervix, vulva, vaginal squamous cell carcinoma, oesophageal carcinoma		Overexpression is seen in cervical adenocarcinoma

vitro inhibition of breast and ovarian carcinoma cell growth and in vivo growth retardation of ovarian carcinoma xenografts in nude mice. Indeed the anti-erbB2 monoclonal antibody, MAb 4D5, has also demonstrated in vivo synergism with the cytotoxic chemotherapeutic agent cisplatinum.

Pentosan polysulphate is a polyanionic compound that binds many extracellular growth factors in a relatively non-specific manner. Despite this wide range of action, pre-clinical studies have demonstrated in vitro and in vivo retardation of mammary carcinoma cell growth, and the compound has recently entered phase 1 clinical trial examination in the USA.

2D-NMR structural resolution of alpha-helical peptides mimicking the transmembrane region in conjunction with computer-generated models of transmembrane receptor packing have recently been used to model pepitomimetic molecules that might selectively interfere with receptor activation by preventing dimerisation.

A number of preliminary reports suggest that antisense phosphorothioate oligonucleotides can also limit the growth of *erb*B2 overexpressing cell lines; however, this approach has not been examined in vivo. In addition, a single report has examined the feasibility of triple helix formation on the *erb*B2 promoter; however, transcriptional deregulation by such an approach will probably require the detailed characterisation of the functional promoter elements in different *erb*B2 overexpressing cell types in order to maximise the efficiency of transcriptional blockade.

References

1   Waterfield MD. Epidermal growth factor and related molecules. Lancet 1989, **1**, 1243–1246.
2   Evans R. The steroid and thyroid hormone receptor superfamily. Science 1988, **240**, 889–895.
3   Aaronson S. Growth factors and cancer. Science 1991, **254**, 1146–1153.
4   Hanks SK, Quinn AM, Hunter T. The protein kinase family: conserved features and deduced phylogeny of the catalytic domains. Science 1988, **241**, 42–52.
5   Todaro GJ, Rose TM, Spooner CE, Shoyab M, Plowman GD. Cellular and viral ligands that interact with the EGF receptor. Semin Cancer Biol 1990, **1**, 257–263.
6   Higashiyama S, Abraham JA, Miller J, Fiddes JC, Klagsbrun M. A heparin binding growth factor secreted by macrophage-like cells that is related to EGF. Science 1991, **25**, 936–939.
7   Holmes WE, Sliwkowski MX, Akita RW et al. Identification of Heregulin, a specific activator of p185[erbB-2]. Science 1992, **256**, 1205–1210.
8   Peles E, Bacus S, Koski R et al. Isolation of the neu/HER-2 stimulatory ligand: a 44 kd glycoprotein that induces differentiation of mammary tumour cells. Cell 1992, **69**, 209–216.

9   Lupu R, Colomber R, Kannan B, Lippman ME. Characterisation of a growth factor that binds exclusively to the *erb*B-2 receptor and induces cellular responses. Proc Natl Acad Sci USA 1992, **89**, 2287–2291.

10 Wen D, Peles E, Cupples R et al. Neu differentiation factor: a transmembrane glycoprotein containing an EGF domain and an immunoglobulin homology unit. Cell 1992, **69**, 559–572.

11 Dobashi K, Davis JG, Mikami Y, Freeman JK, Hamuro JK, Greene M. Characterisation of a neu/*erb*B-2 protein specific activating factor. Proc Natl Acad Sci USA 1991, **88**, 8562–8596.

12 Lupu R, Wellstein A, Sheridan J et al. Purification and characterisation of a novel growth factor from human breast cancer cells. Biochemistry 1992, **31**, 7330–7340.

13 Ullrich A, Schlessinger J. Signal transduction by receptors with tyrosine kinase activity. Cell 1990, **61**, 203–212.

14 Sternberg MJE, Gullick WJ. Neu receptor dimerisation. Nature 1989, **339**, 587.

15. Lin CR, Chen WS, Lazar CW, et al. Protein kinase C phosphorylation at Thr 654 of the unoccupied EGF receptor and GEF binding regulate functional receptor loss by independent mechanisms. Cell 1986;**44**, :839–848.

16 Cantley LC, Auger KR, Carpenter C et al. Oncogenes and signal transduction. Cell 1991, **64**, 281–302.

17 Koch CA, Anderson D, Moran MF, Ellis C, Pawson T. SH2 and SH3 domains: elements that control interactions of cytoplasmic signalling proteins. Science 1991, **252**, 668–674.

18 Sistonen L, Holtta E, Lehvaslaiho H, Lehtola L, Alitalo K. Activation of the neu tyrosine kinase induces the Fos/Jun transcription factor complex, the glucose transporter and ornithine decarboxylase. J. Cell Biol. 1989, **109**, 1911–1919.

19 Bravo R. Genes induced during the Go/G1 transition in mouse fibroblasts. Semin Cancer Biol 1990, **1**, 37–46.

20 Chao MV. Growth factor signalling where is the specificity? Cell 1992, **68**, 995–997.

21 Sporn MB, Todaro GJ. Autocrine secretion and malignant transformation of cells. N Engl J Med 1980, **303**, 878–880.

22 Wong ST, Winchell LF, McCune BK et al. The TGFα precursor expressed on the cell surface binds to the EGF receptor on adjacent cells, leading to signal transduction. Cell 1989, **56**, 495–506.

23 Kudlow JE, Bjorge JD. TGFα in normal physiology. Semin Cancer Biol 1990, **1**, 293–302.

24 Prigent SA, Lemoine NR, Hughes CM, Plowman GD, Selden C, Gullick WJ. Expression of the c-erbB-3 protein in normal and fetal tissues. Oncogene 1992, **7**, 1273–1278.

25 Wright NA, Pike G, Elia E. Induction of a novel growth factor secreting lineage by mucosal ulceration in human gastrointestinal stem cells. Nature 1990, **343**, 82–85.

26 Gullick WJ. Prevalence of abberrant expression of the epidermal growth factor receptor in human cancers. Br Med Bull 1991, **47**, 87–98.

27 Di Fiore PP, Pierce JH, Fleming TP. Overexpression of the human EGF receptor confers a EGF dependent transformed phenotype to NIH 3T3 cells. Cell 1987, **51**, 1063–1070.

28 Bouchard L, Lamarre L, Tremblay PJ, Jolicoeur P. Stochastic appearance of mammary tumours in transgenic mice carrying the MMTV/c-neu oncogene. Cell 1989, **57**, 931–936.

29 Mendelsohn J. The epidermal growth factor receptor as a target for therapy for anti-receptor monoclonal antibodies. Semin Cancer Biol 1990, **1**, 339–344.

30 Kraus MH, Popescu NC, Amsbaugh SC, King CR. Overexpression of the EGF receptor-related proto-oncogene *erbB-2* in human mammary tumour cell lines by different molecular mechanisms. EMBO J 1987, **6**, 605–610.
31 Yin Y, Tainsky MA, Bischoff FZ, Strong LC, Wahl GM. Wild-type p53 restores cell cycle control and inhibits gene amplification in cells with mutant p53 alleles. Cell 1992, **70**, 937–948.
32 Livingstone LR, White A, Sprouse J, Livanos E, Jacks T, Tisty TD. Altered cell cycle arrest and gene amplification potential accompany loss of wild type p53. Cell 1992, **70**, 923–935.
33 Kawasaki K, Kudoh J, Omoto K, Shimizu N. Megabase map of the epidermal growth factor (EGF) receptor gene flanking regions and structure of the amplification units in EGF receptor hyperproducing squamous carcinoma cells. Jpn J Cancer Res 1988, **79**, 1174–1183.
34 Kraus MH, Issing W, Miki T, Popescu NC, Aaronson SA. Isolation and characterisation of ERBB-3, a third member of the ERBB/epidermal growth factor receptor family: Evidence for overexpression in a subset of human mammary tumours. Proc Natl Acad Sci USA 1989, **86**, 9193–9197.
35 Lemoine NR, Barnes DM, Hollywood DP et al. Expression of the erbB-3 gene product in breast cancer. Br J Cancer 1992, **66**, 1116–1121.
36 Perren TJ. c-erbB-2 oncogene as a prognostic marker in breast cancer. Br J Cancer 1991, **63**, 328–332.
37 Gullick WJ, Love SB, Wright C et al. c-erbB-2 protein overexpression in breast cancer is a risk factor in patients with involved and uninvolved nodes. Br J Cancer 1991, **63**, 434–438.
38 Lovekin C, Ellis IO, Locker A et al. c-erbB-2 oncoprotein expression in primary and advanced breast cancer. Br J Cancer 1991, **63**, 439–443.
39 O'Reilly SM, Barnes DM, Camplejohn RS, Bartkova J, Gregory WM, Richards MA. The relationship between c-erbB-2 expression, S-phase fraction and prognosis in breast cancer. Br J Cancer 1991, **63**, 444–446.
40 Nicholson S, Sainsbury JRC, Halcrow P, Chambers P, Farndon JR, Harris AL. Expression of epidermal growth factor receptors associated with lack of response to endocrine therapy in recurrent breast cancer. Lancet 1989, **1**, 182–185.
41 Gusterson BA, Gelber RD, Goldhirsch KN et al. Prognostic importance of c-erbB-2 expression in breast cancer. J Clin Oncol 1992, **10**, 1049–1056.
42 Gullick WJ. Inhibitors of growth factor receptors. In: Carney D, Sikora K eds. Genes and Cancer. Wiley, Chichester, 1990, 263–263.
43 Lofts FJ, Gullick WJ. Growth factor receptors as targets. In: Kerr DJ, Workman P eds. DCISnNew Targets for Cancer Chemotherapy. CRC Press, Boca Raton, FL, 1993, in press.
44 Maguire HC, Greene MI. The neu (c-erbB-2) oncogene. Semin Oncol 1989, **16**, 148–155.

Molecular Biology for Oncologists
Edited by John Yarnold, Michael Stratton, Trevor McMillan
© *1993, Elsevier Science Publishers B.V. All rights reserved*

CHAPTER 7

GTP-binding proteins and cancer

Julian Downward

Signal Transduction Laboratory, Imperial Cancer Research Fund,
44 Lincoln's Inn Fields, London WC2A 3PX, UK

Introduction

The interpretation of extracellular signals and their translation into intracellular effects is central to all biological systems and is of particular importance in complex multicellular organisms. Information is gathered from outside the cell by the binding of signalling molecules such as hormones or growth factors to cell surface receptors that span the plasma membrane. The protein molecules responsible for transferring information from these receptors onto intracellular target enzymes fall into a number of functional families: one of the largest is the protein kinases (discussed in Chapter 6), while another is the guanine nucleotide-binding proteins, the subject of this chapter.

Guanine nucleotide-binding proteins are found in many different forms. They all display very limited sequence homology, which in most cases is restricted to the sites of contact of the protein with the guanine nucleotide. They all share the ability to bind guanosine triphosphate (GTP) and catalyse its hydrolysis to guanosine diphosphate (GDP). The GTP-bound form of the protein has different properties and a different conformation than the GDP-bound form. Not all guanine nucleotide-binding proteins are involved in signalling pathways; a number are structural, such as tubulin, while others are involved in macromolecular synthesis, such as the elongation factors of protein synthesis.

The signal transducing guanine nucleotide-binding proteins are made up of two principal groups: the heterotrimeric GTP-binding proteins, generally referred to as G proteins, and the monomeric ras superfamily proteins. For all these proteins the

GTP-bound form represents an active state that is capable of interacting with some cellular target enzyme system, while the GDP-bound form is without biological activity. The activity of the proteins is regulated by modulating the rate at which bound GDP is replaced by GTP and the rate at which bound GTP is hydrolysed. GTP-binding proteins are often compared to a timer switch; they have an on (GTP-bound) and an off (GDP-bound) state and an inbuilt clock (the rate of the GTP hydrolysis reaction).

The normal function of G proteins

G proteins are a diverse family of heterotrimers

The heterotrimeric G proteins are usually located at the inner surface of the plasma membrane, held in place by covalently attached lipids (palmitylations or myristylations). They are made up of three different subunits: the α subunit, which contains the nucleotide binding site, and the β and γ subunits (reviewed in [1]). So far, some 20 distinct mammalian α subunits have been characterised, along with five β and four γ subunits. These could be combined in different permutations to give perhaps 400 distinct G proteins. The α subunits characterised so far are at least 50% identical to each other at the level of amino acid sequence, while the β and γ subunits are even more conserved at 80% or more identicalness. Remarkably similar G proteins are found in all eukaryotic organisms including yeast.

G proteins couple cell surface receptors to intracellular enzymes

G proteins all interact directly with members of a class of cell surface receptor proteins that have seven membrane spanning regions (see Figure 1). These share a degree of amino acid sequence similarity and include the adrenergic receptors, the muscarinic acetylcholine receptors, rhodopsins, neuropeptide receptors, purinergic receptors and many more. In all, more than 100 distinct receptors have been identified that are coupled to G proteins. The G proteins only interact with activated receptors, that is receptors that have bound to their specific ligand. This interaction stimulates the exchange of GDP bound to the α subunit of the inactive heterotrimer for GTP. The α subunit now acquires the active, GTP-bound conformation and dissociates from the βγ subunits. The GTP-bound α subunit then goes on to interact with and activate its target enzyme, often referred to as an effector. In an interesting economy of design, the free βγ subunits can also interact with and regulate several different effector enzymes. The signal is terminated when the α subunit hydrolyses GTP to GDP, an event in some cases catalysed by interaction with the effector, and reassociates with βγ. Since each ligand-bound receptor can

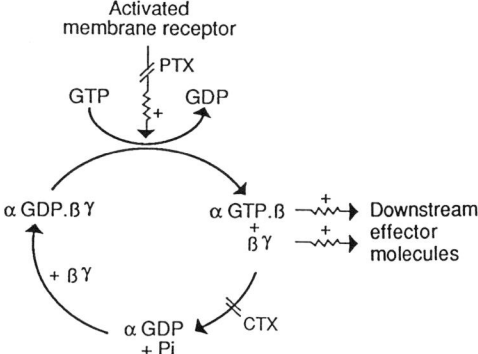

Figure 1. The G protein cycle. Activated receptor (R*) causes αβγ trimer to release GDP and take up GTP. This complex is unstable and breaks up to give α.GTP and βγ, both of which can control various effector enzymes. GTP on α is hydrolysed to GDP resulting in rebinding of α.GDP to βγ and signal termination. Cholera toxin (CTX) blocks GTP hydrolysis on α, while pertussis toxin (PTX) inhibits the interaction of activated receptor with G protein trimer.

activate several molecules of G protein and each G protein can activate several molecules of effector, there is an inbuilt signal amplification. This is particularly marked in the detection of light in the retina. The effector systems controlled by G proteins are almost as diverse as the receptors involved. They include several types of adenylyl cyclases, phospholipases C and A_2, cyclic GMP phosphodiesterase and many different ion channels. These enzymes are all capable of causing changes in the intracellular concentration of small molecules, or ions, involved in cellular signalling.

Disruption of γ protein function in cancer and other diseases

G proteins are targets for bacterial toxins

Recently it has been established that G proteins play an important role in a number of human tumours. Their original identification, however, was due to their involvement in two well-known infectious diseases. *Vibrio cholerae* produces severe disease through the production of cholera toxin; this multisubunit toxin delivers an ADP ribosylase to target cells. This ADP ribosylase is specific for the α subunit of G_s, a G protein that normally functions to positively regulate the activity of adenylyl cyclase and hence cyclic AMP (cAMP) levels. Cholera toxin ADP-ribosylates $G_{s\alpha}$ at arginine 201 and causes a very large reduction in the rate at which bound GTP is hydrolysed. The G protein therefore remains locked in the active, GTP-bound state, leading to the continuous activation of its effector, adenylyl cyclase, and the generation of abnormally high levels of cAMP. The uncontrolled production of cAMP in the cells lining the gut leads to a large efflux of sodium ions and water, causing the characteristic symptoms of cholera. Another organism, *Bordetella pertussis*, produces a G protein-specific ADP ribosylase. In this case the target is $G_{i\alpha}$, a closely related subgroup of G proteins that also regulates adenylyl cyclase, but in an inhibitory manner. Pertussis toxin ADP-ribosylates $G_{i\alpha}$ on cysteine 350: this results in an inhibition of the interaction of the G protein with its activating receptor. The G protein cannot therefore be stimulated to a GTP-bound state which would inhibit adenylyl cyclase, but instead remains inactive. As with cholera, a rise in the intracellular levels of cAMP results.

G_s can act as an oncogene

In 1989, the discovery was made that heterotrimeric G proteins can act as oncogenes and be involved in the development of tumours (see Table 1). Analysis of $G_{s\alpha}$ in eight pituitary tumours [2] revealed that half had undergone point mutations, either at codon 201 (the site of ADP ribosylation by cholera toxin) or at codon 227. Both mutations caused strongly reduced GTPase activity and the tumour cells contained elevated cAMP levels; intracellular cAMP concentration is known to correlate with growth in this cell type. More recently, similar mutations have been found in $G_{s\alpha}$ in a number of thyroid tumours [3]. As with all known oncogenes, the mutation of Gsα is likely to represent only one of several steps required to achieve a full malignant phenotype. The limited number of tumour types in which mutations in Gsα have been found may reflect the fact that cAMP is only growth stimulatory in a limited number of cell types.

G_i can act as an oncogene

A second type of heterotrimeric G protein, $G_{i2\alpha}$, has also been shown to be capable of acting as an oncogene [3]. A subset of adrenal cortical and ovarian tumours carry mutations at sites cognate to those mutated in $G_{s\alpha}$. These mutations again

Table 1. Activated G protein oncogenes in human tumours.

Tumour type	Incidence (%)	Predominant oncogene
Pituitary adenoma (growth hormone secreting)	43	$G_{s\alpha}$
Thyroid carcinoma	4	$G_{s\alpha}$
Ovarian stromal tumours	30	$G_{i2\alpha}$
Adrenal cortical tumours	27	$G_{i2\alpha}$

give rise to a protein that is constitutively activated due to decreased GTPase activity. While it has been proven that mutant $G_{i2\alpha}$ can transform Rat-1 fibroblast cells in culture, it is not clear by what mechanism this occurs; the documented targets for $G_{i2\alpha}$, adenylyl cyclase inhibition, potassium channels and phospholipase A_2, have not been correlated with growth stimulation in adrenal cortical or ovarian cells. Given the rapid progress in this field, it is quite likely that mutation of other G proteins will be found to be involved in the genesis of a number of other tumour types.

The normal function of ras proteins

Ras proteins make up a superfamily of related GTP-binding proteins

Like the heterotrimeric G proteins, members of the ras superfamily of proteins bind GTP and hydrolyse it to GDP, being active in the GTP-bound state and inactive in the GDP-bound state. However, they are monomeric and only about half the size of a G protein α subunit, hence the term "low molecular weight GTP binding protein" which is sometimes used to describe them. They possess only limited homology to G proteins; how analogous their functions are is still a matter of debate.

Some 40 distinct mammalian ras superfamily proteins have been identified (reviewed in [4]). They are all at least 30% related to each other and can be grouped into three or more distinct families in which each member is at least 50% homologous to every other member (see Table 2). The main families within the ras superfamily are those named after, and typified by, the ras, rho and rab proteins. Ras proteins were the first to be identified, due to their role in human cancer (see next section); the others were subsequently identified due to their sequence homology to ras. Three very closely related proteins in the ras family, H-ras, K-ras and N-ras, are primary regulators of cell growth and frequently mutationally activated in tumours. The function of other members of the ras family (R-ras, ral, rap) is not known, though there is some evidence that rap may act antagonistically to ras.

Ras proteins control cell growth

The mechanisms by which the three oncogenic ras proteins control cell growth has been the subject of intense study. It is fair to say that, unlike the case of the heterotrimeric G proteins, a clear picture has not yet emerged. A number of growth stimulatory enzymes have been identified that are rapidly activated by active ras proteins, most notably the calcium and phospholipid dependent family of protein kinases (protein kinase C) and the raf family of protein kinases. However, these are

Table 2. The mammalian ras superfamily.

ras family	rho family	rab family
Ha-ras	rhoA	rab1A
Ki-ras (A and B)[a]	rhoB	rab1B
N-ras	rhoC	rab2
	rhoG	rab3A
rap1A[b]	racl	rab3B
rap1B	rac2	rab4
rap12	TC10	rab5
	CDC42 Hs	rab6
R-ras		rab7
		BRL-ras
ralA		rab8
ralB		rab9
		rabs10–17

[a]Products of alternative splicing.
[b]Also known as Krev-l and smg p21.

almost certainly not the direct targets for ras proteins, which may remain to be identified. At least one of the three oncogenic ras proteins is expressed in all eukaryotic cells, and in most cases disruption of the function of these proteins results in an arrest of cell growth. In a few cell types, particularly neuronal lineages, ras proteins appear to regulate a differentiation response rather than cell growth.

GTPase activating proteins (GAPs) negatively regulate ras proteins

Much progress has been made in understanding the regulation and function of ras proteins recently through the study of proteins that interact with them (see [5] for review). Unlike G proteins, ras proteins have very slow endogenous rates of GTP hydrolysis; the first protein, p120^{GAP}, found to interact with ras was a GTPase activating protein (GAP) that speeds up the hydrolysis reaction and hence the inactivation of the ras proteins (see Figure 2). At least one other GAP exists for the oncogenic ras proteins: neurofibromin, the product of the type 1 neurofibromatosis gene (see next section). While these GAPs act as negative regulators of ras, there is some evidence that they may be downstream targets for ras as well [6]. In other words, as the GAP acts on GTP-bound ras to switch it off, it may also receive a growth promoting signal from the ras protein to pass on into the cell. Alternatively, a ternary complex of ras, GAP and another protein could be involved. It is not known with what target enzymes GAPs or other ras effectors might interact.

Guanine nucleotide exchange factors positively regulate ras proteins

Ras proteins are also regulated by control of the rate of guanine nucleotide exchange. Various exchange proteins have been identified which are likely to be key activators of ras [7]. These, called CDC25 and sos ("son of sevenless"), were named after the homologous genes in yeast and fruit fly which were discovered first. Ras is activated when cells are treated with any of a number of growth factors, especially those that regulate tyrosine kinases. This is achieved in many cases by the activation of the exchange factors, but also in some cases by inhibition of the GAPs. The mechanisms by which the activity of these regulators is controlled is still not fully understood.

Rho proteins control the actin cytoskeleton

The other main families of ras-related proteins have very different, but equally critical, functions. Members of the rho family of proteins are involved in controlling the structure of the actin cytoskeleton. Rho controls actin stress fibre formation [8] while rac controls actin-mediated plasma membrane ruffling [9]. Like ras, they

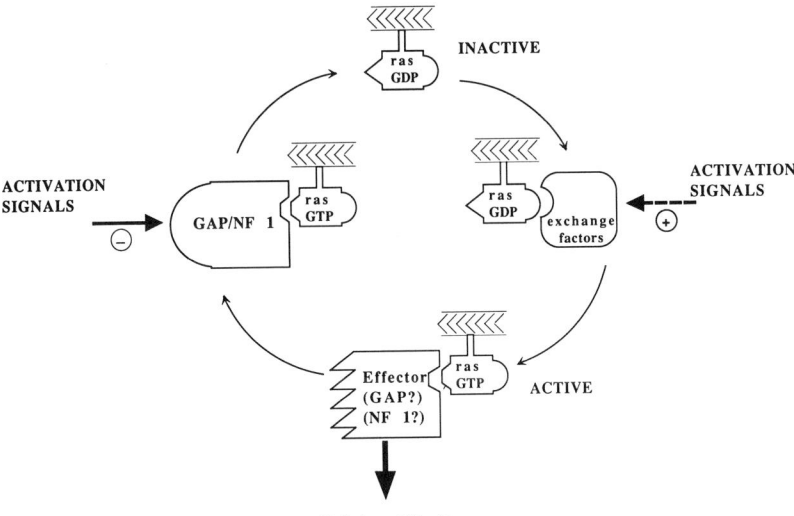

Figure 2. The ras cycle. GDP bound ras at the plasma membrane is stimulated to exchange GDP for GTP by interaction with exchange factors. ras.GTP interacts with unknown effectors (possibly GAP or neurofibromin) to transmit its growth signal to the cell. The signal is terminated when GTP on ras is hydrolysed to GDP, a reaction catalysed by GAP and neurofibromin (NF1). Control of ras is either by inhibition of GAPs or activation of exchange factors. From [2]. Reprinted with permission.

are activated by various mitogenic growth factors. GAPs and exchange factors have also been identified for members of this family; intriguingly, some of these have been previously identified as oncogenes in their own rights (vav, dbl), even though rho proteins are not oncogenic [10]. The rab family of ras-related proteins is involved in the regulation of secretion and intracellular vesicle traffic, with a different rab protein being associated with each of the intracellular vesicular compartments, possibly as a marker or sorting protein [11]. GAPs and exchange factors have also been identified for these proteins.

Disruption of ras protein function in cancer and other diseases

Ras is activated by point mutation in many human tumours

Ras proteins are best known for their involvement in human cancer. The *ras* family members that are capable of causing malignant transformation are H-*ras*, K-*ras* and N-*ras*. Of these, H-*ras* and K-*ras* were first discovered as the products of the retroviral oncogenes of the Harvey and Kirsten murine sarcoma viruses. In the late 1970s it was recognised that retroviruses had picked up their transforming oncogenes from the host genome. In a turning point for the understanding of the molecular basis of cancer, a transforming gene was discovered in a human tumour which was almost identical to the retroviral H-*ras* oncogene. The ability to transfer the *ras* gene from one cell to another by calcium phosphate-mediated transfection led to the discovery that a large number of human tumours carried activated forms of the H-*ras* or K-*ras* genes, or the closely related N-*ras* oncogene. The difference between the oncogenic *ras* genes found in tumours and the normal *ras* genes was a point mutation that resulted in a constitutively active form of the protein. Almost all the mutations found in tumours occur at codons 12, 13 or 61; knowledge of the three-dimensional structure of the ras protein shows that these residues all lie at points of contact of the ras protein with the phosphate groups of GTP. The mutations result in a reduced rate of GTP hydrolysis and a resistance to the inhibitory effects of GAPs. The mutant ras proteins are therefore locked into an active GTP-bound state and transmit a constitutive growth signal to the cell. The advent of the polymerase chain reaction has led to the examination of enormous numbers of human tumours for mutations in *ras* and other oncogenes (see Table 3). *Ras* mutations appear in nearly 30% of all tumours examined [12], with extremely high rates in pancreatic carcinoma (95%), thyroid tumours (60%) and adenocarcinoma of the colon (50%). Analysis of the time course of activation of *ras* has shown it to be a relatively early event in the genesis of many colon carcinomas. *Ras* is known to require other cooperating activated oncogenes, such as p53 or *myc*, to cause full transformation in various model systems.

Table 3. Activated *ras* oncogenes in human tumours.

Tumour type	Incidence (%)	Predominant oncogene
Pancreatic carcinoma	95	Ki-*ras*
Thyroid carcinoma	60	N-*ras*
Colon adenocarcinoma	50	Ki-*ras*
Colon carcinoma	50	Ki-*ras*
Seminoma	40	Ki-*ras*, N-*ras*
Lung adenocarcinoma	40	Ki-*ras*
Myelodysplastic syndrome	30	N-*ras*
Acute myelogenous leukemia	30	N-*ras*
Keratinoacanthoma	30	Ha-*ras*
Melanoma	30	N-*ras*
Squamous cell carcinoma	25	Ha-*ras*
Bladder carcinoma	15	Ha-*ras*
Breast carcinoma	<5	
Cervical carcinoma	<5	
Ovarian carcinoma	<5	
Stomach carcinoma	<5	

Neurofibromatosis type 1 involves defective negative regulation of ras

An unexpected involvement of *ras* has recently emerged in a hereditary disease that is characterised by a high incidence of tumours, type 1 neurofibromatosis (NF1). This is caused by inheritance of a defective allele of the NF1 gene which has been shown to encode a GTPase activating protein (GAP) for *ras* [13]. Since GAPs are negative regulators of *ras*, NF1 patients may be especially susceptible to development of certain tumours due to reduced ability to switch off *ras*, especially in cells where the single healthy allele of NF1 has been lost [14,15]. Increased tumour incidence is confined to cells of neural crest origin, with benign neuro-fibromas being found in almost all patients. It has been suggested that the NF1 gene product might also be involved in delivering a differentiation signal from *ras* to neural crest cells; failure to differentiate is often associated with neoplasia.

As yet there is no evidence for involvement of other members of the *ras* super-family in tumorigenesis. As for other diseases, it is possible that the ADP-ribosy-lation of rho protein by *Clostridium botulinum* exoenzyme C3 could contribute to the illness caused by that organism.

Prospects for therapies

The rapid advances in understanding the role of GTP binding proteins in the for-mation and maintenance of the transformed phenotype have not yet led to improvements in therapy or, with rare exceptions, diagnosis. One case in which ras has been used for diagnosis is the use of polymerase chain reaction to identify mutations in ras from stool samples of patients with colorectal tumours [16]. Considerable efforts are being made by a number of pharmaceutical companies to develop drugs that will interfere with the function of ras proteins, with the hope that the removal of the ras dependent growth signal from tumour cells might cause their growth arrest or death. Two aspects of ras function have been targeted for drug development; one is the essential post-translational modification of ras by farnesylation that is required for its localisation to the plasma membrane and the other is its interaction with its cellular effectors.

The enzymes involved in the post-translational modification are now well characterised and it is likely that reasonable pharmacological inhibitors will soon be available. The problems with the use of such agents would be lack of specificity: they would be very likely to effect other farnesylated proteins, such as nuclear lamins, and would inhibit normal ras in healthy tissues.

Drugs that inhibit the interaction of ras with its effectors are likely to be harder to develop, although the known three-dimensional structure of ras is helpful for modelling. The advantage of these reagents would be that they would probably be more specific for ras; indeed, theoretical and experimental considerations indicate that mutant ras would be more sensitive to inhibition by such drugs than would normal ras protein [17].

Recently permission has been given in the United States for clinical trials of experimental gene therapy to treat disseminated lung cancer with vectors that would express *ras* antisense messenger RNA. As with all gene therapy approaches to cancer treatment, the major difficulty is likely to be the delivery of the therapeutic DNA to all the affected cells. However, as gene therapy protocols and, probably more importantly, drug design improve, the prospect of targeting mutant *ras* as an effective way of treating human cancer becomes increasingly promising.

References

1   Simon MI, Strathmann MP, Gautam N. Diversity of G proteins in signal transduction. Science 1991, **252**, 802–808.
2   Landis CA, Masters SB, Spada A, Pace AM, Bourne HR, Vallar L. GTPase inhibiting mutations activate the alpha chain of Gs and stimulate adenylyl cyclase in human pituitary tumours. Nature 1989, **340**, 692–696.
3   Lyons J, Landis CA, Harssh G et al. Two G protein oncogenes in human endocrine tumors. Science 1990, **249**, 655–659.
4   Downward J. The ras superfamily of small GTP-binding proteins. Trends Biochem Sci 1990, **15**, 469–472.
5   Downward J. Regulatory mechanisms for ras proteins. BioEssays 1992, **14**, 177–184.

6   McCormick F. ras GTPase activating protein: signal transmitter and signal terminator. Cell 1989, **56**, 5–8.
7   Downward J. Exchange rate mechanisms. Nature 1992, **358**, 282–283.
8   Ridley A, Hall A. The small GTP-binding protein rho regulates the assembly of focal adhesions and actin stress fibers in response to growth factors. Cell 1992, **70**, 389–399.
9   Ridley A, Paterson HF, Johnston CL, Diekmann D, Hall A. The small GTP-binding protein rac regulates growth factor-induced membrane ruffling. Cell 1992, **70**, 401–410.
10 Hart MJ, Eva A, Evans T, Aaronson SA, Cerione RA. Catalysis of guanine nucleotide exchange on the CDC42Hs protein by the *dbl* oncogene product. Nature 1991, **354**, 311–314.
11 Balch WE. Small GTP-binding proteins in vesicular transport. Trends Biochem Sci 1990, **15**, 473–477.
12 Bos JL. *ras* oncogenes in human cancer: a review. Cancer Res 1989, **49**, 4682–4689.
13 Downward J. Cell signalling: plugging the GAPs. Curr Biol 1991, **1**, 353–355.
14 Basu T, Gutmann DH, Fletcher JA, Glover TW, Collins FS, Downward J. Aberrant regulation of ras proteins in malignant tumour cells from type 1 neurofibromatosis patients. Nature 1992, **356**, 713–715.
15 DeClue JE, Papageorge AG, Fletcher JA et al. Abnormal regulation of mammalian p21ras contributes to malignant tumor growth in von Recklinghausen (type 1) neuro-fibromatosis. Cell 1992, **69**, 265–273.
16 Sidransky D, Tokino T, Hamilton SR et al. Identification of *ras* oncogene mutations in the stool of patients with curable colorectal tumors. Science 1992, **256**, 102–105.
17 Farnsworth CL, Marshall MS, Gibbs JB., Stacey DW, Feig LA. Preferential inhibition of the oncogenic form of *ras* H by mutations in the GAP binding/"effector" domain. Cell 1991, **64**, 625–633.

Molecular Biology for Oncologists
Edited by John Yarnold, Michael Stratton, Trevor McMillan
© 1993, Elsevier Science Publishers B.V. All rights reserved

CHAPTER 8

The p53 gene in human cancer

Michael Stratton

Section of Chemical Carcinogenesis, Institute of Cancer Research, 15 Cotswold Road, Belmont, Sutton, Surrey, SM2 5NG, UK

How the p53 gene was discovered

Certain DNA viruses can induce tumours in experimental models. The transform-ing activity of one of these, the small DNA virus SV40, is attributable to a portion of the viral genome that encodes a protein termed the large T antigen. In order to understand the mechanism by which large T induces oncogenesis, attempts were made in the late 1970s to detect the host cell protein to which large T binds. Immunoprecipitation of large T from cells infected by SV40 revealed a host cell protein of 53 kDa (known as p53) in addition to large T itself. p53 did not appear to be bound directly to the antibody, but was precipitated because it was associated with large T [1,2]

The first indication that mutation of the p53 gene might be a common step in the development of human cancer emerged from investigations into the role and location of tumour suppressor genes. Comparison of germline and tumour DNAs using polymorphic probes revealed that loss of one or other parental allele (loss of heterozygosity) on the short arm of chromosome 17 occurs at high frequency in many types of neoplasm. This type of result is usually interpreted as indicating the presence of a tumour suppressor gene in the vicinity. The p53 gene had previously been localised to this region and fine mapping of heterozygous deletions in colon carcinoma indicated that the common deleted area includes the p53 locus [3]. Prompted by this clue, the p53 gene was sequenced in two colon carcinomas and subsequently in other tumours showing loss of heterozygosity on chromosome 17p. In most cases single base substitutions were detected in the remaining p53 allele [3,4].

Structural alterations in the p53 gene are common somatic events in human cancer

A substantial body of data contributed by several groups now indicates that the p53 gene is the most commonly mutated gene known in human cancer. Mutations are present in all major histogenetic groups and indeed in most subtypes of cancer [5]. Thus they are found in epithelial, mesenchymal, haemopoietic and lymphoid neoplasms and in tumours of the central nervous system. The proportion carrying p53 mutations is high in some types of cancer and low in others, but in no tumour type does it reach 100%. It therefore appears unlikely to be an obligate step in oncogenesis.

Somatic mutation of the p53 gene in human cancer is often not the initiating event in oncogenesis

Mutation of the p53 gene constitutes an intermediate or late step in the sequence of genetic alterations required for the development of many tumours. For example, mutations usually occur around or during the transition between adenoma and carcinoma of the colon [6]. Similarly they are found in blast crisis of CML but not in the earlier chronic phase [7]. However, p53 gene mutations often occur early in the development of gliomas and are sometimes present in the in situ phase of breast cancer. Exceptions to all these patterns exist and therefore it may be that mutation of p53 occurs at different stages in different tumour types and that adherence to a predetermined sequence of genetic events is not as important as the final accumulation of the necessary set of mutations.

The type and pattern of mutation in the p53 gene suggests inactivation of a tumour suppressor gene

The majority of somatic mutations of the p53 gene found in human tumours are single base substitutions (point mutations) resulting in replacement of one amino acid by another in the p53 protein (missense mutations). A minority result in ab-normalities of mRNA processing, frame shifts or premature termination of trans-lation. Less commonly, complete or partial deletions and other gross rearrange-ments of the gene are detected. Point mutations of p53 are scattered through numerous codons but are mainly confined to four regions of the gene (Figure 1) that are conserved through evolution, located in exons 5–8. Within these conserved regions there are four codons which are "hotspots" and which together account for approximately 30% of mutations in the gene [5]. This diversity of mutation,

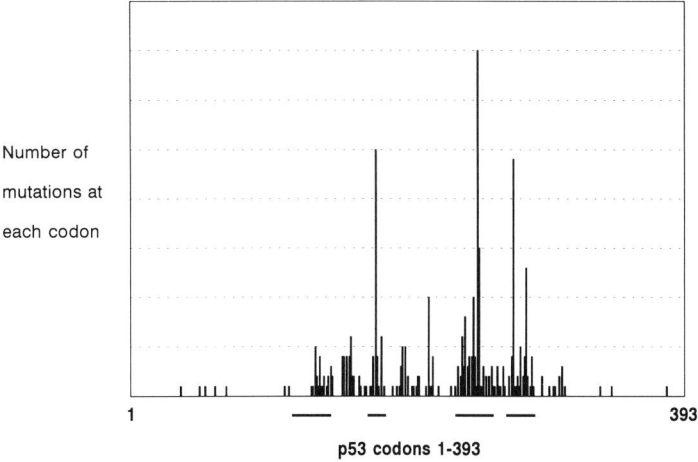

Figure 1. Somatic mutations in the p53 gene are widely scattered over a large number of codons. However, they are preferentially found in regions that show sequence conservation during evolution (black horizontal bars). Moreover, there are hotspots that account for over 30% of mutations. The data are collated from several reports and reflect mutations in many different types of cancer.

coupled with the fact that some must result in absence or truncation of the protein suggest that inactivation is the functional outcome. Comparison of the types of mutation found in different cancers suggests that patterns of mutation differ. It seems plausible that at least part of this variation may reflect exposure to particular carcinogens and hence may generate an insight into aetiological mechanisms (considered in more detail in Chapter 2).

In most normal cells, the level of p53 protein is low and usually undetectable by immunocytochemistry. However, many p53 proteins with single amino acid substitutions have a much longer half-life than the wild type and are consequently present at higher levels in the cell. Thus detection of the p53 protein in tumours by immunohistochemical methods, is often used to provide indirect evidence of a mutated gene [8].

Germline mutations of the p53 gene predispose to the development of several types of cancer

The Li–Fraumeni syndrome (LFS) is a rare familial predisposition to cancer that is transmitted in an autosomal dominant manner [9]. The syndrome is characterised clinically by sarcomas in children (of bone and soft tissue) associated with a high incidence of early onset breast cancer in female relatives. Leukaemia, brain, lung and adrenocortical tumours constitute less common features of the disease.

It transpires that some LFS patients carry one mutated and one wild type p53 allele in their germline and that the predisposition to cancer is transmitted with the mutant allele [10,11]. Many of these mutations are at the same sites as those documented as somatic events in sporadic tumours, including some of the "hotspots" described above. It is therefore thought that LFS patients are predisposed to cancer because one p53 allele is inactivated in the germline and therefore only the remaining allele needs to be altered by somatic mutation before a cell can escape from the tumour suppressor activity of the p53 protein. In normal individuals developing a sporadic tumour, both p53 alleles must be inactivated in the same cell by somatic mutation. In this model, the genetics of p53 are rather similar to those of the prototype tumour suppressor, the retinoblastoma gene (however, see below for complications).

Further studies now indicate that germline p53 mutations account for only a proportion of LFS families, and that the range of tumours developing in individuals carrying such mutations is wider than that classically associated with LFS. Germline mutations in p53 account for approximately 5% of cases of osteosarcoma and of children developing two independent primary cancers [12,13]. A much smaller proportion of early onset breast cancer (<1%) is attributable to germline p53 mutations. In contrast to familial retinoblastoma, the new mutation rate appears to be low. Clinicians are beginning to consider what they may be able to offer such individuals. Cancer screening to provide early diagnosis and prophylactic treatment may be appropriate for some. However, the limited beneficial effects of these may have to be weighed against detrimental effects of testing upon lifestyle and economic status. Some of the problems associated with counselling are exemplified in the kindred shown in Figure 2.

Is p53 a dominant transforming gene or a tumour suppressor gene?

Early in the 1980s, several groups demonstrated that overexpression of p53 can immortalise or transform cells either on its own or in collaboration with other oncogenes such as *ras* [14–16]. This pattern of activity was regarded as similar to that of genes such as c-*myc* which are conventionally regarded as dominantly acting oncogenes (i.e. they can transform cells in the presence of a normal allele and the proto-oncogene usually requires activation to contribute to oncogenesis). Doubt was cast upon this interpretation when it emerged that the p53 cDNA clones used in all these experiments were mutated rather than wild type. Moreover, the pattern of structural alterations found in tumours (widely scattered point mutations and various types of homozygous deletion) is suggestive of a tumour suppressor gene

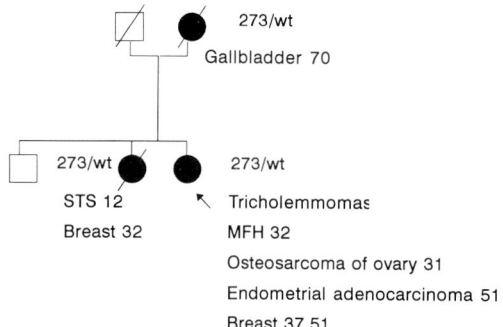

Figure 2. A cancer family due to a germline p53 mutation at codon 273 (Arg → Cys). The pedigree illustrates the wide spectrum of tumours that can be induced and how variable the penetrance can be in individuals with germline p53 mutations. The mother of the index case lived to age 70 before developing a carcinoma of the gallbladder, whilst the index case (arrowed) suffered multiple cancers before the age of 50. Both women and the sister of the index case who developed breast cancer and a sarcoma carry the mutation.

(or recessive oncogene in which both alleles require inactivation) than of a dominantly acting gene. Further experiments have recently revealed that wild type p53 does indeed suppress the transformed phenotype in a number of cell systems [17–19].

However, if p53 were to behave as an orthodox tumour suppressor gene, then mutated versions should theoretically be inactivated and thus not affect the phenotype when introduced into normal cells. Nevertheless p53 clones with several different mutations stubbornly act to transform cells [20,21]. Reconciling the respective transforming and suppressor activities of mutated and wild type p53 has been problematic. One possibility is that p53 protein functions in the form of protein oligomers. Oligomers composed exclusively of wild type protein function normally. Oligomers composed of mutant protein are inactivated. However, hybrid oligomers of mutant and wild type protein take the mutated (i.e. inactivated) phenotype [22] (Figure 3). On this model, a mutation in a single p53 allele may provide the cell with a proliferative advantage without alterations of the remaining allele. Nevertheless, the overall effect is inactivation of the protein (a dominant negative or dominant loss of function effect).

Normal p53 protein regulates the expression of other genes by binding to a specific DNA sequence

p53 is a nuclear phosphoprotein of short half-life that is modified by or interacts with other proteins that are known to be involved in regulation of the cell cycle

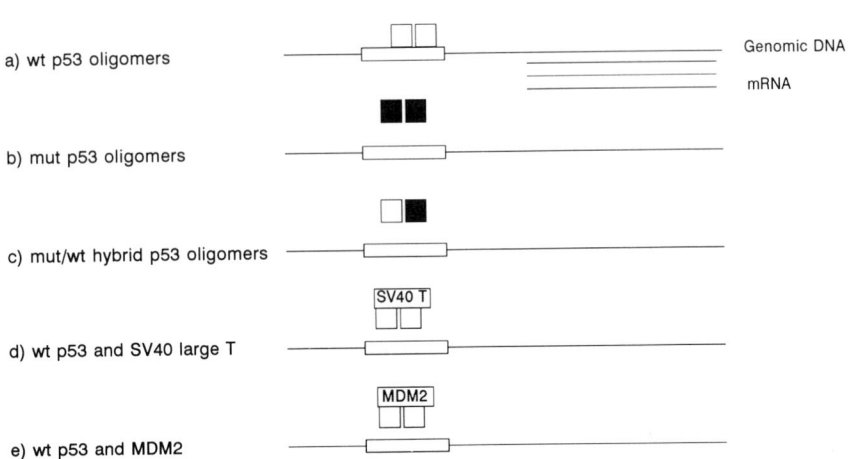

Figure 3. Schematic representation of p53 function and its regulation by other proteins. (a) wt p53 protein binds to specific DNA sequences and regulates the expression (transactivates) of other genes. (b) Mutated p53 cannot bind and transactivate. (c) Hybrid oligomers of mutated and wt p53 cannot bind and transactivate. (d) SV40 large T antigen associates with p53 and prevents it binding or transactivating. (e) MDM2 protein associates with p53 and prevents it binding or transactivating.

[23]. Many mutated versions of p53 protein exhibit a much longer half-life and bind to heat shock proteins [24]. Normal p53 protein binds to a specific DNA sequence [25,26] and can regulate the transcription of (transactivate) genes in the vicinity of such sequences. Mutated forms of p53 found in human cancers, however, cannot bind or regulate transcription through this DNA element [27]. Moreover, mutated p53 (presumably through the formation of hybrid p53 protein oligomers described in the preceding section) and SV40 large T antigen can inhibit the transactivation function of normal p53 protein. Together, these studies support the contention that p53 exerts its biological effects (at least partially) through regulation of transcription and that abrogation of this activity by SV40 large T and mutated p53 may (at least partially) account for their transforming activities.

Wild type p53 can induce apoptosis and has a role in the regulation of the cell cycle

Investigation of the tumour suppressor effect of p53 suggest that at least two biological effects play a role. Firstly, wild type, but not mutated forms of p53 can induce the form of programmed cell death known as apoptosis [28]. Secondly, p53 may have a role in regulating the transition from G1 to S phase of the cell cycle

[29,30]. In the past it has been suggested that G1 arrest may be important in optimising the extent of repair by delaying the onset of DNA synthesis after exposure to DNA damaging agents [30]. If so, inactivation of p53 may render tumour cells more prone to acquire mutations elsewhere in the genome. Indeed, there is now evidence that introduction of mutated p53 or SV40 large T can allow gene amplification (a rather specialised form of mutation) in cell types that were previously resistant to this type of change. Thus p53 may act directly to suppress cell growth and also to make tumour cells more susceptible to the acquisition of mutations.

Interactions of p53 and RB proteins with DNA virus antigens

p53 was discovered through its association with SV40 large T. It also binds to other DNA virus transforming proteins including the E1b of adenovirus and E6 of papillomavirus [31,32]. Interestingly enough it is now known that the protein encoded by the retinoblastoma tumour suppressor gene also binds to SV40 large T, to the E1a protein of adenovirus and the E7 of papilloma virus. There is therefore a remarkable symmetry between p53 and RB1. RB was detected and isolated as a gene involved in a familial cancer syndrome, was subsequently found to be a target for somatic mutation in many tumours and finally was shown to be bound to DNA virus transforming proteins. Conversely p53 was originally discovered as a protein bound to DNA virus products, was subsequently shown to be a target for somatic mutations and finally was implicated in a familial cancer syndrome.

p53 binds to the product of the MDM2 gene which is amplified in certain human cancers

The large T antigen and other viral oncoproteins play a critical role in the normal replication and transforming activity of DNA viruses. The observation that they bind to p53 and to the product of the RB1 gene inevitably rekindled interest in putative eukaryotic homologues of these oncoproteins which might also interact with tumour suppressor proteins. Direct isolation and characterisation of proteins associated with p53 revealed that one of these is the product of the MDM2 gene [33]. MDM2 was originally identified as a gene amplified in spontaneously transformed BALB/c cells and has been shown to have transforming activity when overexpressed in NIH3T3 cells. Interestingly, MDM2 protein appears to inhibit the transcriptional regulatory activity of p53 in a similar way to SV40 large T and mutated p53 [33].

The MDM2 gene is located on the long arm of chromosome 12 close to a locus that is cytogenetically altered in many human sarcomas. This observation led to the discovery that MDM2 is frequently amplified in sarcomas. Future work will no doubt reveal how MDM2 interacts with and regulates p53 activity, its overall role in cell proliferation and may confirm the preliminary suggestion that tumours with MDM2 amplification do not have p53 mutations and that tumours with p53 mutations do not have MDM2 amplification. A tempting speculation that might arise from this observation would be that p53 and MDM2 are components of a regulatory pathway that may be universally deregulated in cancer cells.

References

1   Lane DP, Crawford LV. T antigen is bound to a host protein in SV40 transformed cells. Nature 1979, **278**, 261–263.
2   Linzer DIH, Levine AJ. Characterization of a 54k Dalton cellular SV40 tumor antigen present in SV40-transformed cells and uninfected embryonal carcinoma cells. Cell 1979, **17**, 43–52.
3   Baker SJ, Fearon E, Nigro JM et al. Chromosome 17 deletions and p53 mutations in colorectal carcinomas. Science 1989, **244**, 217–221.
4   Nigro JM, Baker SJ, Preisinger AC et al. Mutations in the p53 gene occur in diverse human tumour types. Nature 1989, **342**, 705–708.
5   Hollstein M, Sidransky D, Vogelstein B, Harris CC. p53 mutations in human cancers. Science 1991, **253**, 49–53.
6   Baker SJ, Preisinger AC, Jessup M et al. p53 gene mutations occur in combination with 17p allelic deletions as late events in colorectal tumorigenesis. Cancer Res 1990, **50**, 7717–7722.
7   Ahuja H, Bar-Eli M, Advani SH et al. Alterations in the p53 gene and the clonal evolution of the blast crisis of chronic myelogenous leukaemia. Proc Natl Acad Sci USA 1989, **86**, 6783–6787.
8   Iggo R, Gatter K, Bartek J et al. Increased expression of mutant forms of p53 oncogene in primary lung cancer. Lancet 1990, **335**, 675–679.
9   Li FP, Fraumeni JF, Mulvihill JJ et al. A cancer family syndrome in twenty-four kindreds. Cancer Res 1988, **48**, 5358–5362.
10 Malkin D, Li FP, Strong LC et al. Germ line p53 mutations in a familial syndrome of breast cancer, sarcomas and other neoplasms. Science 1990, **250**, 1233–1238.
11 Srivastava S, Zou Z, Pirollo K, Blattner W, Chang E. Germline transmission of a mutated p53 gene in a cancer-prone family with Li–Fraumeni syndrome. Nature 1990, **348**, 747–749.
12 Toguchida J, Yamaguchi T, Dayton S et al. Prevalence and spectrum of germ line p53 gene mutations among patients with sarcoma. N Engl J Med 1992, **326**, 1301–1308.
13 Malkin D, Jolly KW, Barbier W et al. Germline mutations of the p53 tumor suppressor gene in children and young adults with second malignant neoplasms. N Engl J Med 1992, **326**, 1309–1315..
14 Parada LF, Land H, Weinberg RA et al. Cooperation between gene encoding p53 tumour antigen and ras in cellular transformation. Nature 1984, **312**, 649–651.
15 Jenkins JR, Rudge K, Currie GA. Cellular immortalisation by a cDNA clone encoding the transformation associated phosphoprotein p53. Nature 1984, **312**, 651–654.

16 Eliyahu D, Raz A, Gruss P et al. Participation of p53 cellular tumour antigen in transformation of normal embryonic cells. Nature 1984, **312**, 646–649.

17 Baker SJ, Markowitz S, Fearon ER et al. Suppression of human colorectal carcinoma cell growth by wild-type p53. Science 1990, **249**, 912–915.

18 Eliyahu D, Michalowitz D, Eliyahu S et al. Wild-type p53 can inhibit oncogene-mediated focus formation. Proc Natl Acad Sci USA 1989, **86**, 8763–8767.

19 Finlay CA, Hinds P, Levine AJ. The p53 proto-oncogene can act as a suppressor of transformation. Cell 1989, **57**, 1083–1093.

20 Hinds P, Finlay CA, Levine AJ. Mutation is required to activate the p53 gene for cooperation with the ras oncogene and transformation. J Virol 1989, **63**, 739–746.

21 Hinds PW, Finlay CA, Quartin RS et al. Mutant p53 DNA clones from human colon carcinomas cooperate with ras in transforming primary rat cells: a comparison of the hot spot phenotypes. Cell Growth Diff 1990, **1**, 571–580.

22 Milner J, Medcalf EA. Cotranslation of activated mutant p53 with wild type drives the wild type p53 protein into the mutant conformation. Cell 1991, **65**, 765–774.

23 Bischoff JR, Friedman PN, Marshak DR et al. Human p53 is phosphorylated by p60-cdc2 and cyclin B-cdc2. Proc Natl Acad Sci USA 1990, **87**, 4766–4770.

24 Finlay CA, Hinds PW, Tan TH et al. Activating mutations for transformation by p53 produce a gene product that forms an hsc70-p53 complex with an altered half life. Mol Cell Biol 1988, **8**, 531–539.

25 Kern SE, Kinzler K, Bruskin A et al. Identification of p53 as a sequence specific DNA binding protein. Science 1991, **252**, 1708–1711.

26 Kern SE, Kinzler KW, Baker SJ et al. Mutant p53 proteins bind DNA abnormally in vitro. Oncogene 1991, **6**, 131–136.

27 Kern SE, Pietenpol JA, Thiagalinam S et al. Oncogenic forms of p53 inhibit p53-regulated gene expression. Science 1992, **256**, 827–829.

28 Yonish-Rouach E, Resnitzky D, Lotem J et al. wild-type p53 induces apoptosis of myeloid leukaemic cells that is inhibited by interleukin 6. Nature 1991, **352**, 345–347.

29 Diller L, Kassel J, Nelson CE et al. p53 functions as a cell cycle control protein in osteosarcomas. Mol Cell Biol 1990, **10**, 5772–5781.

30 Kastan MB, Onyewekwere O, Sidransky D et al. Participation of the p53 protein in the cellular response to DNA damage. Cancer Res 1991, **51**, 6304–6311.

31 Sarnow P, Ho YS, Williams J, Levine AJ. Adenovirus E1b-58kd tumor antigen and SV40 large tumor antigen are physically associated with the same 54kd protein in transformed cells. Cell 1982, **28**, 387–394.

32 Werness BA, Levine AJ, Howley PM. Association of human papillomavirus types 16 and 18 E6 proteins with p53. Science 1990, **248**, 76–78.

33 Momand J, Zambetti GP, Olson DC et al. The mdm-2 oncogene product forms a complex with the p53 protein and inhibits p53 mediated transactivation. Cell 1992, **69**, 1237–1245.

Molecular Biology for Oncologists
Edited by John Yarnold, Michael Stratton, Trevor McMillan
© 1993, Elsevier Science Publishers B.V. All rights reserved

CHAPTER 9

Inherited predispositions to cancer: applications of molecular biology

Victoria A. Murday

Department of Medical Genetics, St. George's Hospital Medical School,
Cranmer Terrace, London, SW17 0RE, UK

Introduction

Predisposition to cancer occurs in a number of different forms. The rare single gene disorders which give rise to an identifiable phenotype with a high risk of malignancy fall mainly into two groups. The dominant cancer susceptibilities, in which the biological defect is unknown, appear to involve genes important in control of cell growth and differentiation. These susceptibilities often give rise to a sufficiently high risk of malignancy to produce familial clustering of the associated cancers. The recessive types are mainly disorders of DNA repair. The clinical features of these syndromes are summarised in Table 1.

Family studies frequently indicate an increase in risk to first degree relatives of cancer patients. This is the case in relatives of colorectal cancer patients, for example, even when the high risk dominant conditions are excluded. The increase in risk is not sufficient to cause significant clustering of affected individuals in families, and can be due to less penetrant single gene variations or shared environmental factors. Variations in the ability of individuals to deal with carcinogens has been one suggested and investigated mechanism.

Linkage studies help to identify individuals at increased cancer risk

In order to establish linkage of an inherited disease with a DNA marker, both affected and unaffected family members are analysed, and co-inheritance for the

Table 1.

Disease	Inheritance	Clinical features	Sensitivity	Cancer risk	Laboratory tests	Mapping
Xeroderma pigmentosa	AR	Freckle-like lesions, sun exposed areas of skin from early childhood; tissue paper scars; microcephaly and developmental delay in severe type	UV	Basal cell and squamous carcinomas; malignant melanomas	Low level unscheduled DNA synthesis following UV damage; several complementation groups A–F; ↑ urinary hyaluronic acid	A–9q22; F–chr15
Werner	AR	Short stature, sparse hair, beaked nose, scleroderma, high pitched voice, cataracts, premature ageing and greying, diabetes		10% malignancy particularly sarcoma and meningioma	Variegated translocation mosaicism in fibroblast cultures; chromatid gaps and breaks with diepoxybutane and 4-nitroquinoline oxide	8p
Secmanora (Nijmegen breakage syndrome)	AR	Microcephaly with normal or mild mental retardation; immune deficiency	X-ray; bleomycin	Leukaemia. lymphoma	Spontaneous translocations; X-irradiation as for AT; immunoglobulins (variable)	Unmapped
Chediak-Higashi	AR	Partial albinism; neutropenia, anaemia, thrombocytopenia, immunodeficiency, hepatosplenomegaly, may be mental retardation, seizures and peripheral neuropathy		Unusual malignant lymphoma	Cytoplasmic inclusions in leukocytes	Unmapped

Ataxia telangiectasia	AR	Progressive cerebellar ataxia; oculomotor apraxia; telangiectasia best seen in conjunctiva; immune defects, hypoplasia thymus; T cell deficiency; late puberty	X-rays	Lymphoma; lymphocytic leukaemia; heterozygotes raised risk of breast cancer	Several complementation groups. AFP ↓, IgA ↓, IgE ↓; chromatid breakage following X-irradiation or alkalating agents; chromosome rearrangements especially 14q12 translocations; prenatal possible	Complimentation groups A, C and D mapped to 11q22-23
Blooms	AR	Short stature, prenatal onset, <3rd centile; sun sensitive, butterfly rash on face by 1st year; spotty hypo- and hyperpigmented lesions of skin; mild dysmorphism with large nose; severe immune deficiency; common in Ashkenasis; survival to adulthood rare	UV	Acute leukaemia; other malignancies including Wilm's	IgA, IgM ↓; raised spontaneous sister chromatid exchange; prenatal possible	Unmapped
Fanconi	AR	Pancytopenia, anaemia, leukopenia, thrombocytopenia; radial ray anomalies, congenital heart disease, renal abnormalities, pigmentary abnormalities	Alkylating agents	Leukaemia	Spontaneous breaks exchanges or raised levels with DNA cross linking agents	Chromosome 20

[a]AR, autosomal recessive.

V.A. Murday

DNA marker and of the disease is looked for. The frequency of co-inheritance depends upon the distance between the two genetic loci and can be used as a measure of genetic length [1]. The frequency with which the marker locus and disease are inherited (segregate) differently is expressed as the recombination rate. A recombination rate of 0.05 means the marker recombines with the disease 5% of the time (i.e. the marker is inherited with the disease 95% of the time).

Tightly linked markers can be used diagnostically within families, but there are limitations to these DNA tests (Figure 1). Firstly, there must be the right family structure to enable the marker of high risk to be identified. Secondly, enough individuals from the family must be available to give samples. Thirdly, the affected parent of the individual being tested must be heterozygous at the marker locus to be informative, otherwise the chromosome carrying the disease cannot be identified.

Identification of the gene itself offers a solution to this problem. In particular, this helps with a family in which a new mutation appears. The mutation can be sought directly in the affected individual and in other family members. Originally this required sequencing of the gene in its entirety. However, there are now a number of techniques which permit screening for mutations. These are still relatively labour intensive but are becoming more routine.

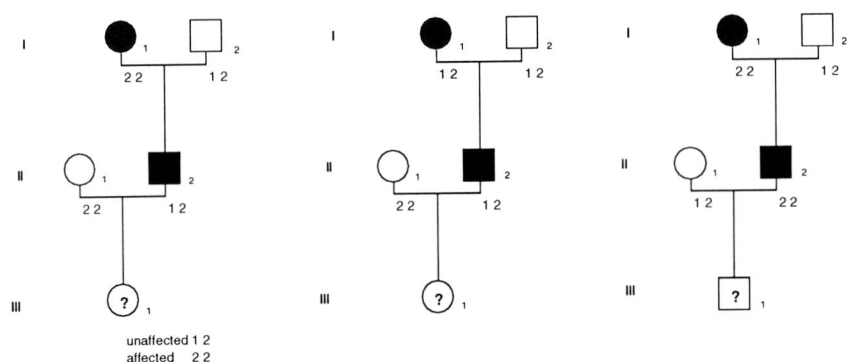

The disease must be linked to allele 2 since I.1 is homozygous 22. If III.1 receives allele 1 from II.2 she will be unaffected. If she receives allele 2 she will be affected. The accuracy will depend on the recombination fraction i.e. if = 0.5 the test will be 95% accurate.

In this family both grandparents are 12. it is therefore impossible to deduce if the 1 or 2 have been passed from I.1 to II.2 with the disease. A test for III.1 is impossible.

Here the disease must be passed with allele 2 to II.2, but he has also received allele 2 from I.2 and is not heterozygote for the marker. A test for III.1 is impossible.

Figure 1. Use and difficulties of diagnosis using linked markers.

Familial polyposis coli

Several well-recognised dominant cancer syndromes have been successfully mapped. Probably the commonest and best known is familial adenomatous polyposis (FAP). The gene for this condition has been mapped to 5q21 [2]. Affected individuals develop a few hundred to over a thousand adenomatous polyps in their large intestine. Adenomas are the benign precursors of malignancies, and an untreated individual with adenomatous polyposis will develop one or more colorectal carcinomas.

Screening should start at about 14 years of age. Many children already have polyps when they are first examined, although the age of onset may be as late as the mid-thirties (Figure 2). Age-specific risks can be used to modify the management of individual patients [3]. Combining information from clinical examination with results of linked markers can further modify age-specific risks. This reduces the need for such frequent examinations for those at low risk, and provides the information necessary for genetic counselling. Age-specific risks are combined with information from the linked markers at 2cM and 5cM, as shown in Figure 3 [3].

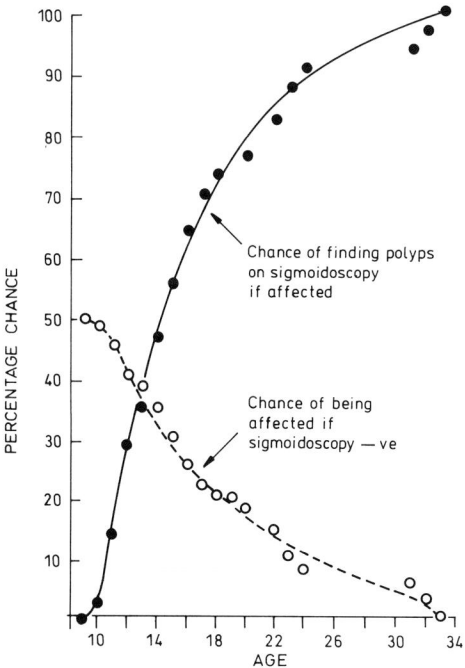

Figure 2. Age of onset of polyps in familial adenomatous polyposis.

As markers are identified closer to the gene, the need for these calculations becomes less important. This is because the linked polymorphic marker will establish the genotype without the risk of recombination, particularly if the marker is within the gene. It is then only necessary to follow up and examine those at high risk.

The identification of the gene itself has meant that even those families that were previously unsuitable for linkage analysis can now be studied by direct mutation analysis.

Prenatal diagnosis in familial polyposis

Another useful application of DNA technology is prenatal diagnosis. Some individuals known to have FAP may feel that a 50% risk to their offspring of developing the disease is too high. A family's experience of the disease will often influence their attitude. For instance, if a couple have lost a child with hepatoblastoma, a

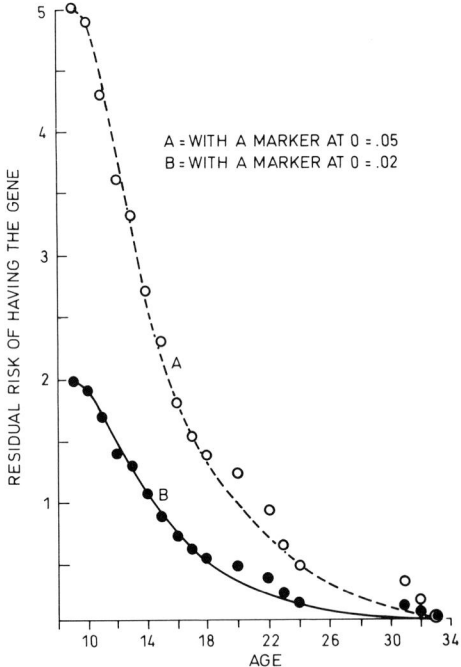

Figure 3. Age specific risks in familial adenomatous polyposis using sigmoidoscopy and DNA marker studies. Combining the results of a negative sigmoidoscopy and inheriting a low risk marker in an individual with a 1 in 2 risk of familial adenomatous polyposis.

malignancy occasionally occurring in FAP, they may feel unable to proceed with another pregnancy without prenatal screening. This procedure can be carried out between 8 and 12 weeks by chorionic villous biopsy or early amniocentesis, allowing parents the option of termination in the first trimester.

Familial breast cancer

Epidemiological and family studies in breast cancer suggest that the genetic contribution to the aetiology of breast cancer is relatively low. However, there is strong evidence for a rare dominant gene that causes susceptibility to breast cancer at a young age, often premenopausal, that is highly penetrant. It is therefore possible to identify families in which there appears to be dominant susceptibility. The individuals in these families have no other known phenotype except the susceptibility to breast cancer and in some families ovarian cancer. Linkage studies have been carried out and it appears that at least half the families are linked to markers on chromosome 17q [4]. However, if these linked markers are to be used diagnostically for susceptible individuals, the family has to be sufficiently large to show co-segregation with 17q markers. Otherwise the family may be unlinked to 17q and the wrong prediction made. Identifying the gene itself will solve this difficulty of genetic heterogeneity, since the mutation in the gene can then be looked for directly in a single affected individual.

Li–Fraumeni syndrome

A second rare genetic form of breast cancer is seen in Li–Fraumeni syndrome. In these families, there is dominant inheritance of susceptibility to early onset breast cancer and other cancers including sarcomas, CNS tumours and leukaemias. Recently, Li–Fraumeni syndrome has been found to be due to an inherited mutation of p53 [5]. In this condition, it is now possible to have a genetic diagnosis by mutation analysis. This type of hereditary breast cancer can now be clearly distinguished from other types at the DNA level. The gene is amplified in sections by PCR. The generated fragments are then either screened for mutations by one of the methods now available, or directly sequenced. The relatives can be screened and counselled. Individuals who are susceptible can then be offered clinical follow-up and screening, whilst those not at risk can be reassured.

p53 and population surveys of cancer susceptibility

When a cancer susceptibility gene is identified, like p53, it provides an opportunity not only to identify those at risk within families but also to ask the question whether cancers arise in the general population as a result of that particular susceptibility gene. At present it is possible to ask how many sarcomas, early breast cancers or second malignancies occur because of germline p53 mutations. A number of population studies have now been done. One hundred eighty-one random sarcoma patients (mostly osteosarcomas) were tested and two were found to have germline mutations. Both had a family history of cancer, in other words about 1% of the sarcoma population [6]. On the other hand, 15 sarcoma patients were selected because of a family history, or a history of multiple cancers, and 6 had germline mutations (3 of these had no family history) [12]. This shows that the population rate is relatively low but that family history or a history of multiple tumours in an individual are good indicators for testing for the mutation.

Extending the phenotype using DNA testing

In the past, the phenotypes of cancer families have been defined clinically. The identification of cancer susceptibility genes enable families to be examined which do not fit the classical phenotype. This has been done using p53, and the polyposis gene on chromosome 5, allowing the clinical phenotype to be expanded. For instance, a hereditary form of bowel cancer similar to polyposis, with a later age of onset and fewer adenomas, has now been linked to the locus on 5q [7].

Identification of germline mutations in children at risk of malignancy

Retinoblastoma

The child of an individual who once had bilateral retinoblastoma has almost a 50% risk of developing retinoblastoma. Individuals presenting with bilateral disease must therefore have germline mutations. In affected children with a positive family history, 68% have bilateral tumours and 32% have unilateral disease, whereas in the general population, 25–30% of children present with bilateral tumours. It is therefore assumed that 10–15% of patients with unilateral tumours and no family history carry germline mutations. However, there is a positive family history in only 10% of children with retinoblastoma, so new mutations are common.

Penetrance for the gene is 90%, so single cases may receive the mutation from an unaffected parent [8]. The risks to siblings are therefore 5% for the siblings of bilaterally affected children, and 1% for the siblings of unilaterally affected individuals. Identification of a germline mutation in affected individuals will allow parents and siblings to be screened. Those truly at risk in families can then be identified and followed-up. In addition, couples who are at risk of having another affected child may be offered prenatal diagnosis.

Of those with germline mutations for retinoblastoma 15–20% develop second tumours, especially osteosarcoma. There is thought to be no increase in risk to those without germline mutations. Again, this illustrates the importance of finding those with a germline mutation, allowing those at risk of a second tumour to be carefully followed-up [8].

Radiation and other forms of treatment are often implicated in the development of second cancers. In the case of retinoblastoma, it appears that radiotherapy will shorten the period between the malignancies. In a case with a germline mutation where a second cancer is likely, it may be possible to modify therapy to take this into account [3].

Cancer genes and development

From the developmental point of view, paediatric oncology is of interest because many of the genes involved may be expressed during development. The emerging understanding of cancer susceptibility in childhood has interesting possibilities for learning about developmental genes. There are several syndromes with both developmental abnormalities and cancer susceptibility. Drash, WAGR, and Beckwith–Wiedemann are associated with a high risk of Wilm's tumour (see Table 2). Turcot's syndrome is associated with astrocytomas, Gorlin syndrome is associated with medulloblastoma, and in addition, there are isolated anomalies such as hemihypertrophy that can be associated with malignancy [9].

The possibility that genes expressed during development may also be involved in the development of cancer is perhaps best illustrated at the current time by the work done on the genetics of Wilm's tumour. It may be inherited as an autosomal dominant but can also be associated with developmental abnormalities. Two to three percent of children with Wilm's tumours have a chromosome deletion at 11p13. Beckwith–Wiedemann syndrome, with a high risk of Wilm's tumour, is sometimes associated with a duplication at 11p15. The deletion at 11p13 is associated with a chromosomal deletion syndrome with the acronym WAGR, standing for Wilm's, Aniridia, Genital anomalies and Gonadoblastoma, and mental Retardation (see Table 3). Some males may be undermasculinised or have complete sex reversal. The Drash syndrome is a related syndrome, but aniridia does not occur. In addition, children with Drash have an early onset neuropathy with mesangial sclerosis.

Molecular studies of the locus at 11p13 have identified the gene responsible, known as WT1. This is a zinc finger protein which functions as a transcription factor. This has enabled the identification of mutations in the Drash syndrome, all of which have so far been located in the important DNA-binding zinc finger motif. It is therefore very likely that the risk of Wilm's tumour and the developmental

Table 2. Developmental anomalies and risk of Wilm's tumour.

Clinical syndrome	Risk of Wilm's tumour (%)
Dominant families	60
Sporadic aniridia	30–50
WAGR	50–60
Drash	50–60
Beckmann Wiedemann with hemihypertrophy	20
Beckmann Wiedemann without hemihypertrophy	5
Hemihypertrophy	1

abnormalities in the genital tract in this condition, and in WAGR, are a direct result of mutations or deletions of this gene [10].

The cloning of the aniridia gene [16], which is close but separate to the Wilm's locus, explains the association of this abnormality with WAGR. It also explains the existence of families with dominant aniridia without evidence of Wilm's susceptibility, and the lower risk of Wilm's with sporadic aniridia than is seen in children with WAGR. In the past, all children with aniridia needed screening for Wilm's. Now, identification in these children of a mutation limited to the aniridia gene means that they are not at increased risk of Wilm's tumour and that they do not require screening.

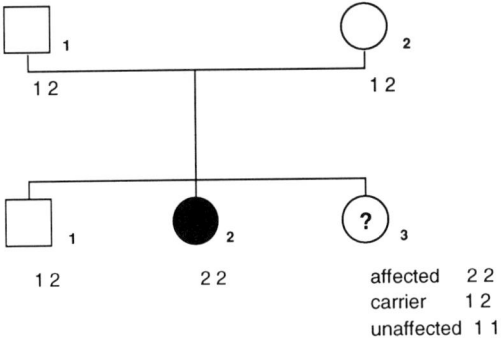

Figure 4. Use of linked markers in diagnosis in a recessive disease. Since the affected offspring is 22 the mutation in both parents is on the chromosome with 2 allelle. Diagnosis is therefore possible for III.3.

Table 3. Clinical features of mapped dominant cancer syndromes.

Disease	Inheritance	Clinical features	Associated tumours	Mapping
Familial polyposis coli	Autosomal dominant	Adenomatous polyposis; epidermoid cysts; supernumery teeth; osteomas (especially jaw); congenital hypertrophy retinal pigment epithelium (CHRPE)	Colorectal cancer; 10% upper gastrointestinal malignancy; thyroid cancer; benign fibrous desmoid tumours (10%)	5q21; gene identified
Von Hippel–Lindau	Autosomal dominant	Retinal haemangiomas; cystic kidneys; pancreatic cysts	Cerebellar haemangioblastomas; cord haemangioblastomas; renal cell carcinoma; phaeochromocytoma	3p25-26
Men.type II	Autosomal dominant	Type II(a) no associated features; type II(b) gastrointestinal tract mucosal neuromata; prominent lips; thickened anti-everted eyelids; marfanoid cubitus; nerve fibres in cornea; cutaneous neurofibromata; developmental delay	Medullary carcinoma thyroid; phaeochromocytoma; parathyroid tumours	10q11.2
Neurofibromatosis type II	Autosomal dominant	Often no cafe au lait spots or peripheral neurofibromata; cataracts	Bilateral acoustic neuromas; meningiomas; Schwannomas of dorsal roots; gliomas	22q11.21 –q13.1
Gorlins	Autosomal dominant	Hypertelorism; macrocephaly; jaw cysts; basal cell nevae; pits in skin especially palms and soles; linear calcification of falx; rib abnormalities; mild developmental delay; ovarian cysts	Medulloblastoma; basal cell carcinomas	9q22.3–q31

DNA repair deficiencies may be associated with cancer susceptibility

Several DNA repair deficiency genes are now mapped, and in some cases the gene itself has been isolated. This allows some families with an inherited repair defect to undertake prenatal or presymptomatic diagnosis. There are limitations to these tests, just as in the dominant cancer susceptibility diseases. In the family, both parents must be heterozygous for the linked marker if the test is to be informative. In addition, they must already have an affected child to study in order to establish the carrier alleles in the parents. Most repair defects present during childhood with associated clinical factors rather than the development of a malignancy [9]. Inherited repair deficiency syndromes are summarised in Table 1.

There are other causes of an inherited cancer susceptibility

In addition to investigating single gene disorders, other types of susceptibility can now be investigated using DNA technology. This has been made possible by the cloning of other candidate susceptibility genes, for instance enzymes involved in carcinogen metabolism, another source of inherited cancer susceptibility. The proportion of the population at increased risk is large and the effects are small. Family trees are not helpful in identifying the population at risk because the effects are not large enough to produce familial clustering. This population can only be identified by testing affected individuals for a specific variation.

Cytochrome P-450

Amongst the best known of such enzymes are the mono-oxygenases, or cytochrome P-450 enzyme system. These enzymes metabolise many substances from inert into reactive, or carcinogenic, compounds. A controlled study of cigarette smokers with and without lung cancer has shown a sixfold reduction in frequency of slow metabolisers of debrisoquine, a drug metabolised by P-450, amongst patients with lung cancer [11]. Most of the P-450 enzymes have now been cloned. This means that it is now possible to study genetic variation in these enzymes at the DNA level. As yet there has not been any convincing confirmation of the lung cancer study, demonstrating the importance of repeating these studies at the DNA level.

Acetylation phenotype

Variation in the ability to acetylate is implicated in susceptibility to malignancies [12]. This applies in particular to individuals exposed to arylamine chemicals in industry. An excess of patients with fast acetylator status is found in breast and colorectal cancer [13,14]. The *N*-acetyl transferase gene involved in this polymorphism has been cloned. It is now possible to type individuals directly, avoiding the problem of enzyme induction. One of the criticisms of the enzyme studies has been that the disease process may alter the activity of the enzyme. Studying the polymorphism at the DNA level avoids this problem.

References

1   Botstein D, White RL, Skolnick M, Davis RW. Construction of a linkage map using restriction fragment length polymorphisms. Am J Hum Genet 1980, **32**, 314–331.
2   Bodmer WF, Bailey CJ, Bodmer J, et al. Localization of the gene for familial adenomatous polyposis on chromosome 5. Nature 1987, **328**, 614–616.
3   Murday V, Slack J. Inherited disorders associated with colorectal cancer. Cancer Surv 1989, **8**; 139.
4   Hall JM, Lee MK, Newman B et al. Linkage of early onset breast cancer to chromosome 17q21. Science 1990, **250**, 1684–1689.
5   Malkin MD, Li FP, Strong LC et al. Germ line p53 mutations in a familial syndrome of breast cancer, sarcoma, and other neoplasms. Science 1990, **250**, 1233–1238.
6   Toguchida J, Yamaguchi T, Dayton SH et al. Prevalence and spectrum of germ line mutations of the p53 gene among patients with sarcoma. N Engl J Med 1992, **326**, 1301–1308.
7   Leppert M, Burt R, Hughes JP et al. Genetic analysis of an inherited predisposition to colon cancer in a family with a variable number of adenomatous polyps. N Engl J Med 1990, **322**, 904–908.
8   Goodrich DW, Lee W-H. Molecular genetics of retinoblastoma. Cancer Surv 1990, **9**, 529–554.
9   Jones KL ed. Smith's Recognizable Patterns of Human Malformation, 4th edition. WB Saunders, 1988.
10 Pritchard-Jones K, Hastie ND. Wilm's tumour as a paradigm for the relationship of cancer to development. Cancer Surv 1990, **9**, 555–578.
11 Ayesh R, Idle JR, Ritchie JC, Crothers MJ, Hetzel MR. Metabolic oxidation phenotypes as markers for susceptibility to lung cancer. Nature 1984, **312**,169–170.
12. Cartwright RA. Historical and modern epidemiological studies on populations exposed to N-substituted aryl compounds. Environ Health Perspect 1983, **49**, 13–19.
13 Bulovskaya LN, Krupkin RG, Bochina AA, Shipkova ??, Pavlova MV. Acetylator phenotype in patients with breast cancer. Oncology 1978, **35**, 185–188.
14 Illett KF, David BM, Detchon P, Castleden WM, Kwa R. Acetylation phenotype in colorectal carcinoma. Cancer Res 1987, **47**, 1466–1467.

Molecular Biology for Oncologists
Edited by John Yarnold, Michael Stratton, Trevor McMillan
© 1993, Elsevier Science Publishers B.V. All rights reserved

CHAPTER 10

Multistage carcinogenesis

Andrew H. Wyllie

CRC Laboratories, Medical School, Teviot Place, Edinburgh, EH8 9AG, UK

Introduction

Carcinogenesis means the acquisition of those cellular properties that render a cell competent to grow into a tumour. Once tumours have formed, further new properties continue to appear in the cells within them, some carrying selective growth advantage and hence of sinister import for the patient. Properly speaking, this evolution of phenotype within tumours is referred to as *tumour progression*, whilst the term *carcinogenesis* refers to the events leading to the appearance of a tumour in the first place [1]. Both sets of events, however, represent changes that lead from normality to aggressive tumour growth, and they are considered together in this chapter.

Evidence from many different sources indicates that the events involved in carcinogenesis and tumour progression are multiple. This chapter describes this evidence, and gives an account of some of the different events known to participate in the development of common human tumours. Finally, certain significant implications of multistage carcinogenesis are discussed.

Evidence for multiplicity of carcinogenic events

Examination of tumour pathology has often suggested that tumours evolve by a series of discrete transitions. Thus there are clearly recognisable morphological distinctions between benign and malignant tumours. Similarly, the terms atypical hyperplasia, intra-epithelial neoplasia and invasive carcinoma reflect different degrees of cytological aberration in structure and behaviour. Often these lesions occur in isolation, but there are also many observations of, for example, invasive carcinoma co-existing and contiguous with intra-epithelial carcinoma, atypical hyperplasia apparently merging with areas of intra-epithelial neoplasia or carcinoma arising within a pre-existing benign tumour. The most obvious interpretation of these histological patterns is that new behaviour is acquired in a sequential, stepwise fashion. There is now molecular evidence in support of this view, some of which is discussed later, but it is much more difficult to prove that a similar stepwise process is a feature of all tumour development.

Multistage carcinogenesis was also recognised in experimental reconstructions of the carcinogenic process. In experiments involving chemical carcinogens, Berenblum distinguished between tumour *initiators* (substances capable of causing tumour growth on their own) and tumour *promoters* that do not cause tumour growth on their own but enhance the effect of initiators [2]. In a series of classical experiments, he showed that initiators altered cells in a permanent manner, that might nonetheless remain undetected by morphological observation until proliferation of the altered cells was stimulated by the promoter. Promoters had no observable effect if applied before initiators, but could dramatically increase the yield of tumours if applied afterwards, even after an interval of many months. Essentially similar observations have since been made with a great many carcinogens and a wide variety of tissues (see [1,3] for reviews). Most initiators are now known to be mutagens, whereas the most widely used promoters, extracted from croton oil, contain tetradecanoyl phorbol 13-acetate (TPA), an analogue of diacylglycerol and a potent stimulator of protein kinase C.

Experiments of the same design as Berenblum's have been repeated with a specific focus on mutations in critical cancer-related genes. In experimental mammary and skin cancer of rodents (see Figure 1), the tumours were shown to be clonal expansions of cells bearing *ras* oncogenes activated by mutations. The mutations were of the specific types expected from cellular exposure to the initiating agents that had been applied long before [4,5]. Moreover, it could be shown that the initiator induces its characteristic mutations in cells that contribute only to hyperplasia, requiring further events before a subclone from them emerges as a carcinoma [6].

From an entirely different angle, epidemiological data have shown that the age-related increase in human tumour incidence, with few exceptions, follows a steeply rising curve. The shape of this curve is amenable to mathematical modelling, and one model interprets the slope, following log–log transformation, to indicate the number of separate events involved in carcinogenesis. For the great majority of adult-onset carcinomas, this slope lies close to five or six. For some childhood tumours it is less, but it is always greater than unity, suggesting that carcinogenesis always involves more than one cellular event [7,8].

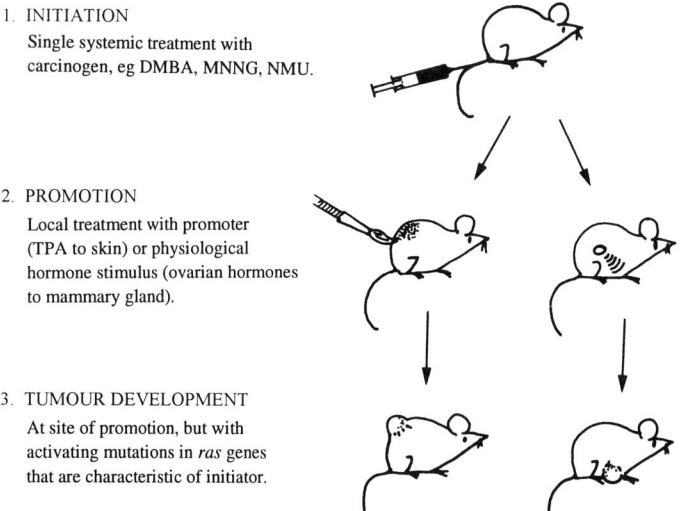

1. INITIATION
   Single systemic treatment with
   carcinogen, eg DMBA, MNNG, NMU.

2. PROMOTION
   Local treatment with promoter
   (TPA to skin) or physiological
   hormone stimulus (ovarian hormones
   to mammary gland).

3. TUMOUR DEVELOPMENT
   At site of promotion, but with
   activating mutations in *ras* genes
   that are characteristic of initiator.

Figure 1. Diagram of experiments that demonstrate initiation and promotion at the molecular level (see [4,5]).

Most of this chapter is devoted to the unravelling of what these multiple carcinogenic events might be, and how they interact with each other. As a cautionary note, it should be pointed out that it is always easier to discuss areas of science where good data exist than to peer into the darker regions where information is not available. This chapter links genetic events strongly to multistage carcinogenesis and tumour progression because there is now a solid factual basis for doing so. It is possible that other types of event also occur, such as alteration in the level of transcription of important genes without requiring structural changes in the genes themselves. An example might be the expression of the cell attachment molecule E-cadherin, which appears to be depressed preferentially in metastatic tumours. As much less is known of such modifications in the cellular levels of gene products, or of how they might occur in a fixed manner in cancer, they are not further discussed here.

Genetic changes in human cancers

Many genetic abnormalities are known to occur in tumours. Some of these are not site-specific. Thus a generalised *hypomethylation* has been described in many tumours, both benign and malignant [9]. Very little is known of the regulation of this change in the methylation status of the genome, but it does not appear to be specific to particular sets of genes. It has been postulated that this hypomethylated DNA in premalignant cells is more susceptible to mutagenesis, perhaps because of less tight chromatin packing and correspondingly enhanced carcinogen access.

Similarly, a high proportion of all malignant tumours are aneuploid or show other evidence of *karyotypic instability*, such as the presence of bizarre marker chromosomes, double minutes or truncated telomeric sequences. Unlike hypomethylation, aneuploidy is seldom a feature of benign tumours. A superficial assessment of the karyotypes of aneuploid tumour cells reveals a wide variety of abnormalities and a broad range of DNA content. However, two large groups can often be discerned: "near diploid" tumours, that differ from normal to a relatively small degree, perhaps by virtue of a single extra chromosome, or part of a chromosome arm; and "near tetraploid" tumours that contain double sets of most chromosomes. These patterns suggest that different mechanisms may be at work in the genesis of tumour aneuploidy, some based on damage to or non-disjunction of individual chromosomes, and others involving an endoreduplicative event in which two rounds of chromosome replication occur without intervening cell division [10]. Despite the fact that karyotypic instability in tumours has a large number of possible outcomes in terms of gene number, integrity and expression, it now appears probable that it may result from disorder in only a few critical genes.

Specific genetic alterations are acquired in carcinogenesis in *oncogenes* and *oncosuppressor genes*. Human colorectal tumours [11], and to a lesser extent tumours at other sites such as the breast, bladder, cervix, ovary and lung, have been studied for such changes in some detail. About 40% of all colorectal tumours include clonal expansion of cells that carry a Ki-*ras* gene activated by point mutation. Around 50–60% bear a mutated p53 gene, and between 60 and 80% have nonsense or stop codon mutations in *APC* [12], the gene responsible for transmission of familial adenomatous polyposis. Allele deletions that involve the p53 locus on chromosome 17p are found in around 70%, the *DCC* locus on chromosome 18q in 70–80%, and the *APC* locus on chromosome 5q in about 50%. Other allele deletions, presumably indicating the location of further, but as yet undiscovered oncosuppressor genes, are observed in chromosomes 8p, 1p and 22q [13]. Very few colorectal carcinomas fail to show clonally expanded alterations at any of these sites, and most have multiple defects. However, adenomas and carcinomas with proven better prognosis tend to carry fewer genetic defects than carcinomas with poor prognosis [14].

Rather similar principles appear to apply to carcinomas at other sites. Thus breast carcinoma, in addition to mutation and allele deletion of p53, frequently shows amplification of a region in chromosome 17q that includes the proto-oncogene c-*erb*B2, and allele deletions at a wide variety of other sites, including loci on chromosome arms 1q, 3p, 6q, 16q, 17q and 18q. In lung carcinomas, the L-*myc* oncogene is often amplified whilst a proportion show activating Ki-*ras* mutations. Allele deletions are also common at many loci, including p53, the RB (retino-blastoma susceptibility) oncosuppressor gene at 13q14, and still anonymous loci in 3p and 11p. Ovarian cancers show a different spectrum of allele loss (4p, 6p, 7p, 12p, 19q as well as 17p and 17q). Interestingly, cervical carcinoma, in the majority of which oncogenic human papillomavirus (HPV) genomes are present, seldom show alterations at the loci that are commonly involved in other tumours, including mutation and deletion of p53. These examples show how different tumours tend to show differing but overlapping combinations of frequently affected genes. This pattern of multiple gene abnormality in tumours raises several questions. Does each abnormality represent a distinct stage in tumour evolution, or is the critical factor in tumour progression merely the total number of defects? Do the genetic events become added sequentially in time, or do they occur simultaneously in those cells that are the targets of carcinogen attack? Do the events interact with each other? Are there alternative routes to the same end-point? The answers to these questions bear on the whole concept of multistage carcinogenesis and tumour progression, and are discussed below.

Multiple genetic events: stage-specific or cumulative?

Early anecdotal observations in colorectal tumours suggested that some genetic lesions could be acquired at almost any stage of tumour evolution, and the statement is often reiterated that it is the total number of defects rather than their nature that determines tumour behaviour. Much subsequently gathered data, however, indicate that certain genetic defects preferentially appear at specific stages of tumour progression, presumably because they initiate or permit the clonal expansion of cells with characteristic neoplastic behaviour (see Figure 2) [15]. Thus Ki-*ras* mutations in colorectal cancer appear in the larger and more villous adenomas with the same incidence as in carcinomas, but are relatively rare in small, tubular

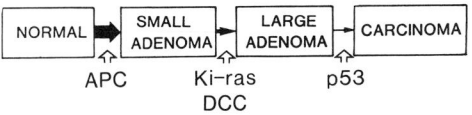

Figure 2. Sequence of pathological changes in colorectal carcinogenesis with stages at which several acquired gene abnormalities probably exert their major effects.

adenomas. Where a Ki-*ras* mutation appears in a carcinoma that bears a contiguous fragment of adenoma (from which, presumably, it had derived), the same mutation is often found in the adenoma. Thus the evidence supports the view that presence of a mutationally activated Ki-*ras* gene confers growth advantage on cells within the larger adenomas, but offers little to the cells of established carcinomas.

Point mutations in *APC* also occur with the same frequency in adenomas and carcinomas, but here even small adenomas are involved [12]. Indeed, point mutations in *APC* appear to be a major determinant for adenoma growth: where such mutations are inherited (as in the Mendelian dominant condition of familial adenomatous polyposis), mucosa that is heterozygotic for the mutation engages in adenoma formation, usually to a dramatic extent.

Mutations in p53 show an entirely different pattern in colorectal carcinogenesis. They are rare in adenomas of all kinds, even those that are undergoing transition to carcinoma, but are relatively common in carcinomas [16]. Thus, p53 mutation appears to confer no growth advantage on cells evolving from normal to adenoma, but provides a potent advantage to carcinoma cells. Moreover, the advantage seems to be limited to cells early in the process of growing as carcinoma: divergent stemlines of carcinomas that bear no p53 mutation tend not to acquire one [17]. The reason for this distinctive relationship is becoming clearer. One normal function of p53 appears to be to arrest entry to DNA synthesis as a response to DNA damage, probably as part of a mechanism that facilitates repair before the damaged DNA perpetuates its errors through replication [18,19]. Cells that lack a functional p53 gene therefore tend both to accumulate DNA damage and to engage in premature DNA replication. This has the effect of increasing the frequency of gene amplification, a known stimulus for abnormal recombinational and deletional chromosomal events [20–22]. Cells with defective p53 function might also be expected to become tetraploid through endoreduplicative mitosis. In these ways, mutation in p53 may facilitate the development of aneuploidy, although it is probably not the only genetic abnormality that does so.

Multiple genetic events: sequential or simultaneous?

There is much to support the view that genetic abnormalities are acquired in presumptive tumour cells in a sequential fashion with time. Foremost is evidence from persons who inherit a tumour susceptibility gene, such as a mutated copy of *APC*, *RB*, p53 or the genes responsible for ataxia telangiectasia. Such individuals are usually born tumour-free, but tumours develop during childhood or early adulthood, as further genetic lesions are acquired in the cells of the tissues at particular risk of carcinogenesis.

Nonetheless, there are difficulties in accepting that each genetic event in a carcinoma occurs sequentially in time and independent of the others. For example, it is not unusual to find carcinomas of the colorectum that bear five or six of the defects mentioned earlier (mutation of Ki-*ras*, mutation and deletion at the p53, *DCC* and *APC* loci, deletion of 8p alleles, aneuploidy). Tumours with only two or three of these lesions are also common. The question therefore arises how sufficient concentrations of carcinogen can be made available to colorectal stem cells so as to deliver five consecutive hits to one, without ensuring that many more cells receive two or three. Yet five-defect carcinomas can occur as single tumours; colorectal specimens bearing multiple carcinomas are relatively rare.

In theory, there are several ways to rationalise this situation. It is possible that all cancers require five or six hits (as suggested by the epidemiological studies mentioned above) and those in which smaller numbers are recorded merely represent those bearing genetic alterations at hitherto unidentified chromosomal loci. In view of the negative correlation between number of identified genetic events per tumour and patient prognosis, this seems unlikely. Alternatively, mutagenic events may be clustered within certain regions, and even within certain cells of the mucosa. Careful colonoscopic studies have shown that adenomas do indeed tend to occur in clusters. An extreme version of this perspective would be that some cells may receive multiple simultaneous hits in the course of a single exposure to carcinogen. Finally, the mutagenic events may not be entirely independent of each other, so that the progeny of cells that have undergone one or two events may be much more susceptible to undergo more. Certainly some acquired genetic lesions in tumours occur together more frequently than would be predicted from chance association, suggesting potential interactions between them. Some of these are explored in the next section.

Potential interactions between genetic events

Experiments with cultured rodent fibroblasts reveal that carcinogenesis requires cooperation between certain groups of oncogenes [23]. Transformation in vitro and tumorigenicity in vivo occur in diploid embryo fibroblasts when they are forced (by gene transfer) to express a mutated *ras* oncogene together with mutated p53, c-*myc* or the adenoviral oncogene E1b. None of these genes on its own produces a transformed phenotype in such cells. In contrast, the *ras* oncogenes are highly effective in transforming rodent fibroblasts that have been immortalised through prolonged culture in vitro. Presumably the process of prolonged culture permits or selects for changes that have equivalent effects to expression of the cooperating genes described above. It has been pointed out that this cooperation between oncogenes often involves interaction between a membrane-associated oncogene protein (e.g. p21^{ras}) and one that enters the nucleus (e.g. p62^{c-myc}, p53), but this association with cell compartments may be fortuitous.

Further evidence for gene cooperation in carcinogenesis comes from studies with transgenic animals in which foreign genes are inserted into the embryonic stem cells and so appear in cells that colonise all tissues including the germline. In the progeny of such animals, it is possible to ensure that entire organs are populated with cells that both contain and express a potentially oncogenic transgene. For example, mice have been produced in which all breast epithelial cells express an activated *ras* oncogene, introduced in a progesterone-sensitive construct and hence specifically expressed in progesterone target tissues [24]. Despite the universal expression of the oncogene in the target tissue, only a few cells there undergo malignant change. The remainder sometimes show hyperplasia. Hence there must be additional carcinogenic factors that cooperate with the inserted oncogene, but are specific to only a few cells in the gland. Rather similar results have been obtained from other experiments of this design, aimed at expressing different oncogenes in a variety of tissues. Only in the case of *erb*B$_2$ expression in mammary epithelium did widespread malignant transformation follow the expression of a single oncogene [25], and it is possible that this could be attributed to unusual, genetically determined susceptibility in the animals used.

Further examples of gene cooperation in carcinogenesis are suggested by observations in human tumours. In lymphomas, a frequent association is c-*myc* dysregulation with overexpression of the oncogene *BCL*-2. c-*myc* expression appears to induce a "high-turnover" state in which cells may either proceed around the replication cycle or die by apoptosis, depending on their circumstances at the time [26], whereas *BCL*-2 facilitates survival and inhibits induction of apoptosis [27]. Hence the co-existence of these two oncogene abnormalities may have a ready explanation.

In tumour progression, loss of heterozygosity at the p53 locus frequently co-exists with mutation of the residual p53 allele. This can be interpreted to mean that residual wild-type p53 still exerts some tumour-suppressing effect, despite the "dominant negative " nature of the mutation. For rather less obvious reasons, loss of the residual wild-type *ras* alleles appears to be associated with tumour progression in cells of animal tumours that bear activated *ras* oncogenes [4].

Some tumours bear specific genetic defects at unexpectedly low frequency. These may indicate that one genetic event can supply the tumorigenic stimulus more usually associated with another. Attention to such absences of involvement may therefore lead towards the definition of critical pathways in carcinogenesis, dependent on multiple gene products, only one of which need be abnormal to subvert the whole (Figure 3). Thus sarcomas seldom show mutations in p53, but commonly amplify the *mdm* 2 gene, which codes for a protein that binds, and per-

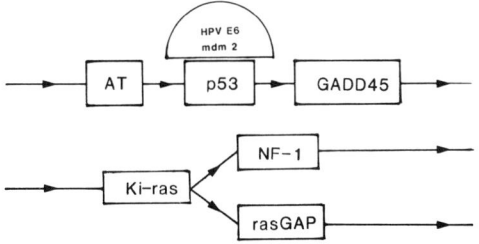

Figure 3. Diagrammatic summary of two pathways that may cooperate in carcinogenesis. The upper line shows the DNA repair pathway involving p53, and the lower line a portion of the *ras* signalling pathway. Abnormality in any one element in either pathway may influence the efficiency of the pathway as a whole. If carcinogenesis requires both pathways to be defective together, it is clear that a wide variety of possible combinations of abnormality could be responsible. In more detail, AT represents the products of the ataxia telangiectasia genes. These may be responsible for detecting DNA damage, and are abnormal in the recessive disorder of ataxia telangiectasia, a cancer-prone condition in which chromosome breaks and other manifestations of karyotypic instability are seen. The half-moon shape represents proteins that bind p53 and prevent it from interacting with its downstream effector sites. Two such proteins are known: the papillomavirus oncoprotein E6 and the endogenous proto-oncogene product, MDM2. GADD45 is a protein induced by p53 after DNA damage, and may be the effector molecule responsible for cell cycle arrest in G_1 phase (see [20,28]). The activity of the Ki-*ras* protein is regulated by at least two other proteins that activate its GTPase and so eventually terminate the *ras* signal, rasGAP (GTPase activating protein and NF-1). Oncogenic mutations in *ras* genes, such as occur in many tumours, produce proteins in which the GTPase activity is sustained. NF-1 is abnormal in some tumours, and is inherited in defective form in the tumour-susceptibility syndrome, neurofibromatosis.

haps limits the effectiveness of cellular p53 [28]. Similarly p53 mutations are seldom seen in cervical carcinoma cells in which there is infection with HPV 16 or 18, but do appear in the minority in which HPV is not involved [29]. Presumably altered p53 function is critical for the malignant transformation of these cells, but may arise either through mutation of the p53 gene itself, or sequestration of its product through binding to the viral E6 oncoprotein. Mutation of p53 already inactivated through binding to HPV E6 would not be expected, as it would convey no further growth advantage to the cell in which it had occurred.

These observations invite further speculation. Although many different combinations of genetic events appear in tumours at different sites, it is possible that the essence of multistage carcinogenesis is to alter a rather small number of critical pathways within the cell, by affecting one element within each. Clear knowledge of these pathways, and how to re-instate them by pharmacological means, might then lead to new means of treatment of rather widespread applicability. Some encouragement for this hope comes from experiments in which intact normal onco-suppressor genes, or fragments of chromosomes that contain them, are introduced into cancer cells, with resulting loss of tumorigenicity [30]. Thus, despite the fact that many steps may be needed to lead from normal to cancer, reversal of even one of them may have profound effects on cellular behaviour.

References

1   Farber E. The multistep nature of cancer development. Cancer Res 1984, **44**, 4217–4223.
2   Berenblum I, Shubik P. An experimental study of the initiating stage of carcinogenesis, and a re-examination of the somatic cell mutation theory of cancer. Br J Cancer 1949, **3**, 109–118.
3   Pitot HC, Sirica AE. The stages of initiation and promotion in hepatic carcinogenesis. Biochim Biophys Acta 1980, **605**, 191–215.
4   Balmain A, Brown K. Oncogene activation in chemical carcinogenesis. Adv Cancer Res 1988, **51**, 147–182.
5   Zarbl HS, Sukumar S, Arthur AV et al. Direct mutagenesis of Ha-*ras*-1 oncogenes by N-nitroso-N-methyl urea during initiation of mammary carcinogenesis in rats. Nature 1985, **315**, 382–385.
6   Miyamoto S, Sukumar S, Guzman RC et al. Transforming c-Ki-*ras* mutation is a preneoplastic event in mouse mammary carcinogenesis induced in vitro by N-methyl-N-nitrosourea. Mol Cell Biol 1990, **10**, 1593–1599.
7   Armitage P, Doll R The age distribution of cancer and a multistage theory of carcinogenesis. Br J Cancer 1954, **8**, 1–12.
8   Stein WD Analysis of cancer incidence data on the basis of multistage and clonal growth models. Adv Cancer Res 1991, **56**, 161–182.
9   Jones PA, Buckley JD The role of DNA methylation in cancer. Adv Cancer Res 1990, **54**, 1–23.
10 Shackney SE, Smith C, Miller BW et al. Model for genetic evolution of human solid tumors. Cancer Res 1989, **49**, 3344–3354.
11 Hamilton SR. Molecular genetics of colorectal cancer. Cancer 1992, **70**, 1216–1221.
12 Powell SM, Zilz N, Beazer-Barclay Y et al. APC mutations occur early during colorectal tumorigenesis. Nature 1992, **395**, 235–237.
13 Vogelstein B, Fearon ER, Kern SF et al. Allelotype of colorectal carcinomas. Science 1989, **244**, 207–211.
14 Offerhaus GJA, De Feyter EP, Cornelisse CJ et al. The relationship of DNA aneuploidy to molecular alterations in colorectal carcinoma. Gastroenterology 1992, **102**, 1612–1619.
15 Fearon ER, Jones PA. Progressing towards a molecular description of colorectal cancer development. FASEB J 1992, **6**, 2783–2790.
16 Purdie CA, O'Grady J, Piris J et al. p53 expression in colorectal tumors. Am J Pathol 1991, **138**, 807–813.
17 Carder PJ, Wyllie AH, Purdie CA et al. Aneuploid clonal divergence in colorectal carcinoma is associated with abnormal p53. Unpublished.
18 Kastan MB, Onyekwere O, Sidransky D et al. Participation of p53 protein in the cellular response to DNA damage. Cancer Res 1991, **51**, 6304–6311.
19 Lane DP. p53: the guardian of the genome. Nature 1992, **358**, 15–16.

20 Kastan MB, Zhan Q, El-Deiry WS et al. A mammalian cell checkpoint pathway utilizing p53 and GADD45 is defective in ataxia telangiectasia. Cell 1992, **71,** 587–597.

21 Livingstone LR, White A, Sprouse J et al. Altered cell cycle arrest and gene amplification potential accompany loss of wild type p53. Cell 1992, **70,** 923–936.

22 Yin Y, Tainsky MA, Bischoff FZ et al. Wild type p53 restores cell cycle control and inhibits gene amplification in cells with mutant p53 alleles. Cell 1992, **70,** 937–948.

23 Weinberg RA. The action of oncogenes in the cytoplasm and nucleus. Science 1985, **230,** 770–776.

24 Sinn E, Muller W, Pattengale P et al. Coexpression of MMTV/ v-Ha-*ras* and MMTV/c-*myc* genes in transgenic mice: synergistic actions of oncogenes in vivo. Cell 1987, **49,** 465–475.

25 Muller WJ, Sinn E, Pattengale P et al. Single step induction of mammary adenocarcinoma in transgenic mice bearing the activated c-*neu* oncogene. Cell 1988, **54,** 105–115.

26 Evan G, Wyllie AH, Gilbert C et al. Induction of apoptosis in fibroblasts by c-*myc* protein. Cell 1992, **69,** 119–128.

27 Fanidi A, Harrington EA, Evan GI. Cooperative interaction between c-*myc* and *bcl*-2 proto-oncogenes. Nature 1992, **359,**554–556.

28 Oliner JD, Kinzler KW, Meltzer PS et al. Amplification of a gene encoding a p53-associated protein in human sarcomas. Nature 1992, **358,** 80–83.

29 Crook T, Wrede D, Tidy JA et al. Clonal p53 mutation in primary cervical cancer, association with human papilloma virus-negative tumours. Lancet 1992, **339,** 1070–1073.

30 Goyette ML, Cho K, Fasching CL et al. Progression of colorectal cancer is associated with multiple tumor suppressor defects but inhibition of tumorigenicity is accomplished by connection of any single defect via chromosome transfer. Mol Cell Biol 1992, **12,** 1387–1395.

Molecular Biology for Oncologists
Edited by John Yarnold, Michael Stratton, Trevor McMillan
© *1993, Elsevier Science Publishers B.V. All rights reserved*

CHAPTER 11

Human papillomavirus (HPV) and cervical cancer

Rachel C. Davies and Karen H. Vousden

Ludwig Institute for Cancer Research, St. Mary's Hospital Medical School,
Norfolk Place, London W2 1PG, UK

Epidemiological data first linked HPV infection with cancer of the cervix

Cervical cancer was recognised as a sexually transmitted disease more than a century ago and since then numerous infectious agents have been suggested to play a causative role. The link between human papillomaviruses (HPVs) and cervical carcinoma was first suggested in 1976 [1]. Subsequently, a large body of epidemiological and experimental evidence has accumulated to support a role for HPVs in the development of these malignancies [2]. HPV DNA can be detected in about 90% of tumours, although the HPVs associated with such malignancies are of specific types (most frequently HPV 16 and 18), designated the "high risk" HPV types. Other common genital HPV types, such as HPV 6 and 11, almost exclusively give rise to benign lesions or warts and are therefore considered "low risk" HPV types. Despite the strong association between HPV infection and cervical carcinoma, it is difficult to assign an absolute role to the virus, since infection with a high risk papillomavirus type alone is not sufficient to induce malignant growth [3]. This is indicated both by the long phases between initial viral infection and tumour formation and the observation that not all women carrying a high risk HPV infection develop cervical cancer. Whilst it seems clear that HPV infection represents only one stage in a multistep carcinogenic process, there is evidence that expression of the HPV-encoded oncoproteins is necessary for the maintenance, as well as the development, of the malignant phenotype. The prospect of targeting viral products for therapeutic intervention has prompted an enthusiastic research effort aimed at understanding the mechanisms by which the viral proteins function. Although still not fully resolved, it seems that one of the mechanisms by which the HPVs contribute to malignant development is by interference with some of the important negative growth signals through which the cell normally regulates proliferation.

Papillomaviruses are small DNA viruses exhibiting strong host and tissue tropism making them difficult to study in experimental systems

HPVs are small DNA viruses, consisting of 8 kb of double stranded DNA enclosed within a 55 nm viral capsid. The viral genome encodes 9 major open reading frames (Figure 1) which are expressed as proteins, although the complexity of expression is increased by multiple splicing patterns. Studies of both human and animal papillomaviruses have allowed the assignment of at least one role to most of the major protein products (Table 1), although it is clear that the function of each protein may not be identical in different papillomavirus types [4]. HPVs are epitheliotropic viruses and viral gene expression is dependent on the host cell growth and differentiation, late gene expression and viral particle production only occurring in terminally differentiated keratinocytes. This stage of differentiation is extremely difficult to mimic in culture and no simple experimental system for viral growth is available, resulting in a very poor understanding of the natural life cycle of the virus. A recent report of virion production in a highly specialised system involving the grafting of cell layers onto athymic mice highlights the difficulties of

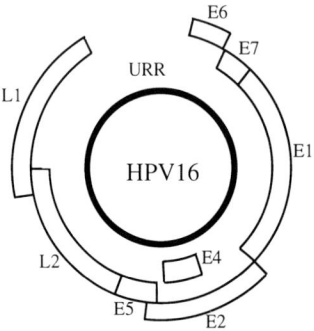

Figure 1. Genome of HPV 16 showing the position of major open reading frames with the potential to encode proteins and the upstream regulatory region containing transcription and replication control elements. Functions of the proteins encoded by the various open reading frames are described in Table 1.

Table 1. Some functions of high risk genital HPV encoded proteins.

Early proteins
E1	Control of replication
E2	Control of transcription
E5	Transformation (not expressed in cancers)
E6	Immortalisation/transcriptional control
E7	Immortalisation/transcriptional control

Late proteins
L1	Structural protein
L2	Structural protein
E4	Disruption of cell keratin cytosketeton

Non-coding region
URR	*Cis* acting transcription and replication control

studying the virus [5]. Consequently, most of the work on HPV has used viral sequences cloned into bacterial plasmid vectors. Using such techniques, the entire HPV 16 genome can be introduced into normal human genital keratinocytes, resulting in the immortalisation of these primary cells [6]. Keratinocytes infected and immortalised with HPV 16 are unable to differentiate normally and under appropriate culture conditions closely resembled pre-malignant cervical lesions, supporting the association between infection with a high risk HPV type and the development of a pre-cancerous lesion [7].

Integration of viral sequences into host DNA may be an important step in carcinogenesis

In benign productive lesions, such as warts, HPV DNA is maintained episomally in the nucleus of the host cell, often at a high copy number. In tumour cells, however, viral DNA is frequently integrated, resulting in the loss or lack of expression of most of the viral genome [8]. The only genes which are consistently retained and expressed under these circumstances are E6 and E7, leading to the hypothesis that these two proteins are involved in the development of the malignancy. Although the consequences of integration are not fully understood, it may be of some significance that the E2 and E1 open reading frames are frequently interrupted. The E2 and E1 proteins are involved in control of transcription and replication and a recent study has shown that mutations resulting in loss of expression of either E2 or E1 can increase the immortalising activity of HPV 16 DNA [9]. It therefore seems likely that integration is an important step in malignant progression, resulting, at least in part, in the de-regulated expression of the E6 and E7 oncoproteins.

Viral E6 and E7 proteins are responsible for cellular transformation and immortalisation

The hypothesis that E6 and E7 are the major transforming proteins gained further credence with the demonstration that both proteins have transforming and immortalising activities in cultured cells. Primary human genital epithelial cells, the natural target cell for the virus, can be fully immortalised by cooperation between E6 and E7, and E7 alone shows both transforming and immortalising activities in rodent cells [10]. Although immortalisation of cells in culture is a rather artificial assay, the relevance of this activity of E6 and E7 to the development of human cancers is supported by two observations. Firstly, the E6 and E7 proteins encoded by the high risk HPV types function much more efficiently in all these assays than E6 and E7 encoded by the low risk HPV types, providing a satisfying correlation between in vitro transforming activity and apparent malignant potential. Secondly, human keratinocytes immortalised by E6 and E7 remain phenotypically untransformed and do not become tumorigenic. Immortalisation in culture therefore seems to represent only a step in the malignant conversion of these cells, mimicking the contribution of HPV infection to the development of cervical cancer.

E6 and E7 certainly play a role in the normal life cycle of the virus and their ability to activate transcription from heterologous promoters may be a reflection of such activity. However, these proteins are also clearly involved in the oncogenic activity of the HPVs and this property has attracted a great deal of research effort over the last few years, aimed at elucidating the mechanisms by which E6 and E7 function within the cell. Interestingly, another HPV 16 protein, E5, has also recently been shown to encode transforming activity in rodent cells, although the significance of this activity to the malignant development in vivo is not clear since expression of this region is frequently lost in cervical carcinomas [11].

E6 and E7 are zinc binding proteins with some similarity to each other

E6 and E7 are small (100–150 amino acids) nuclear proteins which bind zinc via a C-terminal cysteine motif (Figure 2). The zinc binding regions of E6 and E7 show limited conservation and it has been proposed that the proteins may have arisen by amplification and divergence of a 33 amino acid peptide [12]. The zinc binding

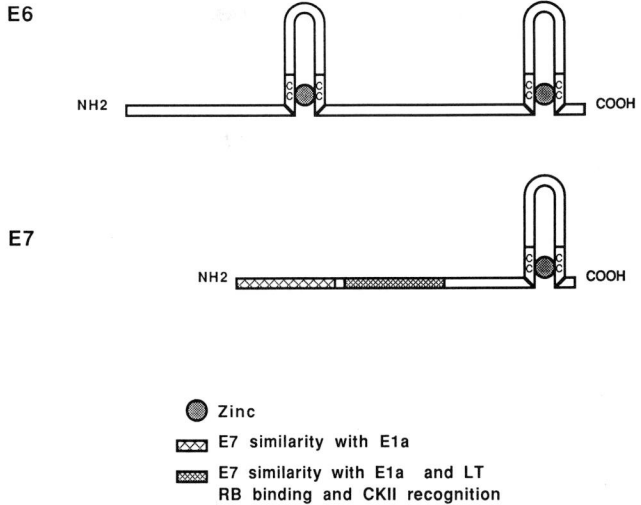

Figure 2. Structure of HPV E6 and E7 proteins showing similarity between them in the cysteine motifs which form zinc binding domains. Regions of homology between E7, E1a and LT are also indicated, as is the region of E7 which is responsible for Rb binding and contains recognition sequences for phosphorylation by CKII. No sequence similarity between E6 and any other protein has yet been described although functional similarities between E7, E1b and LT in their ability to bind p53 are described in the text. From [10]. Reproduced with permission.

cysteine motifs appear to be important for the stability of both proteins and are therefore essential for E6 and E7 function. Recent studies, however, have shown that other regions of these small proteins play a much more direct role in the activity of E6 and E7 by contacting other cell encoded proteins.

The E7 protein disrupts cell cycle control by binding Rb and other cell proteins

The amino terminal half of E7 contains regions of similarity with transforming proteins of other DNA tumour viruses, namely adenovirus E1a and SV40 large T antigen (Figure 2) stimulating the idea that these proteins act, at least in part, by a common mechanism [13]. This region of similarity has now been shown to play a role in the ability of all three proteins to form a complex with the cell encoded retinoblastoma gene product (Rb) (see Figure 3, p. 153), and also provide the recognition signals for phosphorylation by casein kinase II. Mutations within this region of E7 have shown that both Rb binding and phosphorylation are important for efficient transforming activity in rodent cells [14] and the ability to target the Rb protein probably represents at least one important function of E7 in the development of a malignant cell.

The significance of the ability of viral oncoproteins to associate with Rb has only been fully appreciated with a more detailed understanding of the normal function of Rb. The retinoblastoma gene was originally identified in children suffering from hereditary retinal tumours and is described in more detail in Chapter 1. In these individuals, one mutant *Rb* allele is inherited in the germline from one parent resulting in a predisposition to disease. Subsequent tumour development is associated with a second mutation in the other allele, resulting in the complete loss of Rb protein. The correlation between loss of Rb and malignancies suggests that the normal function of Rb is to negatively regulate cell growth and it is now clear that *Rb* belongs to a family of tumour suppressor genes, encoding proteins that play a role in preventing cell division and tumour formation [15]. Recent studies have shown that Rb inactivation occurs in many other types of tumours, as perhaps might be anticipated for a protein with a proposed role in tumour suppression. Further evidence of a negative role for Rb in cell growth has been provided by studies in which the reversion of malignant phenotype of several Rb negative tumour cell lines was achieved by re-introducing a wild-type *Rb* gene.

The mechanisms by which Rb negatively regulates cell growth are not fully understood although recent evidence has indicated that one important function may be to regulate the activity of various cell transcription factors. Rb has been shown to be associated with transcription factors such as E2F, DRTF1 and c-*myc* during certain stages of the cell cycle and a simple hypothesis is that the association with Rb inactivates the transcription factors, thereby preventing expression of genes important for progress through the cell cycle. Cell cycle-dependent sequential phosphorylation of Rb indicates that the activity of Rb may be regulated by a kinase which is activated as the cell proceeds through division. Phosphorylation of Rb correlates with the release the bound transcription factors, allowing progress through the cell cycle (Figure 3). Interestingly, E7 has recently been shown to interfere with the formation of complexes between Rb and transcription factors [16]. It is tempting to speculate that expression of E7 overcomes the suppression of growth normally mediated by Rb at certain stages of the cell cycle by preventing the normal association between Rb and other cell proteins (Figure 3) [17].

Despite the apparent importance of Rb binding to E7 function, it is clear that the ability to bind Rb is, by itself, not sufficient for efficient transforming activity. Another potentially important function of E7 is the ability to bind an Rb related protein, p107, which is itself able to bind to the transcription factor E2F in a similar manner to Rb. The p107/E2F complex forms later during the cell cycle than the Rb/E2F complex, once cells have begun to replicate their DNA, and it is

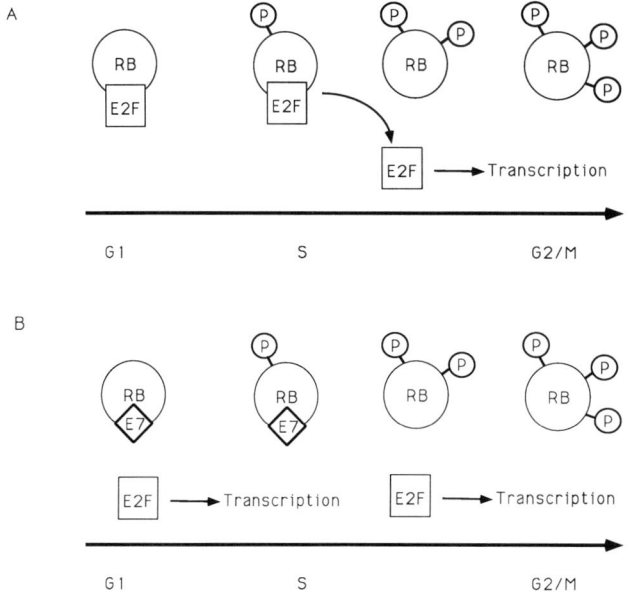

Figure 3. Model for Rb function. During the G1 stage of the cell cycle, Rb is hypophosphorylated and binds transcription factors such as E2F or *myc*, preventing their function. In a normal cell (A), sequential phosphorylation of Rb results in the release of transcription factors, allowing expression of genes necessary for progress through the cell cycle. In HPV infected cells (B), interference with the Rb/transcription factor complex by E7 allows inappropriate expression of these genes and loss of normal regulation of cell division.

anticipated that E7 would also be able to disrupt the p107/E2F complex, again freeing transcription factors and hence driving the proliferation signals.

The E6 protein targets cell encoded p53 for rapid proteolytic degradation

Whilst E7 alone is able to transform and immortalise rodent cells, both E6 and E7 are needed for efficient immortalisation of primary human keratinocytes and E6, like E7, is almost always retained and expressed in cervical carcinoma cells. Unlike E7, E6 does not display sequence similarity with the transforming proteins of other DNA tumour viruses. A clear functional similarity is seen, however, in that HPV E6, adenovirus E1b and SV40 LT all target the same cell protein, p53.

p53 was originally identified as a protein which complexes with SV40 large T antigen and initially classified as the product of a dominant oncogene. Subsequent studies, however, revealed that many tumours contain deletions, rearrangements or mutations in both alleles of the p53 gene and that Li–Fraumeni patients, who show a predisposition to a diverse range of tumours, transmit a p53 mutation in the germline, leaving them particularly susceptible to subsequent mutation. This suggests that p53 has a similar role to Rb in protecting against tumour formation and allows the classification of p53 as a tumour suppressor gene [18]. Consistent with a role for p53 in the negative regulation of cell growth are the observations that the wild-type protein can inhibit the transformation by various oncogenes, including E7, and is able to specifically suppress the growth of p53 negative human carcinoma cells. The contribution of p53 to malignant development appears to be more complicated than simple loss of function, however, since many p53 mutations result in the expression of a mutant protein which can display dominant transforming activity. It has been suggested that mutant p53 can inactivate wild-type p53 protein following oligomerisation and numerous tumours are, in fact, heterozygous in that they retain one mutant and one wild-type p53 allele. Although mutant p53 proteins almost certainly do function in this dominant negative manner, there is also evidence that at least some of these mutants acquire further transforming activity beyond interference with wild-type p53 function leading to the possibility that expression of mutant p53 can contribute more strongly to malignant development than straightforward loss of wild-type p53 function.

The normal function of p53 is not fully understood although the frequency with which mutations within this gene occur during the development of cancers suggests a critical role in maintaining normal growth control. Studies with SV40 have suggested that p53 may play a role in the control of DNA replication and more re-cent studies show that p53 may play a specific role in the control of transcription. The potential activities of this protein are discussed in detail in Chapter 8.

The mechanism by which HPV E6 proteins contribute to abnormal cell growth appears to be by binding and rapidly targeting the p53 protein for degradation via the ubiquitin pathway [19]. This presumably results in the loss of wild-type p53 function and so the release from p53-mediated negative growth control. The im-portance of this E6/p53 interaction has been highlighted by studies examining the state of genomic p53 sequences in HPV positive and negative ano-genital cancers. Whilst HPV infection is implicated in many of these tumours, some malignancies do develop without evidence for infection by HPVs and these HPV negative tumours frequently show a worse prognosis than HPV positive cancers. Analyses of these two groups of cancers revealed an inverse relationship between the presence of HPV sequences and the presence of a somatic p53 mutation (Table 2). This strongly suggests that perturbation of wild-type p53 function is important for the development of ano-genital cancers but that this can occur in one of two ways,

Table 2. Inverse relationship between HPV infection and somatic p53 mutation.

Cervical carcinoma cell lines [22, 23]	
6 HPV positive	Wild-type p53
2 HPV negative	Mutant p53
Primary anal carcinomas [24]	
6 HPV positive	Wild-type p53
3 HPV negative	Mutant p53
Primary cervical carcinomas [25]	
28 HPV positive	Wild-type p53
3 HPV negative	Mutant p53

either by mutation of the p53 gene or by expression of the HPV E6 protein. The worse prognosis of HPV negative tumours might be a reflection of the expression of mutant p53 proteins, which may result in more severe transformation than the simple loss of wild-type p53 mediated by E6 expression.

Therapeutic approaches include HPV vaccines and drugs that target the E6 and E7 proteins

The accumulating epidemiological and experimental evidence implicating certain HPV types as causative agents in the development of most cervical cancers has led to the hope of using anti-viral agents in both the prevention and treatment of this disease. Vaccination against the high risk HPV types is an exciting prospect which might dramatically reduce the incidence of cervical cancer and may also have therapeutic uses. Studies in a number of animal model systems have demonstrated the efficacy of prophylactic and therapeutic vaccination for treatment of papillomavirus-induced lesions and much effort is currently being expended to extend these studies into humans [20].

An alternative strategy to the treatment of HPV induced pre-malignancies and cancers is to target the activity of the viral oncoproteins, E6 and E7. The consistency of E6 and E7 expression in both cervical cancers and cervical carcinoma cell lines, combined with some experimental evidence showing reversion of the tumorigenic phenotype following inhibition of E6 and E7 expression, suggest that the continued presence of the E6 and E7 proteins is necessary for the maintenance of malignancy. This is an important concept, since it implies that drugs designed to specifically interfere with E6 and E7 function could be used in the treatment of HPV-induced tumours. Preliminary experiments have shown that small peptides identical to the Rb binding region of E7 can specifically block the ability of E7 to bind Rb in an in vitro assay [21]. The activity of such a peptide in cells has not been assessed yet, although there is evidence that such a peptide can prevent association between Rb and its normal partners such as c-*myc*. This casts some doubt over the use of such a peptide for therapy since it is quite possible that the peptide will mimic E7 function rather than prevent it. The question whether potential therapeutic agents designed will act as agonists or antagonists of E7 function remains to be resolved, and potentially a more successful strategy will be to target the E7 protein itself to prevent interactions with cell factors.

References

1   zur Hausen H. Condylomata acuminata and human genital cancer. Cancer Res 1976, **36**, 794.
2   Vousden KH. Human papillomaviruses and cervical cancer. Cancer Cells 1989, **1**, 43–49.
3. zur Hausen H. Papillomaviruses as carcinoma viruses. In: Advances in Viral Oncology, G Klein, ed. Raven Press, New York, 1989, pp 1–26.
4. Spalholz BA, Howley PM. Papillomavirus-host cell interactions. In: Advances in Viral Oncology 8: Tumorigenic DNA Viruses, G Klein, ed. Raven Press, New York, 1989, pp 27–53.
5   Sterling J, Stanley M, Gatward G, Minson T. Production of human papillomavirus type 16 virions in a keratinocyte cell line. J Virol 1990, **64**, 6305–6307.
6   Dürst M, Dzarlieva-Petrusevska RT et al. Molecular and cytogenetic analysis of immortalised human primary keratinocytes obtained after transfection with human papillomavirus type 16 DNA. Oncogene 1987, **1**, 251–256.
7   McCance DJ, Kopan R, Fuchs E, Laimins LA. Human papillomavirus type 16 alters human epithelial cell differentiation *in vitro*. Proc Natl Acad Sci USA 1988, **85**, 7169–7173.
8   Cullen AP, Reid R, Campion M, Lörincz AT. Analysis of the physical state of different human papillomavirus DNAs in intraepithelial and invasive cervical neoplasm. J Virol 1991, **65**, 606–612.
9   Romanczuk H, Howley PM. Disruption of either the E1 or the E2 regulatory gene of HPV 16 increases viral immortalization capacity. Proc Natl Acad Sci USA 1992, **89**, 3159–3163.
10 Vousden KH. Human papillomavirus oncoproteins. Semin Cancer Biol 1990, **1**, 415–424.
11 Leechanachai P, Banks L, Moreau F, Matlashewski G. The E5 gene from human papillomavirus type 16 is an oncogene which enhances growth factor-mediated signal transduction to the nucleus. Oncogene 1992, **7**, 19–25.
12 Danos O, Yaniv M. E6 and E7 gene products evolved by amplification of a 33 amino acid peptide with a potential nucleic acid-binding structure. In: Cancer Cells, BM Steinberg, JL Brandsma, LB Taichman eds. Cold Spring Harbor Laboratory, New York, 1987, pp 145–149.
13 Phelps WC, Yee CL, Münger K, Howley PM. The human papillomavirus type 16 E7 gene encodes transactivation and transformation functions similar to those of adenovirus E1A. Cell 1988, **53**, 539–547.

14 Barbosa MS, Edmonds C, Fisher C et al. The region of the HPV E7 oncoprotein homologous to adenovirus E1a and SV40 large T antigen contains separate domains for Rb bindings and casein kinase II phosphorylation. EMBO J 1990, **9**, 153–160.

15 Marshall CJ. Tumor suppressor genes. Cell 1991, **64**, 313–326.

16 Chellappan S, Kraus V, Kroger B et al. Adenovirus E1A, simian virus 40 tumor antigen, and human papillomavirus E7 protein share the capacity to disrupt the interaction between transcription factor E2F and the retinoblastoma gene product. Proc Natl Acad Sci USA 1992, **89**, 4549–4553.

17 Vousden KH. Human papillomavirus transforming genes. Semin Virol 1992, **2**, 307–317.

18 Levine AJ, Momand J, Finlay CA. The p53 tumour suppressor gene. Nature 1991, **351**, 453–456.

19 Scheffner M, Werness BA, Huibregtse JM et al. The E6 oncoprotein encoded by human papillomavirus types 16 and 18 promotes the degradation of p53. Cell 1990, **63**, 1129–1136.

20 Campo MS. Vaccination against papillomaviruses. Cancer Cells 1991, **3**, 421–426.

21 Jones RE, Wegrzyn RJ, Patrick DR et al. Identification of HPV-16 E7 peptides that are potent antagonists of E7 binding to the retinoblastoma suppressor protein. J Biol Chem 1990, **265**, 12782–12785.

22 Crook T, Wrede D, Vousden KH. p53 point mutation in human papillomavirus negative cervical carcinoma cell lines. Oncogene 1991, **6**, 873–875.

23 Scheffner M, Münger K, Byrne JC et al. The state of the p53 and retinoblastoma genes in human cervical carcinoma cell lines. Proc Natl Acad Sci USA 1991, **88**, 5523–5527.

24 Crook T, Wrede D, Tidy JA et al. Status of c-*myc*, p53 and retinoblastoma genes in human papillomavirus positive and negative squamous cell carcinomas of the anus. Oncogene 1991, **6**, 1251–1257.

25 Crook T, Wrede D, Tidy JA et al. Clonal p53 mutation in primary cervical cancer: association with human-papillomavirus-negative tumours. Lancet 1992, **339**, 1070–1073.

Molecular Biology for Oncologists
Edited by John Yarnold, Michael Stratton, Trevor McMillan
© *1993, Elsevier Science Publishers B.V. All rights reserved*

CHAPTER 12

Cell cycle control and cancer

Antony M. Carr

MRC Cell Mutation Unit, Sussex University, Falmer, Brighton, BN1 9RR, UK

Introduction

In the past 20 years much progress has been made in understanding the processes by which cells regulate their growth [1]. Much of this progress has resulted from studies concentrating on the cell cycle of single cellular eukaryotes such as *Saccharomyces cerevisiae* (budding yeast) and *Schizosaccharomyces pombe* (fission yeast). One of the reasons that these studies have been so informative is the fact that the two species of yeast are evolutionarily as distinct from each other as they are from mammalian species such as man. This has enabled the identification of conserved protein structures (amino acid sequences) and functions essential for the cell cycle.

In complex multicellular organisms, a defect in growth regulation in a population of cells can lead to the formation of tumours and ultimately the death of the organism. An understanding of the cellular decisions leading to cell division is fundamental to our understanding of cancer biology and cancer genetics. In this chapter, the concepts involved in the cell cycle are introduced and linked to our current understanding of cancer biology.

Cellular duplication involves several events that must occur in a predetermined order: the cell cycle

Initial studies of the cell cycle identified two distinct marker events, DNA synthesis ("S" phase) and mitosis/nuclear division ("M" phase). These two key events are

147

separated by gaps ("G_1" and "G_2") where no marker events occurred. Thus the cell cycle is considered, in its simplest form, to be composed of four stages (Figure 1); G_1–S–G_2–M which occur in a cyclical manner in rapidly dividing cells. The cell cycle (or cell division cycle) is thus a description for the events that are required for cellular duplication.

There are two major control points in the cell cycle: "start" and the G_2/M transition

Cell cycle events need to be controlled. Even in the simple unicellular organisms such as yeast, decisions on whether to proceed with a new round of division are required before each cell cycle is initiated. These decision points are fundamental to the control of the cell cycle. There are two major decision points in the cell cycle. The first occurs at the transition between the G_1 and S phases and is known as "start". Transition of this control point usually indicates that the entire cycle will be completed (resulting in two cells). A second decision point exists immediately preceding mitosis; this is known as the G_2/M transition. At this point cells are able to arrest the cell cycle for a prolonged period of time. However, when cells cease proliferation, they usually "rest" in what can be considered as a specialised G_1 state that is often known as G_0.

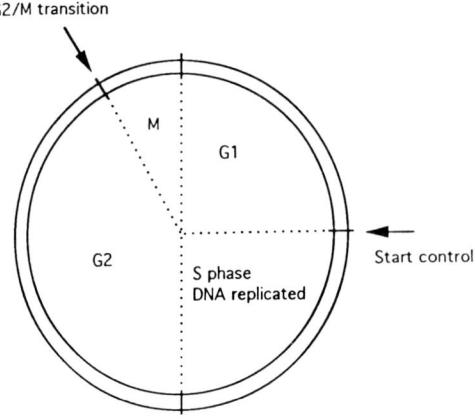

Figure 1. The cell cycle. "Start" is the point at which cells make a decision whether to divide or enter an alternative developmental state (such as differentiation). The G_2/M transition is a point at which cells are able to arrest should circumstances (DNA damage for example) change during the cell cycle. The passage through both start and the G_2/M transition is controlled by the p34^{cdc2} protein kinase.

Progression through the cell cycle is controlled by p34^{cdc2} kinase and a number of associated proteins

In the unicellular yeasts, genetic studies have identified a number of gene products that are involved in controlling transition of the G$_2$/M boundary. This entry of cells into mitosis is tightly controlled, and this control appears to be mainly channelled through the p34^{cdc2} gene product (this notation indicates a 34 kDa protein, equivalent to the product of the *cdc*2 gene of fission yeast and is often used as a universal label for equivalent conserved proteins in other organisms). The p34^{cdc2} protein has been genetically identified independently in both budding and fission yeast. The protein was also found to be a component of maturation promotion factor (MPF), a biochemical activity identified and purified by virtue of its ability to promote the maturation of amphibian oocytes, a process requiring a nuclear division akin to mitosis. p34^{cdc2} protein has been found in essentially all eukaryotic organisms. The p34^{cdc2} gene product is a protein kinase, the activity of which peaks during mitosis.

Genetic and biochemical studies have also identified a number of highly evolutionarily conserved proteins that interact directly with p34^{cdc2} to control the entry of cells into mitosis [2]. The key proteins identified in yeast are: p80^{cdc25}, p108^{wee1}, p13^{suc1} and p56^{cdc13}. The important properties of these proteins are indicated in Table 1. In brief, p80^{cdc25} activates p34^{cdc2} by dephosphorylation of a specific amino acid residue (tyrosine 15, Y15) and p108^{wee1} inactivates p34^{cdc2} by phosphorylation of the same residue (see Figure 2). The balance of dephosphorylation (activating) and phosphorylation (inactivating) of Y15 dictates the precise timing of mitosis. The action of the p34^{cdc2} protein also requires association with two further proteins, p13^{suc1} and p56^{cdc13}. While little is known about the precise role of p13^{suc1}, the role of p56^{cdc13} is better understood. p56^{cdc13} has been recognised as a

Table 1. Properties of cell cycle proteins.

p34^{cdc2}	Protein kinase controls entry into mitosis	The workhorse that acts to phosphorylate proteins (such as nuclear lamins) and signal the onset of mitosis
p108^{wee1}	Inactivates p34^{cdc2} by phosphorylation of Y15	This protein kinase serves to inactivate p34; in yeast, at least one other protein (*mik*1) acts to phosphorylate p34
p80^{cdc25}	Activates p34^{cdc2} by dephosphorylation of Y15	In human cells there are several p80 related proteins
p56^{cdc13}	Associates with p34^{cdc2}, an essential part of active kinase complex	More than five cyclin like proteins interact with p34 related proteins at specific and distinct points in the cell cycle
p13^{suc1}	Capable of binding to p34^{cdc2}	Probably required for exit from mitosis

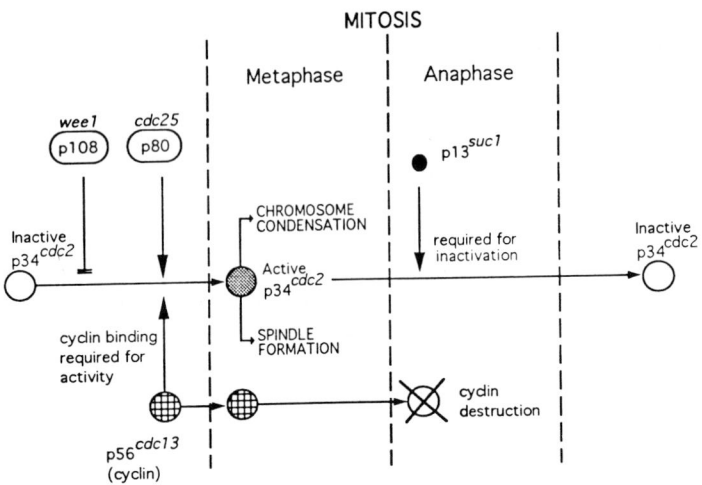

Figure 2. A model for the control of mitosis by the p34^{cdc2} protein kinase in fission yeast. The control of the p34^{cdc2} protein kinase activity by the inhibitor p108^{wee1} and the activator p80^{cdc25} is shown, as are the required associations with cyclin related protein p56^{cdc25} and with p13^{suc1}. The existence of p80^{cdc25}, p56^{cdc13} and p34^{cdc2} protein families greatly complicates the process. From [2]. Reprinted with permission.

member of the cyclin family of proteins. These proteins were identified in amphibian oocytes as proteins whose gross levels fluctuated, or cycled, in tight association with the passage of mitosis. Essentially, p56^{cdc13} is required to bind to p34^{cdc2} for entry into mitosis and is then destroyed. p56^{cdc13} degradation is thought to be necessary for successful exit from mitosis and may serve as a molecular marker event signalling the completion of mitosis.

The passage of "start" in yeast cells also requires the p34^{cdc2} protein. Furthermore, proteins related in structure to p56^{cdc13} (cyclin) are found associated with p34^{cdc2} during passage of "start" and serve to control the activity of the p34^{cdc2} protein. These observations show that the transition of the two major control points in the cell cycle is dependent on the activation of p34^{cdc2}. The control of the activation of p34^{cdc2} is thus of considerable importance in the understanding of cell cycle regulation. The situation is, however, greatly complicated by the observation that four or more p56^{cdc13} related proteins (G$_1$ cyclins) interact with p34^{cdc2} during the control of the "start" event. Furthermore, these proteins are functionally able to substitute for each other's essential role in the passage of "start" (this is known as functional redundancy). The precise role of each G$_1$ cyclin has yet to be elucidated in yeast. The presence of multiple functionally redundant proteins not only serves to underline the complexity of the molecular events underlying cell cycle control, but also makes studying them more difficult.

Cell cycle progression is influenced by extracellular signals

In the unicellular eukaryotic yeasts, cell cycle transition points can be controlled in response to extracellular signals. Yeast cells in G_1 are capable of several developmental fates; they can mate, enter a G_0-like stationary state or they can commit themselves to another cell cycle. The fate of an individual cell is determined by the presence of nutrients such as carbon (sugars) and nitrogen sources and the presence of mating pheromones (released by adjacent cells of opposite mating type). Complex signalling pathways exist that sense the presence of such agents and process this information. Ultimately, the extracellular environment dictates whether the cell will traverse "start" and commit itself to a complete round of cell division. This process is channelled through the ubiquitous p34^{cdc2} protein kinase.

If cell cycle events are interfered with, checkpoints postpone cell cycle progression until normal processes are restored

Once a cell has embarked on the cell cycle, it will usually complete the cycle unless the temporal sequence of events (i.e. DNA synthesis, spindle formation, chromosome alignment etc.) is interrupted. In the event of such an interruption, feedback mechanisms delay the execution of the next event until the preceding ones are completed.

Two examples serve to illustrate the existence of checkpoints. If DNA synthesis is interrupted (for example by chemical inhibitors) then cells do not proceed into mitosis until the block to synthesis is removed and the DNA is completely replicated. Mutations in a number of genes in fission yeast have been shown to abolish this checkpoint. In such mutant cells, a block to DNA synthesis results in a catastrophic mitosis of unreplicated chromosomes and cell death [3]. Cell proliferation is not adversely affected if replication proceeds normally. A similar relationship is observed in budding yeast between mitotic spindle formation and mitosis [4]. A chemical block to spindle formation produces a delay in mitosis. This feedback mechanism can also be disrupted by mutations. Such checkpoints are examples of the complexity of control over cell cycle events. Although the molecular nature of checkpoints is poorly understood, it is clear that the DNA replication checkpoint can inhibit the activity of the p34^{cdc2} protein kinase to prevent the onset of mitosis.

Human cell cycle controls are very similar to those in yeast

In human cells there are a minimum of two p34^{cdc2} related proteins, three p80^{cdc25} related proteins and five p56^{cdc13} related proteins. These are likely to be underestimates. The number of potential combinations and interactions of these three protein families is very large. The $p56^{cdc13}$ cyclin family of proteins is the most extensive. Different members of this protein family associate with the $p34^{cdc2}$ related proteins during the transition of "start", progress through S phase and transition of the G_2/M boundary. This serves to illustrate the complexity of interactions between the $p34^{cdc2}$ protein kinase and its regulatory molecules.

Interference/disruption of normal cell cycle control mechanisms are potent factors in cancer development

Many cancers are thought to be dependent on activating mutations in individual oncogenes. Many of these oncogenes have subsequently been found to be integral to growth factor detection and/or the signal transduction mechanisms that respond specifically to the presence of external growth factors. Thus an "activating" mutation in an oncogene protein may signal to the cell, the presence of a growth factor that is not in fact present. This can cause cells to grow in apparently inappropriate circumstances. Of the many different cell types present in humans, most can respond to appropriate growth factor stimulation under specific circumstances. Thus an artificially activated growth factor receptor/signalling system can be one of the key events leading to the cancerous state.

A second class of proteins have been found to be involved in some cancers. These have been called the tumour suppressors. These proteins are apparently required to prevent uncontrolled cellular proliferation. Examples of tumour suppressors are the retinoblastoma Rb antigen, the p53 protein and the Wilm's tumour protein. In each case, mutations are thought to inactivate the protein function, which appears to be involved in the suppression of cellular proliferation. p53 was initially found because it binds SV40 large T antigen (this protein can transform cells to immortality). It is now thought that the binding of SV40 large T antigen to the p53 protein may effectively inactivate p53, thus removing growth suppression. A similar association has recently been established between adenovirus E1A and papillomavirus E7 proteins and the Rb protein. Interestingly, the human papillomavirus E6 protein and the E1B protein of adenovirus both bind p53. These interactions are summarised in Figure 3.

A direct link between tumour suppressors and the cell cycle control mechanisms has recently been demonstrated in a number of laboratories. The Rb tumour suppressor is associated with a complex of proteins that are bound to specific sites on the DNA and which are thought to control the transcription of adjacent genes [5]. Biochemical studies have identified a member or members of the $p34^{cdc2}$ and $p56^{cdc13}$ (cyclin) families to be involved in this DNA binding protein complex and the Rb protein is known to be phosphorylated by $p34^{cdc2}$/cyclin complexes [6]. It has been suggested that the Rb tumour suppressor may block transcription unless it

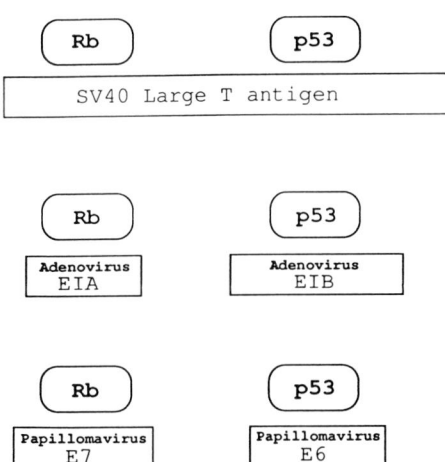

Figure 3. The interaction of viral oncogenes with tumour suppressors.

is phosphorylated by p34^{cdc2} kinase. Once phosphorylated, Rb antigen is inactivated and transcription of adjacent genes can occur. Repression of transcription at these loci may be one of the functions by which Rb suppresses cell growth and division.

Cyclins have recently been directly implicated in cancer by several observations showing that changes in cyclin behaviour are linked to cellular transformation. Specifically, in human hepatocellular carcinoma, the hepatitis B virus is integrated at the site of one of the cyclin genes, and cyclins are also found to be associated with proteins linked to cellular transformation by adenovirus. For a review, see [7].

Summary

The level of structural (amino acid sequence) conservation observed between the cell cycle control proteins of yeasts and humans is variable but significant. What has been surprising has been the level of functional homology within a single class of proteins from distantly related organisms such as yeast and humans. These observations testify to the ubiquitous nature of cell cycle control systems in all eukaryotic cells. The conservation of the cell cycle proteins indicates that the basic mechanism by which all cells control progression through the cell cycle has been conserved throughout the eukaryotic kingdom.

A yeast cell has only a very limited set of developmental fates available to it. These are controlled by relatively simple external stimuli that are monitored by mechanisms ultimately interacting with the p34^{cdc2} protein kinase. During evolution into multicellular organisms, the developmental fate of individual cells has become more complex. Differentiation is often irreversible and almost always ultimately static. Thus, a host of control systems have evolved to tightly regulate cellular proliferation in the context of the whole organism. It appears that these control systems have been superimposed on, and interact with, a highly conserved p34^{cdc2} based mechanism controlling the passage of "start" and the timing of mitosis.

In mammalian cells, the control of cell proliferation is complex and paramount to the ordered construction and maintenance of multicellular and differentiated structures. When the control of proliferation fails, tumours are often formed that can be fatal. In order to understand the development of the cancerous state, and ultimately to either prevent or treat it, we need to understand the control of cell cycle initiation and cell cycle progression.

References

1   Morris NR. Lower eukaryotic cell cycle: perspectives on mitosis from the fungi. Curr Opinions Cell Biol 1990, **2**, 252–257.
2   MacNeill SA, Nurse P. Genetic interactions in the control of mitosis in fission yeast. Curr Genet 1989, **16**, 1–6.
3   Al-Khodairy F, Carr AM. DNA repair mutants defining G$_2$ checkpoint pathways in *Schizosaccharomyces pombe*. EMBO J 1992, **11**, 1343–1350.
4   Li R, Murray AW. Feedback controls of mitosis in budding yeast. Cell 1991, **66**, 519–531.
5   Partridge JF, La Thangue NB. A developmentally regulated and tissue-dependent transcription factor complexes with the retinoblastoma gene product. EMBO J 1991, **10**, 3819–3827.
6   Lees JA, Buchkovich KJ, Marshak DR, Anderson CW, Harlow E. The retinoblastoma protein is phosphorylated on multiple sites by human cdc2. EMBO J 1991, **10**, 4279–4290.
7   Hunter T, Pines J. Cyclins and cancer. Cell 1991, **66**, 1071–1074.

Molecular Biology for Oncologists
Edited by John Yarnold, Michael Stratton, Trevor McMillan
© 1993, Elsevier Science Publishers B.V. All rights reserved

CHAPTER 13

Differentiation and cancer

Malcolm D. Mason

Department of Clinical Oncology, Velindre Hospital, Whitchurch, Cardiff, CF4 7XL, UK

Normal tissue renewal requires that mature cells are replaced by cellular differentiation from a pool of undifferentiated "stem cells"

Normal cell renewal is usually not accomplished by the division of mature cells. Instead, normal tissue growth and renewal is accomplished by division and differentiation from a pool of undifferentiated cells called *stem cells*. Stem cells are the origin of the mature cells that characterise an individual tissue. Such a system certainly exists in the haemopoietic system, in most common epithelia, and may even exist in some other non-epithelial tissues, although the latter is not firmly established [1].

Stem cells are unique in that when they divide, their two daughter cells are different (i.e. cell division is asymmetric). One cell identical to the parent stem cell replaces it in the tissue pool to maintain the overall number of stem cells. The other is said to be *committed* as it embarks on a chosen pathway of differentiation and will eventually give rise to a mature cell whose phenotype is characteristic of the tissue in which it arises. What governs the decision by a daughter cell to self-renew or to differentiate is one of the most fascinating questions in cell biology [1,2]. In an organised tissue, the presence of stem cells may only be inferred from their effects on tissue growth and renewal. For example, the existence of a bone marrow multipotent stem cell was suggested many years ago, but its formal identification has been difficult [3,4].

Differentiation is linked to proliferation

It is axiomatic that, in order to renew cells lost by death or injury, cell division must occur and feed a pool of new, differentiating cells into the system. The population undergoing this cell division is initially in the stem cell compartment of a tissue. At the other extreme is the fully mature cell, for which cell division is not only unnecessary for tissue renewal, but is thought to be impossible in most tissues that have a stem cell system of organisation. Such mature, fully differentiated cells are said to be *post-mitotic*, but the mechanisms by which mitosis is blocked are unknown. Also implicit in the stem cell model is that in normal tissues, between the undifferentiated stem cell stage and the post-mitotic end cell stage are a number of intermediate stages, during which a committed cell may have limited proliferative capacity but yet not be completely differentiated. It seems, therefore that as a generalisation, a cell must either be a stem cell, or engaged in the process of differentiation in order to retain proliferative capacity. The phenotypic diversity of the intermediate cells could go some way towards explaining the corresponding diversity of tumours that arise from a single tissue, if such tumours were caricatures of normal differentiation. For the post-mitotic cell, the only destiny is cell death. This is as true of terminally differentiated cells arising from malignant stem cells as it is of post-mitotic cells from normal stem cells (Figure 1).

Under normal circumstances cells remain faithful to one lineage pathway of differentiation, and their destiny is defined by determination during development

The term *determination* refers to a heritable undertaking by a cell, usually during

WHICH CELLS IN A TISSUE ARE DIVIDING?

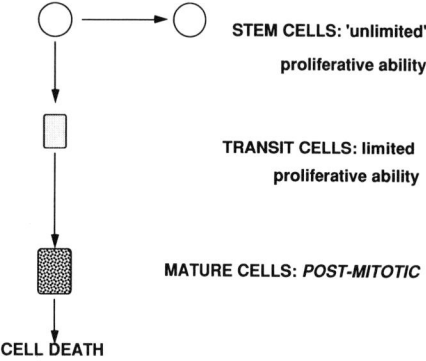

STEM CELLS: 'unlimited' proliferative ability

TRANSIT CELLS: limited proliferative ability

MATURE CELLS: *POST-MITOTIC*

CELL DEATH

Figure 1. Although the characteristic morphology of a tissue depends on the mature cells that make it up, it is usually the most primitive cells that have the ability to divide, while the mature cells themselves have lost this capacity and are said to be *post-mitotic.*

embryonic development, to follow a particular pathway of specialised development at some stage in the future. Once differentiated, a cell acquires certain structural and functional characteristics that endow it with the ability to undertake a specialised task, e.g. to carry oxygen, or to absorb nutrients. Differentiation may involve, for example, the secretion of certain specialised molecules that are not produced by the undifferentiated cell, and therefore determination must by definition precede differentiation. According to classical dogma, once a cell is committed, it will remain faithful to one lineage pathway of differentiation only; a committed intestinal epithelial precursor cell will not turn into a red blood cell, for example. Clearly, some cells, particularly very undifferentiated cells, must have the option of following several alternate pathways, the best example of which is the bone marrow pluripotent stem cell, which may differentiate along erythrocytic, leucocytic or megakaryocytic pathways. However, it is also clear that such options in normal tissues are severely restricted.

Some malignant tumour cells retain the ability to differentiate along alternative lineage pathways

Fundamentally, the dogma of fidelity to one pathway is almost certainly correct. However, certain features of malignant cells imply that early on in differentiation, a cell may keep its options open with regard to which of several possible lineage pathways it intends to follow, and primitive cells may carry markers characteristic of more than one lineage (referred to as lineage promiscuity by Greaves [5]). Indeed, there is evidence that the system can *sometimes* be flexible to the point that "impossible" transitions, such as from epithelial cell to fibroblastoid cell, do in fact, occur [6], but phenomena such as this should be seen as exceptions and not the rule. Such "epithelial-mesenchymal transformations" could be common events in embryogenesis [6], and may yet change our perceptions of a cell's fidelity to one lineage pathway.

Germ cell tumours provide the most dramatic model of cancer cell differentiation. The diversity of cell types present in such tumours can be understood in the light of evidence that the embryonal carcinoma (EC) stem cell in non-seminomas is very similar to a pluripotent embryonic stem cell, i.e. it is capable of differentiating along multiple lineage pathways. In the mouse, EC probably corresponds to cells of the late blastocyst [7]. When human EC cells of the cell line NTERA-2 are treated in vitro with retinoic acid (RA) they undergo somatic differentiation into neuron-like cells. By contrast, cells of the mouse EC cell line F9 treated with RA undergo extra-embryonic differentiation into parietal endoderm. Treatment of NTERA-2 cells with another differentiation-inducing agent, hexamethylene bisacetamide (HMBA) induces them to differentiate into a large, flat cell type with entirely different phenotypic markers to the RA-treated cells [8]. The outcome of an interaction between a cell and a differentiation-promoting agent clearly varies according to the cell and according to the stimulus.

Normal differentiation is often blocked in malignancy, but "dedifferentiation" is usually a misnomer

The central theme of this chapter is the concept that cancers are merely caricatures of normal differentiation, and that the primary characteristic of a malignant cell is that the process of normal differentiation has somehow been blocked, leading to a cell whose maturation has been "frozen". Human lymphoid malignancies provide the most comprehensively understood model of the relationship between normal differentiation and the phenotype of cancer cells. The phenotypic characteristics of normal lymphoid cells along their pathways of differentiation have been elaborated to a high degree of sophistication. What is striking about lymphoid malignant cells is that their phenotype is often remarkably similar to that of normal cells somewhere in the pathway between primordial stem cell and mature cell [5]. The real problem in haemopoietic malignancies and perhaps in some other cancers may not be so much that *abnormal* differentiation is taking place, but, rather, that the freezing of normal differentiation at an intermediate stage leads to the clonal expansion of a cell type which should be transient and therefore extremely rare [5]. The motive of altered gene expression in this setting would be the stabilization of this transient cell type. Some specific genomic abnormalities may merely reflect the method whereby this is achieved. For example, the t(14;18) translocation in lymphomas involving the *BCL*-2 gene may be important because *BCL*-2 codes for a protein that is an inhibitor of programmed cell death [9]. The t(15;17) translocation seen in acute promyelocytic leukaemia results in the fusion of a retinoic acid receptor gene to a different site, which is interesting given that RA promotes the differentiation of acute promyelocytic leukaemia (APML) cells [10,11].

In haemopoietic malignancies, the block in normal differentiation can sometimes be overcome; injection of leukaemia cells into the placentae of mouse embryos sometimes results in normal mice whose haemopoietic tissues are chimeric. Therefore, all of the leukaemia cells injected into these mice were able to participate in normal differentiation [12]. Differentiation of erythroleukaemia cells can also be induced by HMBA [13].

Germ cell tumours are caricatures of normal development in which *somatic differentiation* gives rise to the embryo and subsequently to the animal itself, and *extra-embryonic* differentiation gives rise to supporting tissues such as yolk sac and trophoblast, which do not form the animal proper. By understanding what sort of cell the malignant teratoma stem cell thinks it is supposed to be, we can begin to understand certain aspects of its behaviour. It was the experiments of Kleinsmith and Pierce [14] that demonstrated that malignant teratoma cells could differentiate into "benign" (i.e. post-mitotic) tissues. They found that it was possible to inject a single teratocarcinoma cell into a mouse and generate a tumour comprised of both malignant teratoma and benign differentiated tissue. Even more dramatic were the experiments by Brinster, in which a malignant teratoma cell was introduced into a normal mouse blastocyst which was then transferred to a foster mother. The healthy offspring resulting from this blastocyst was a chimeric mouse, in which its tissues were derived partly from normal mouse embryonic cells and partly from teratocarcinoma cells (see [15] for review). The molecular processes favouring the survival of malignant teratocarcinoma stem cells are, indeed, "subservient to the normal differentiation programme" [5]. In one chimera, the cells derived from the teratoma included those of the germinal epithelium, and it was possible to propagate a second generation of apparently normal mice whose "father" was, in a sense, a teratocarcinoma cell! "Once a cancer cell, always a cancer cell?": clearly not, as these experiments dramatically illustrate. Interestingly, the transplantation of a normal mouse blastocyst into an extra-uterine site led to the formation of a malignant teratocarcinoma [15]. Although it is not possible to perform such an experiment in human teratomas, it is clear that, as in the mouse, human EC cells are pluripotent [7]. Whether the differentiated cells produced in this context are truly benign and post-mitotic is open to question; the clinical phenomenon of sarcomatous degeneration of differentiated testicular teratoma deposits argues that such tissues are not always stable.

Malignant tumours, far from being in a state of total anarchy as is sometimes portrayed, may, therefore, be subject to a "tissue" system of organisation based on a stem cell compartment [16]. In some tumours, the stem cell population will pre-dominate and no differentiated elements will be apparent to the light microscopist. In others, mixtures of undifferentiated cells and more mature cells give a malignant tumour its characteristic morphology. It was Pierce, working on teratocarcinomas, who postulated that malignant tumours arose from stem cells, that they were the target for carcinogenic stimuli and not the mature end-cells that give a tissue its morphology [17].

In this view of cancer as a caricature of normal differentiation, malignant stem cells give rise to a larger pool of similar malignant stem cells by self-renewal, and also to committed cells that have a variable capacity for differentiation, including terminal differentiation. It is implicit from what has been said that differentiation is a one-way process. Usually, normal mature post-mitotic cells do not revert to undifferentiated stem cells and give rise to cancer by "dedifferentiation". This term is used loosely by clinicians without any regard for the enormous biological implications that such a label carries. Cell biologists now believe that limited dedifferentiation may occur in some specific cell types and in special circumstances, but the use of the term in the clinicians' ordinary vocabulary should be banned. The evolution of a low-grade tumour into a high-grade tumour can be viewed as resulting from a shift in emphasis from differentiation to self-renewal, and not from the supposed "dedifferentiation" of well-differentiated tumour cells [16].

The genetic control of normal differentiation is better understood in simpler organisms like *Drosophila* than in mammals

Much of our understanding of normal differentiation comes from studies of embryonic development in lower animals, and comparatively little is known about normal adult cell differentiation. It may seem surprising to the oncologist that non-mammalian systems are relevant to human cancer, but such is indeed the case. Mutations of *Drosophila* have added to the studies on mammalian cells in identifying genes that are of crucial importance in normal development and differentiation. These include homeotic genes, encoding DNA-binding proteins that are responsible for implementation of the body "plan" in *Drosophila* development, instruct cells on their spatial position in the body, and hence implement their proper pathway of differentiation. Homeotic genes include the highly conserved homeobox genes [18], a class of gene which has been found in other organisms, including mammals and even plants. It is noteworthy in this context that the action of retinoic acid (RA), a potent inducer of differentiation in several tumour systems in vitro, appears to involve the action of several homeobox genes [19].

Table 1. Conserved genes involved in differentiation/development in non-mammalian systems that are also found in mammalian cells and in malignant tumours.

Mammalian/cancer cell	Non-mammalian cell	Reference
gli	*kruppel*	20
Int-1	*wingless*	21
TGF-β	*decapentaplegic*	22
EGF	*notch lin*-12	23, 24
EGF receptor	*torpedo*	25
RA receptor	*knirps*	26

Some highly conserved genes that control normal differentiation are disrupted in malignancy

Several homologues of non-mammalian genes from *Drosophila* and from the nematode *Caenorhabditis elegans* are involved in growth and differentiation of both normal and malignant mammalian cells (Table 1). At first sight these examples of homology are confusing because they do not present a clear picture of the relationship between normal differentiation and malignancy. Our understanding will be enhanced in the future as the functions of these genes in normal development are elaborated, and already their patterns of expression are being described. For example, the *int*-1 (*Wnt*-1) gene, expressed in mouse mammary tumours, is not expressed in the normal mouse mammary gland, although it is in the developing central nervous system. However, several members of the *Wnt* family of genes, which have substantial homology to *Wnt*-1, are expressed in mouse mammary glands during pregnancy and lactation [27]. As another example, the candidate Wilm's tumour gene, *WT1*, is normally expressed in the early stages of nephron de-

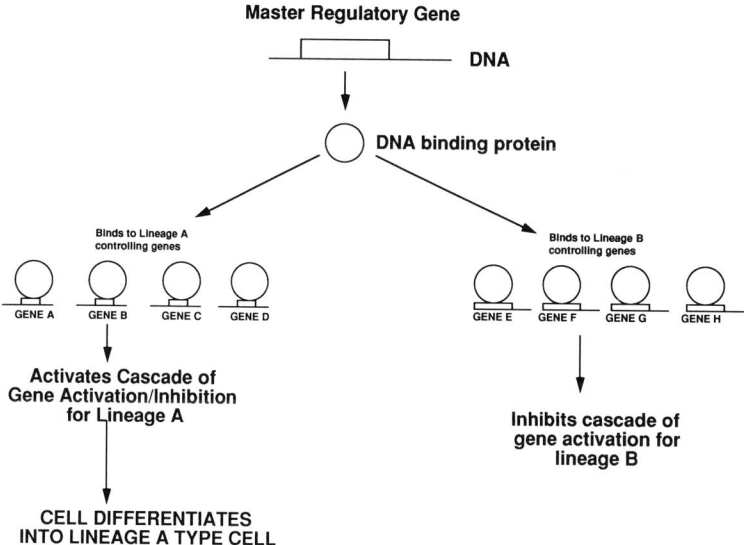

Figure 2. A schematic representation of how a master regulatory gene might control cellular differentiation. In this example a cell has two alternative lineage pathways, A and B. The product of the master regulatory gene is a protein that binds to many genes, in this example activating those reponsible for the features of lineage A while inactivating those responsible for lineage B. Different master regulatory genes might use different combinations of activation and inactivation to achieve their objective.

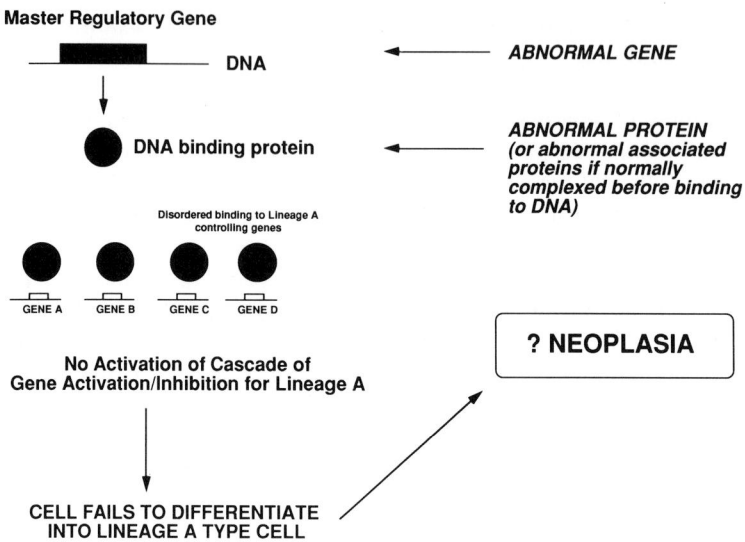

Master Regulatory Gene

DNA

ABNORMAL GENE

DNA binding protein

ABNORMAL PROTEIN
(or abnormal associated
proteins if normally
complexed before binding
to DNA)

Disordered binding to Lineage A
controlling genes

GENE A   GENE B   GENE C   GENE D

No Activation of Cascade of
Gene Activation/Inhibition for Lineage A

? NEOPLASIA

CELL FAILS TO DIFFERENTIATE
INTO LINEAGE A TYPE CELL

Figure 3. Hypothetical diagram to show how disorders of the master regulatory gene system might produce tumours. The gene itself might be mutated, producing an abnormality in the DNA-binding protein that interferes with its ability to regulate genes. Alternatively, if the DNA-binding protein complexes with other proteins, mutations in these associated proteins could also interfere with master regulatory gene function. The result could be to "freeze" differentiation at a primitive stage. If such mechanisms could be found, novel strategies might be developed to reverse the malignant phenotype.

velopment, in the foetal gonad and foetal mesothelium [28]. Other cancer-associated genes will undoubtedly be shown to have a role in normal development. The $nm23$ gene, sometimes called the "anti-metastatic gene" because of the inverse correlation between its expression and the incidence of metastatic disease in breast cancer, is expressed in mice at high levels in the non-proliferating and pre-pubertal breast, but at very low levels during pregnancy and lactation, suggesting a possible role of this gene in the regulation of normal breast proliferation and differentiation [29].

Confusing as these observations may be, they serve to illustrate two points. First, that many molecules which regulate growth and differentiation have multiple and possibly diverse functions [30]. Second, that in order to understand how a biologically active molecule influences carcinogenesis or the behaviour of malignant cells, we may first have to understand its functions in normal embryological development.

Tempting as it is to suggest that every important cancer-associated gene will have a role in normal development and differentiation, there will be important exceptions. The tumour suppressor gene coding for p53 has been implicated in a wide variety of tumours. Rather surprisingly, mouse embryos that are genetically engineered so that the individuals carry a deficiency of p53 are developmentally normal, yet they are highly susceptible to the development of malignant tumours [31].

Master regulatory genes control differentiation in mammalian cells and may be implicated in malignancy

Homeobox genes may be an example of a class known as master regulatory genes. Such genes are extremely potent, in that they initiate a clear signal that, in effect, irrevocably switches on a given pathway of differentiation. A master regulatory gene has now been identified in cells of muscle lineage, where it has been named *MyoD* [32]. The protein produced by *MyoD* is a DNA-binding protein of the helix-loop-helix class, with sequence homology to the *myc* family of proteins [33]. The MyoD protein will directly or indirectly regulate all of the other genes whose activation or repression make up the phenotype of skeletal muscle cells (Figure 2). So powerful is the *MyoD* gene that, if it is transfected into non-muscle cells such as amniocytes, and even if it is transfected into some tumour cells, it will over-ride the cells' prevailing phenotype, and induce terminal differentiation into muscle cells [32]!

   Will such master regulatory genes turn out to be important in cancer? It has already been suggested that *MyoD* normally associates with a tumour suppressor gene product, and that defects in this associated protein inhibit differentiation in rhabdomyosarcomas [32] (Figure 3). Perhaps abnormalities of master regulatory gene function will turn out to be a major fundamental defect in many cancers.

References

1   Hall PA, Watt FM. Stem cells: the generation and maintenance of cellular diversity. Development 1989, **106**, 619–633.
2   Horvitz HR, Herskovitz I. Mechanisms of asymmetric cell division: two Bs or not two Bs, that is the question. Cell 1992, **68**:237–255.
3   Till JE, McCulloch EA. A direct measurement of the radiation sensitivity of normal mouse bone marrow cells. Radiat Res 1961, **14**, 213–222.
4   Baum CM, Weissman IL, Tsukamoto AS et al. Isolation of a candidate human haemopoietic stem-cell population. Proc Natl Acad Sci USA 1992, **89**, 2804–2808.
5   Greaves MF. Differentiation-linked leukemogenesis in lymphocytes. Science 1986, **234**, 697–704.
6   Boyer B, Tucker GC, Valles AM et al. Rearrangements of desmosomal and cytoskeletal proteins during the transition from epithelial to fibroblastoid organization in cultured rat bladder carcinoma cells. J Cell Biol 1989, **109**, 1495–1509.

7   Pera MF, Roach S, Elliss C. Comparative biology of mouse and human embryonal carcinoma. Cancer Surv 1990, **9**, 243–262.

8   Andrews PW. Human teratocarcinomas. Biochim Biophys Acta 1988, **948**,17–36.

9   Hockenberry D, Nuñez G, Milliman C et al. Bcl-2 is an inner mitochondrial membrane protein that blocks programmed cell death. Nature 1990, **348**, 334–336.

10 de Thé H, Chomienne C, Lanotte M et al. The t(15;17) translocation of acute promyelocytic leukaemia fuses the retinoic acid receptor alpha gene to a novel transcribed locus. Nature 1990, **347**, 558–561.

11 Warrell RP, Frankel SR, Miller WH et al. Differentiation therapy of acute promyelocytic leukemia with tretinoin (all-*trans*-retinoic acid). N Engl J Med 1991, **324**, 1385–1393.

12 Gootwine E, Webb GC, Sachs L. Participation of myeloid leukaemia cells injected into embryos in hematopoietic differentiation in adult mice. Nature 1982, **299**, 63–65.

13 Melloni E, Pontremoli S, Viotti PL et al. Differential expression of protein kinase C isozymes and erythroleukaemia cell differentiation. J Biol Chem 1989, **264**, 18414–18418.

14 Kleinsmith LJ, Pierce GB. Multipotentiality of embryonal carcinoma cells. Cancer Res 1964, **24**, 1544–1551.

15 Martin GR. Teratocarcinomas and mammalian embryogenesis. Science 1980, **209**, 768–776.

16 Pierce GB, Speers WC. Tumors as caricatures of the process of tissue renewal: prospects for therapy by directing differentiation. Cancer Res 1988, **48**, 1996–2004.

17 Pierce GB, Shikes R, Fink LM. Cancer, a Problem of Developmental Biology. Prentice Hall, Englewood Cliffs, NJ, 1978.

18 Gehring WJ, Hiromi Y. Homeotic genes and the homeobox. Annu Rev Genet 1986, **20**, 147–173.

19 Mavilio F, Simeone A, Boncinelli E, Andrews PW. Activation of four homeobox gene clusters in human embryonal carcinoma cells induced to differentiate by retinoic acid. Differentiation 1988, **37**, 73–79.

20 Kinzler KW, Ruppert JM, Bigner SH, Vogelstein B. The *gli* gene is a member of the *kruppel* family of zinc finger proteins. Nature 1988, **332**, 371–374.

21 Rijsewijk F, Schuermann M, Wagenaar E et al. The *Drosophila* homolog of the mouse mammary oncogene *int*-1 is identical to the segment polarity gene *wingless*. Cell 1987, **50**, 649–657.

22 Gelbart WM. The *decapentaplegic* gene: a TGF-b homologue controlling pattern formation in *Drosophila*. Development 1989, suppl, 65–74.

23 Bender W. Homeotic gene products as growth factors. Cell 1985, **43**, 559–560.

24 Greenwald I. *lin*-12, a nematode homeotic gene, is homologous to a set of mammalian proteins that includes epidermal growth factor. Cell 1985, **43**, 583–590.

25 Price JV, Clifford RJ, Schubach T. The maternal ventralizing locus *torpedo* is analogous to *faint little ball*, an embryonic lethal gene, and encodes the *Drosophila* EGF receptor homolog. Cell 1982, **56**, 1085–1092.

26 Oro AE, Ong ES, Margolis JS et al. The *Drosophila* gene *knirps*-related is a member of the steroid-receptor gene superfamily. Nature 1988, **336**, 493–496.

27 Gavin BJ, McMahon AP. Differential regulation of the *Wnt* gene family during pregnancy and lactation suggests a role in postnatal development of the mammary gland. Mol Cell Biol 1992, **12**, 2418–2423.

28 Pritchard-Jones K, Fleming S, Davidson D et al. The candidate Wilm's tumour gene is involved in genitourinary development. Nature 1990, **346**, 194–197.

29 Nummy KA, Corbier P, Steeg PS, Haslam SZ. Expression of nm23 metastasis suppressor gene in normal mammary gland as a function of development. Proc Am Assoc Cancer Res 1992, **33**, 96 (abstract).

30 Sporn MB, Roberts AB. Peptide growth factors are multifunctional. Nature 1989, **332**, 217–219.

31 Donehower LA, Harvey M, Slagle BL et al. Mice deficient for p53 are developmentally normal but susceptible to spontaneous tumours. Nature 1992, **356**, 215–221.

32 Randall T. Gene scene: a master control switch for myogenesis muscles its way into the clinic. J Am Med Assoc 1992, **267**, 337–338

33 Li L, Olson EN. Regulation of muscle cell growth and differentiation by the MyoD family of helix-loop-helix proteins. Adv Cancer Res 1992, **58**, 95–119.

Molecular Biology for Oncologists
Edited by John Yarnold, Michael Stratton, Trevor McMillan
© *1993, Elsevier Science Publishers B.V. All rights reserved*

CHAPTER 14

Molecular aspects of radiation sensitivity

Catherine Mort and Trevor J. McMillan

Radiotherapy Research Unit, The Institute of Cancer Research,
Cotswold Road, Sutton, Surrey, SM2 5NG, UK

Biochemistry and genetics are increasingly required in the study of radiation sensitivity

In the treatment of cancer, radiotherapy is an important treatment modality which is used for over half of cancer patients. In addition, the effects of ionising radiation on mammalian cells has been well studied in the laboratory. Despite this, we still do not clearly understand the reason why cells differ in their sensitivity to ionising radiation. This question is important both in cancer therapy, where normal cells and tumour cells differ in their sensitivities, and in the field of carcinogenesis where environmental exposure to radiation hazards may have different effects on different individuals.

Measures of DNA damage have suggested a number of possible reasons for differences in radiosensitivity (Figure 1):

(i)   cells may vary in the amount of damage induced by a given dose of radiation;
(ii)   they may differ in the extent of repair of the damage or in accuracy of repair;
(iii) cells may tolerate residual (unrepaired) damage to different degrees.

In order to determine the genes and proteins that are important in the processes that affect radiosensitivity, this research has increasingly involved the specialities of biochemistry and molecular genetics. It is hoped that a greater understanding of the

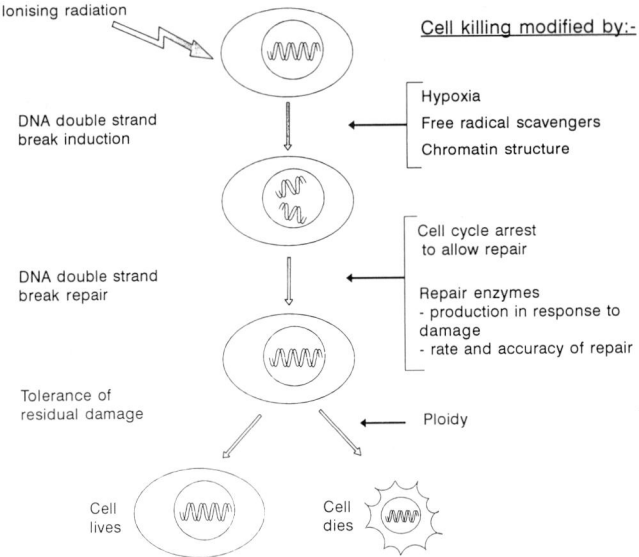

Figure 1. Schematic diagram of a cell's response to ionising radiation and factors that may modify cellular radiosensitivity.

cellular response to radiation will enable us to predict the radiosensitivity of a cell and also to provide a means to manipulate its sensitivity.

The aim of this chapter is to provide a summary of this field of research and to give a framework for the following chapters.

DNA double-strand breaks are likely to be the most significant radiation-induced lesions

Ionising radiation causes cell killing by its action on DNA. A variety of lesions are produced in DNA including base and sugar damage, DNA single-strand and double-strand breaks, DNA–protein cross-links and DNA–DNA cross-links. Of these, the DNA double-strand break is thought to be the lesion that is responsible for cell death. Irradiation performed under conditions that affect the number of double-strand breaks shows that changes in cell survival are seen to closely reflect changes in double-strand break levels [1]. In addition, it has been demonstrated in yeast that a double-strand break leads to cellular inactivation and in mammalian cells, double-strand breaks induced by other methods (e.g. restriction endonucleases) produce similar types of chromosome damage and cytotoxicity to ionising radiation. Other lesions such as single-strand breaks can be shown to be relatively non-cytotoxic when produced by other means. The single-strand breaks produced by hydrogen peroxide, for example, are non-cytotoxic until they interact to form double-strand breaks.

Modification of the number of double-strand breaks induced can modify cell kill

Of the many lesions produced in the DNA following ionising radiation, most are repaired within a few hours at 37°C. If cells are irradiated at low temperature (0–4°C) DNA repair mechanisms are inhibited. Experiments performed under these conditions are able to measure the initial damage induced in the cell.

It is found that the amount of damage induced in the DNA can be affected by chemical processes that occur in the cell during irradiation. The reason for this can be explained by considering the way in which ionising radiation causes damage. Gamma radiation when passing through a cell gives up its energy to produce fast moving electrons. These electrons can act either directly on DNA or indirectly by their interaction with water molecules to produce hydroxyl radicals. It is thought that two-thirds of the damage to DNA in mammalian cells is via the hydroxyl radical. Substances that interact with these free radicals during irradiation can influence the amount of DNA damage produced. Oxygen is able to do this by combining with aqueous electrons, thereby causing the damage induced to be "fixed" and so made irreparable. The majority of the effect of oxygen on cell survival can be explained in terms of this modification of the level of induced damage. Hydrogen donors such as compounds containing sulphydryl groups (–SH) can neutralise the free radicals and therefore have a protective effect. It has been demonstrated that an increase or decrease in the level of cellular sulphydryl compounds leads to parallel changes in clonogenic cell survival.

The importance of these chemical processes in accounting for the observed differences in radiosensitivity between tumours and between tumours and normal tissues is not known. A reduced oxygen tension in parts of poorly vascularised tumours (i.e. hypoxic regions) is believed to be an important cause of treatment failure. Methods to overcome hypoxia (e.g. hyperbaric oxygen, chemical hypoxic cell sensitizers) can improve tumour response in some situations. Therefore, the modification of DNA damage by oxygen is important in radiotherapy.

Whether there may be differences in the level of scavenging compounds such as sulphydryl compounds, between cells in such a way that they lead to differences in sensitivity has been a matter of some debate in recent years. There are now a number of groups who have observed differences in the apparent level of DNA double-strand breaks immediately after irradiation between different cell types [2,3]. The basis of such differences, however, is still not known. Radiation chemistry suggests that absolute levels of scavenging compounds may not be significant due to their inaccessibility to DNA. However, Ward has suggested that if there were differences in chromatin structure between cells, then this could allow a different accessibility of scavenging molecules to damaged DNA [4].

Most double-strand breaks are usually rejoined but cells may vary in the accuracy of repair

DNA repair is widely quoted as being the single most important factor in determining differences in radiation sensitivity. Except in extreme cases, however, there is relatively little evidence underlying this belief. Mutant cell lines isolated from Chinese hamster cells, which are highly radiosensitive, have been demonstrated to have a significant defect in their ability to rejoin DNA double-strand breaks [5]. This is in contrast to most other situations where the vast majority of double-strand breaks appear to be rejoined.

Assays used to measure the number of double-strand breaks can be used to assess double-strand break rejoining but tell us nothing about the accuracy of this process. It is possible that a significant amount of DNA is lost either due to the irradiation or during the repair process. Loss of genetic information can be detected in various ways:

(i)   fragments of chromosomes are often observed in irradiated cells and these can be lost when a cell divides;
(ii) deletions of parts of chromosomes containing specific genetic markers can be identified by Southern analysis or by the polymerase chain reaction;
(iii) small losses have been seen in the DNA sequences of genes from irradiated cells.

No attempts have been made using these techniques to compare cells of differing sensitivity, although repair fidelity has been examined by using a system where artificially damaged foreign DNA is introduced into cells which then attempt repair of double-strand breaks in the foreign DNA [6]. This repair process can be shown to result in significant loss of genetic material and the fidelity of rejoining assessed in this way does vary between cell lines of differing sensitivity [7,8].

Repair functions may be induced by DNA damage

As detailed in Chapter 18, there are a significant number of genes which increase their expression following treatment of cells with DNA damaging agents. In bacteria, it has been well demonstrated that some of these genes are involved in DNA repair processes. Thus, the cell responds to insult by increasing its ability to repair DNA damage. In mammalian cells, the relevance of increases in gene expression to subsequent survival of the cell is much less clear. There are three situations in which it has been suggested that mammalian cells increase their repair ability subsequent to treatment with ionising radiation:

(i)   Treatment of human lymphocytes with a very small dose of radiation (0.01 Gy) has been demonstrated to reduce the number of chromosomal aberrations detected following a second radiation treatment (9). This effect has been termed the adaptive response.

(ii) If viruses are irradiated and then infected into mammalian cells, the damaged virus is repaired by the cell in such a way that survival of viruses is increased. In some situations it has been shown that pretreating the mammalian cell with a small dose of radiation prior to infection with damaged virus can further increase the survival of the virus [10]. This has been inferred to be due to an inducible repair process within the mammalian cells.

(iii) A detailed analysis of clonogenic cell survival of rodent cells following ionising radiation to very low doses has suggested that the initial slope of the cell survival curve is much steeper than is generally thought. Once above these low doses, the curve is more shallow and it has been suggested that this change in slope may be due to inducible repair processes which are activated following the initial small component of the dose given.

Cells may differ in their tolerance of unrepaired DNA damage

It has been suggested by cytogeneticists that in diploid mammalian cells, loss of a single chromosomal fragment leads to cell death. In one extensive study, Revell and his colleagues showed that chromosome fragments were expressed at the first post-irradiation division as micronuclei and that in the diploid Syrian hamster cell system, the presence of acentric fragments correlated on a 1 to 1 basis with inhibited colony formation [11]. Revell also studied two non-diploid cell lines and demonstrated that in these cases more than one fragment needed to be lost before cell death resulted. The suggestion was therefore that once a cell becomes non-diploid it may be able to tolerate fragment loss better than in the diploid cell state. This is obviously relevant in cancer when one considers that a large proportion of human tumour cells are aneuploid.

Genes involved in radiosensitivity may be isolated directly or indirectly

Chapter 17 describes many of the approaches that have been used to isolate genes involved in DNA repair. The most successful approach to date has been the direct reversal of the sensitivity of radiation-sensitive mutants by the introduction of foreign DNA. This has been very fruitful with ultraviolet light, but only one gene involved in X-ray sensitivity has been isolated by this approach. Another approach is used in studies where the influence of genes that have been isolated in other contexts is being sought. The transformation of normal human fibroblasts in culture by infection with the SV40 virus or its transforming molecule (the large T antigen) has been known for some time to increase the radio-resistance of the fibroblasts. Such studies have been refined in recent years by the use of various transforming oncogenes. In rat embryo cells, transformation with the *ras* oncogene plus the v-*myc* oncogene results in cells that are highly resistant compared to the parent cells [12].

There is evidence to suggest that the cell cycle arrest induced by ionising radiation is important for the survival of the cell following irradiation. The theory is that cells respond to DNA damage by slowing down or stopping the cell cycle at specific points allowing time for repair to take place. Anything that reduces this cell cycle delay would reduce the extent of damage repair therefore increasing cell kill. Rad9 mutant yeast cells are more sensitive to radiation than normal yeast cells and also do not undergo a division delay after irradiation. Caffeine reduces or prevents the G2 delay and results in increased radiosensitivity. It has been recently suggested that tumour suppressor genes, in particular p53, can also influence radiosensitivity by regulation of the cell cycle. The increased radiosensitivity in ataxia telangiectasia cells is associated with abnormal cell cycle responses. Thus, genes that play a role in normal cell cycle control may be crucial to a cell's response to irradiation and therefore influence radiosensitivity.

What can the oncologist do with this information?

In the short term, the ability to predict the radiosensitivity of both normal and tumour cells holds great promise for improvements in radiotherapy. Modifications of dose to account for differences in tumour cell radiosensitivity could improve local control rates. This could be combined with measures of normal tissue radiosensitivity to identify those individuals who could tolerate an increased dose and those at high risk of radiation damage to normal tissues where alternative treatment modalities could be considered. In the longer term, it is hoped that identification of critical rate limiting proteins in DNA damage processing may allow the inactivation of those proteins within new therapeutic approaches.

References

1   Radford IR. The level of induced double strand breakage correlates with cell killing after X-irradiation. Int J Radiat Biol 1986, **48,** 45–54.
2   Radford IR. Evidence for a general relationship between the induced level of DNA double strand breakage and cell killing after X irradiation of mammalian cells. Int J Radiat Biol 1986, **49,** 611–620.
3   McMillan TJ, Cassoni AM, Edwards S et al. The relationship of DNA double strand break induction to radiosensitivity in human tumour cell lines. Int J Radiat Biol 1990, **58,** 427–438.
4   Ward JF. The yield of DNA double strand breaks produced intracellularly by ionising radiation: a review. Int J Radiat Biol 1990, **57,** 1141–1150.
5   Jeggo PA, Kemp LM. X-ray sensitive mutants of Chinese hamster ovary cell line. Isolation and cross-sensitivity to other DNA damaging agents. Mutat Res 1983, **112,** 313–327.
6   Eady JJ, Peacock JH, McMillan TJ. Host cell reactivation of gamma-irradiated adenovirus 5 in human cell lines of varying radiosensitivity. Br J Cancer 1992, **66,** 113–118.
7   Debenham PG, Jones NJ, Webb MBT. Vector-mediated DNA double strand break repair in normal and radiation-sensitive, Chinese hamster V79 cells. Mutat Res 1988, **199,** 1–9.
8   Powell S, McMillan TJ. Clonal variation of DNA repair in a human glioma cell line. Radiother Oncol 1991, **21,** 225–232.
9   Shadley JD, Wiencke JK. Induction of the adaptive response by X-rays is dependent on radiation intensity. Int J Radiat Biol 1989, **1,** 107–118.
10 Jeeves WP, Rainbow AJ. Gamma-ray enhanced reactivation of gamma-irradiated adenovirus in human cells. Biochem Biophys Res Commun 1979, **90,** 567–574.
11 Revell SH. Relationships between chromosome damage and cell death. In: Radiation-Induced Chromosome Damage in Man. Alan R. Liss, New York, 1983, pp 215–233.
12 McKenna WG, Iliakis G, Muschel RJ. Mechanism of radioresistance in oncogene transfected cell lines. In: Radiation Research: A Twentieth Century Perspective. ICRR, 1991. Academic Press, New York, 1991, pp 392–397.

Molecular Biology for Oncologists
Edited by John Yarnold, Michael Stratton, Trevor McMillan
© *1993, Elsevier Science Publishers B.V. All rights reserved*

CHAPTER 15

Cells differ in their susceptibility to DNA damage induction and in their ability to repair it

Stephen J. Whitaker

Department of Radiotherapy, The Royal Marsden Hospital, Downs Road,
Sutton, Surrey, SM2 5PT, UK

Damage to the genome underlies the carcinogenic and cell killing effects of many chemical and physical agents

The process of DNA damage and its repair underlies two important aspects of oncology. In carcinogenesis, many malignancies are caused by gene mutations and deletions following environmental exposure to genotoxic agents (such as natural radiation). Equally important, in cancer therapy the cell killing effects of most cytotoxic drugs and radiation is via DNA damage. These two aspects of DNA damage are seen to converge in individuals with radiosensitive syndromes who suffer increased tissue damage from exposure to ultraviolet (UV) or X-rays and are also at increased risk of developing malignancies. This review focuses on differences in DNA damage and repair with reference to the lethal effects of genotoxic agents but many of the points are relevant for mutagenesis and carcinogenesis.

The DNA damage sustained is of many different kinds

While both cytotoxic drugs and ionising radiation share a similar major target for their lethal effects in DNA, certain important differences exist. The chemistry of drug interactions with DNA leads to a narrow range of lesions. Ionising radiation, however, causes a wider range of lesions [1]. Each Gray of radiation absorbed by a cell causes about 10 000 damaged bases, 1000 damaged sugars, 1000 single and 40 double-strand breaks (ssb and dsb), 150 DNA–protein crosslinks (dpc) and 30 DNA–DNA crosslinks [1]. In comparison with drug exposure, these lesions are induced rapidly (at dose rates commonly used in radiotherapy, it takes 1–2 min to deliver this amount of radiation-induced damage to a cell). The majority of these lesions are equally rapidly and effectively repaired, such that the final level of DNA damage after ionising radiation is several orders of magnitude less than the lesions produced by drugs or UV radiation for an equal level of cell kill.

A continuing controversy in understanding cell killing by ionising radiation is whether there exists a particular type of lesion that is lethal for a cell. In addition, it is not clear whether there are a small number of critical targets within the genome where DNA damage is more likely to be lethal. Of the radiation-induced lesions listed above, the dsb appears to be most closely associated with cell lethality [1–3]. This may be because the dsb is a severe lesion for the cell to repair. It is unlikely to be commonly encountered by a cell under normal, physiological situations (unlike ssb and base damage) and requires a multi-step procedure for accurate repair.

Initial DNA damage is modified by chemical scavengers and by enzymatic repair to leave variable amounts of residual damage

The chain of events leading from cytotoxic exposure to cell death involves DNA damage and is shown schematically in Figure 1. DNA damage may be direct, by the drug or radiation itself, or indirect, by free radicals produced by secondary chemical reactions around DNA, often involving the water lying close to DNA [1]. Following damage induction, numerous processes remove and repair the damage in an attempt to restore the genetic sequence to its original state. At the physico-chemical level, damage involves ionisation of atoms within the DNA molecules with consequent disruption of interatomic bonds and the formation of other ab-normal covalent bonds between the constituents of DNA and chromatin. There is a competition between rapid chemical restitution of damage and its fixation by the interaction of O_2. The damage remaining at this stage is termed *initial damage* which may be lethal to a cell if not repaired. The level of initial damage is usually measured in cells treated on ice when enzymatic repair does not occur.

The damage can kill a cell by interfering with replication or gene transcription. Before this the cell attempts to *repair* the damage using a host of general DNA "housekeeping" repair enzymes as well as specific types of damage repair systems. Repair can be observed by exposing cells to a cytotoxic agent, then measuring the level of DNA damage remaining after incubation under physiological conditions. The majority of DNA lesions are rapidly repaired. Following exposure to ionising

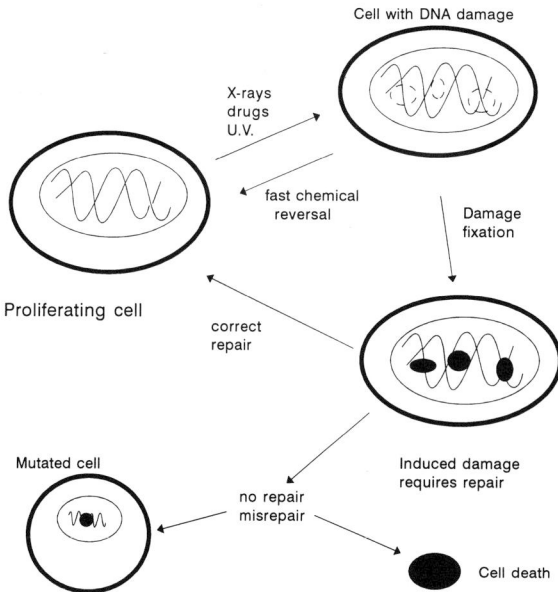

Figure 1. DNA damage leads to cell mutation and death. Chemical reversal of DNA damage and correct enzymatic repair lead to a normal cell which (if proliferating) will show normal growth and will have no detectable genetic changes. Misrepair or no repair may be lethal to a cell or may leave genetic changes (mutations) which are compatible with cell survival. The mutated cell may appear normal and exhibit normal growth. In a non-proliferating cell, the effects of the mutations may not appear until the cell is required to divide at a later date.

radiation, damaged bases are removed and ssb rejoined with half-times of 2–10 min, while dsb are rejoined with half-times of 10 min to several hours (a half-time being the time required for 50% of the lesions to be removed).

DNA damage may be correctly repaired, repaired incorrectly (misrepaired) or completely unrepaired. Any misrepaired or unrepaired damage, i.e. *residual damage* is often lethal to a cell but may be *tolerated* with a greater or lesser effect on cell function and replication.

Residual damage may have no consequences for the cell, it may alter cell behaviour or it may lead to cell death

Not all residual damage may be lethal to a cell. Lesions may be sublethal if the damage resides in a non-coding region of the genome. A striking example of this is the level of UV-induced pyrimidine dimers remaining in the genome of rodent cells compared to human cells even at equivalent levels of survival [4]. Whereas in human cells the majority (80%) of UV-induced dimers are removed by 24 h from both active and inactive regions of the genome, in rodent cells, non-coding regions of DNA still contain 85% of dimers at 24 h. The cells are able to continue DNA replication by a process termed *bypass replication* which is in effect a means of tolerating residual damage. Tolerance to chemotherapeutic gene damage may also occur and underlie some examples of drug resistance in human tumours [5]. Tolerance to ionising radiation is less well established because the low levels of DNA damage remaining have been difficult to detect [6], but newer assays such as pulsed field gel electrophoresis, should make this possible [7,8].

At the molecular level, tolerance is evident following ionising radiation treatment [9,10]. The analysis of cells that have mutated to drug resistance has shown that, generally, very large portions of DNA can be lost, as long as genes essential for cell growth remain intact.

Different kinds of cells vary in their radiosensitivity to the lethal effects of DNA damaging agents

When the cytotoxic effects of ionising radiation and, later, drugs began to be studied in vitro, it was assumed that all mammalian cells were killed with similar efficiency. Survival curves generated by the colony-forming assay of Puck and Marcus in the 1960s demonstrated little difference between the small number of tumour types studied. In the 1980s, however, a substantial review of published data on in vitro sensitivity proved that human tumour cells do differ in their inherent

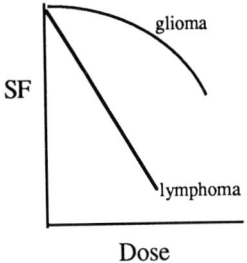

Dose

Figure 2. Inherent cellular radiosensitivity of two human tumour cells. Cells grown from a lymphoma are relatively more sensitive to the lethal effects of radiation than cells from a glioma, and reflect the inherent radiosensitivity of normal lymphocytes and glial tissue. (SF, surviving fraction).

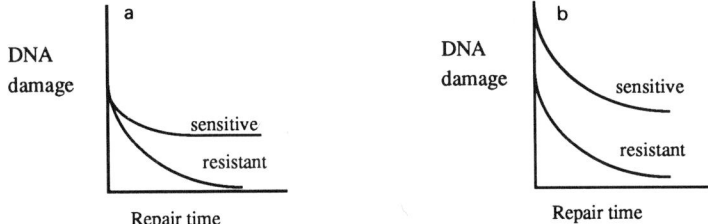

Figure 3. Two models describing variation in sensitivity to ionising radiation. (a) Both radiosensitive and radioresistant cells suffer equal levels of initial DNA damage induced but sensitive cells repair damage slower and to a lesser extent than resistant cells. (b) Radiosensitive cells suffer more initial DNA damage induced but the rate of repair is similar to radioresistant cells.

sensitivity to ionising radiation. This *inherent cellular radiosensitivity* is relevant to variations in radioresponse seen in the clinic and is an additional factor to other determinants of tumour response to X-irradiation such as oxygen tension and cell proliferation kinetics.

The greatest variations are between tumour cells originating from different normal tissues which themselves show variations in cellular sensitivity, such that gliomas are amongst the most resistant and lymphomas are the most sensitive (Figure 2) [1]. Variation is also seen within one tumour type, for example squamous cell carcinomas of the cervix show a range in cellular radiosensitivity [11] and likewise normal cells from different individuals vary in radiosensitivity. These differences could explain the ranges of tumour and normal tissue responsiveness observed clinically.

Cells differ in the amount of initial DNA damage they sustain

DNA damage and repair has often been assumed to be uniform within the genome and damage induction similar between different cells. Differences in inherent cellular sensitivity have classically been attributed to differences in the rate and extent of repair processes throughout the entire genome (shown diagrammatically in Figure 3(a). However, evidence exists that initial DNA damage induction by some genotoxic agents varies between different cells (Figure 3(b)). For example, some human tumour cells have been found to sustain a greater level of initial double-strand breaks induced than radioresistant cells. The molecular basis of this is not clear but presumably is dependent on the post-irradiation chemical processes described above [2,11,12].

The distribution of initial DNA damage is not uniform

Conventionally, the extent of induced DNA damage has been expressed relative to the total quantity of DNA per cell. This follows the theoretical assumption, that damage must be uniform because all parts of the genome are exposed to equal concentrations of agent (drugs or radiation). In addition, it has generally proved difficult to measure damage in specific parts of the genome.

With advances in the techniques of molecular biology, our understanding of the relationship of the structural and functional organization of DNA as chromatin has also increased. In particular, various fractions of cellular DNA can be separated and damage in each specific fraction determined. Measurement of the genomic distribution of DNA damage induced by various cytotoxic agents has become possible.

With supralethal doses of ionising radiation (hundreds of Gray), DNA damage appears to be random and uniform in distribution. After lower doses, however, hot-spots for DNA damage have been identified in both functionally and structurally specific regions of DNA. Using probes for actively transcribed genes, an increase of 5–6-fold in ssb and dpc has been observed in transcribed genes [13]. An increase in ssb has also been found in newly replicated DNA compared to overall DNA ssb [14].

Similar heterogeneity in damage induction has been found for drug-induced damage [5,15]. Strand breakage secondary to treatment with topoisomerase inhibitors, such as etoposide, and direct strand cleavage by bleomycin is greater in transcribed regions [5]. Alkylation is also seen to be greater in GC rich areas, which are associated with coding regions of the genome [5,16].

The increased sensitivity of actively transcribed genes and newly replicated DNA could have a common basis if one considers the relationship of gene function and chromatin structure. Actively transcribed genes are associated with regions of open chromatin conformation which also exist around newly replicated DNA. This chromatin may be more accessible to DNA damaging free radicals and drugs. In support of this is the observation that chromatin, when relaxed in low magnesium concentration, sustains more ssb than in high Mg^{2+} [15].

Cells differ in their ability to repair initial DNA damage

The differences observed in the cellular consequences of cytotoxic exposure have been assumed to arise from variations in the repair of DNA damage. The study of mutant cells selected for increased sensitivity to UV, drugs or ionising radiation has allowed the isolation of cells deficient in DNA repair and, consequently, the steps involved in repair have been identified. Cells sensitive to UV radiation show defects in excision-repair of UV-damaged DNA bases and there are a number of in vitro rodent cell lines whose marked sensitivity to ionising radiation arises from grossly delayed rejoining of dsb as shown in Figure 3(a) [1].

Amongst human cell lines, however, such differences have been difficult to demonstrate. Following ionising radiation, variations in the level of residual dsb after 1-h incubation have been found [11,17]. Cells from ataxia telangiectasia (A-T) individuals, which show increased sensitivity to ionising radiation, do not show any major differences in DNA dsb rejoining.

A major problem with these studies is that they measure rejoining or removal of lesions in DNA, not the accuracy of such repair (see Figure 4). The *fidelity* of repair is as important for the restitution of gene sequence and function as the quanti-

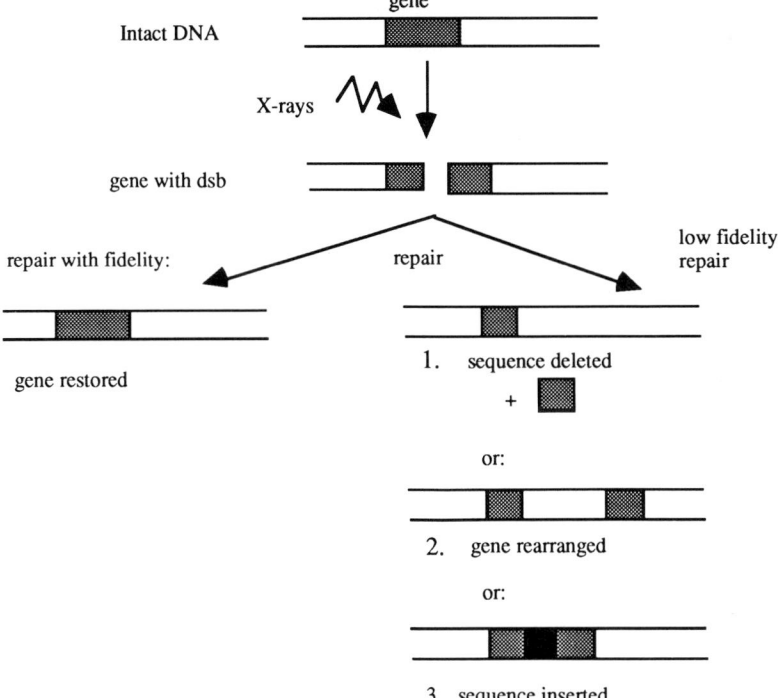

Figure 4. A gene exposed to X-irradiation sustains a double-strand break (dsb). Repair is necessary for gene function to be restored. Faithful repair requires not only rejoining of the strands but accurate restoration of the original sequence. Lack of faithful repair (low fidelity) can be envisaged as sequence loss (by deletion) or inaccurate recombination repair with incorrect sequence insertion.

tative removal of DNA lesions. Low fidelity of repair has been implicated in the radiosensitivity of A-T cells as well as some human tumour cells [1]

DNA damage repair is not uniform across the genome

Like DNA damage induction, the removal and repair of damage was assumed to be uniform across the genome. However, from comparisons between rodent and human cells, it was found that removal of UV-induced pyrimidine dimers was much less complete in the rodent cells (15% at 24 h) compared to the human cells (80% at 24 h) at equi-lethal doses of UV (see Figure 5 and [4,18]). It was found that while a significant difference existed in the overall level of DNA repair, removal of damage from actively transcribing genes was similar in the two cell types. This discrepancy has led to studies of DNA repair in specific regions of the genome. What determines rodent cell death from UV radiation is unrepaired damage in essential, transcribed gene sequences. In proliferating rodent cells, replication can bypass residual dimers. In contrast, in human cells, all dimers must be removed as bypass replication does not appear to occur [18]. In human cells, like rodent cells, dimers are removed more rapidly from active genes (see Figure 5).

Preferential repair of active genes is also observed following drug and ionising radiation damage and is likely to be a fundamental process protecting DNA integrity [5,14]. The biological reason for preferential repair is presumably to ensure that essential sequences required for immediate survival are repaired first and rapidly. The molecular mechanism of this, however, is not completely understood.

One possible explanation is that regions of the genome engaged in gene expression are, by necessity, more accessible; indeed part of the process of control of gene expression is through structural conformation. This is similar to the mechanism invoked for increased sensitivity for damage induction. This may be too simplistic, however. A more sophisticated model of chromatin organization has actively transcribed genes and newly replicated DNA closely associated with the nuclear matrix [19,20].

Repair of UV damage involves excision of damaged DNA and its replacement, with *repair patches*. These are also found associated with the nuclear matrix implying that the DNA repair apparatus may require similar constraints to DNA for its repair or that the repair enzymes are also attached to the matrix [20]. Repair at other points of the DNA loop is slower, either because DNA here requires diffusion of a mobile repair protein, or the DNA loop has to *reel* through the nuclear matrix attachment zone.

An alternative explanation of preferential repair, aside from structural and positional characteristics, is that it is the functional distinction of active genes that

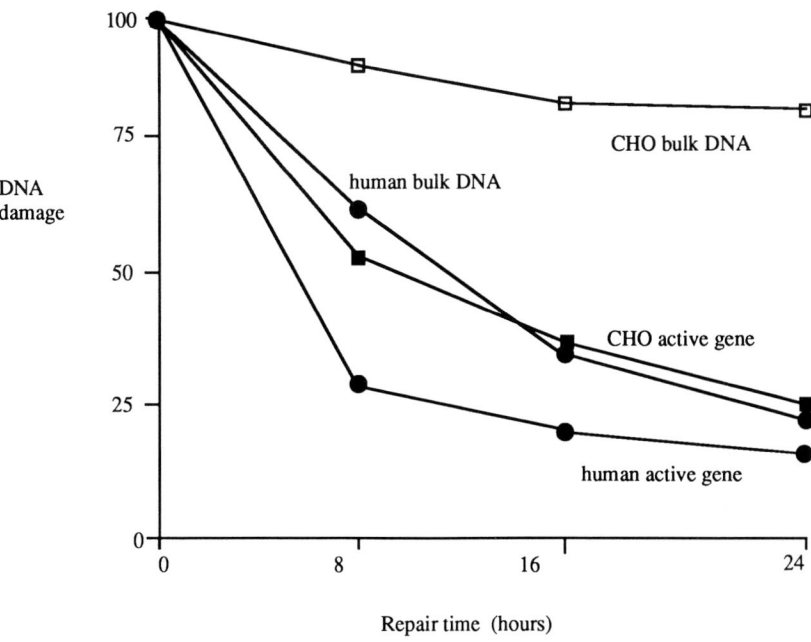

Figure 5. Mammalian cells exposed to UV radiation repair DNA damage in active genes faster than non-coding regions and overall bulk DNA. In CHO cells, significant levels of DNA damage remain at 24 h in bulk regions (which include all the genome, coding and non-coding). In human cells, greater DNA damage repair is seen in active regions but by 24 h residual levels of damage in active and bulk DNA are similar.

facilitates their repair. Evidence for this has come recently with the observation that preferential repair is only for the transcribed strand of the gene, i.e. the sense or coding strand, and not the anti-sense, non-transcribed strand [4,21]. Both strands share the same position and conformation with the only difference being the transcription of the sense strand. This implies that the transcription apparatus RNA polymerases and transcription factors are involved in the repair process. This may be by active participation in repair or by allowing damage recognition and facilitation of repair.

Rare human UV sensitivity syndromes are due to defects in selective DNA repair

There are many different human syndromes which show sensitivity to UV radiation. The best characterized example is xeroderma pigmentosum (XP). This rare, autosomal recessive syndrome is of fundamental interest to the study of both carcinogenesis and cancer therapy as each variant represents a single gene defect responsible for a specific step in DNA repair.

Most of the syndromes are thought to represent defects in the overall extent of DNA repair but one syndrome, Cockayne's, is due to a failure of preferential repair of active genes [18,21,22]. Typical characteristics of individuals with Cockayne's syndrome (CS) include severe photosensitivity, generalized wasting, ataxia, mental retardation and early death from progressive neurological degeneration.

Cells from CS individuals can be useful models for the study of preferential DNA repair [18]. CS fibroblasts and lymphocytes both show marked hypersensitivity to UV radiation, although to a lesser degree than XP cells. CS cells also show a lack of preferential repair of nuclear matrix-associated DNA, indicating again the relationship of active genes with the nuclear matrix.

Interestingly, an important difference between CS individuals and other radiosensitive individuals is the lack of increased susceptibility to malignancy that is so obvious in the other syndromes. Thus, while preferential repair of UV damage in active genes is important in cell survival, it is the overall genomic damage and its repair that is important for carcinogenesis. It is precisely because the residual damage in CS cells is in essential, active genes that the cell is more likely to die than to acquire a mutation giving a proliferation advantage and malignant transformation.

References

1   Powell SN, McMillan TJ. DNA damage and repair following treatment with ionizing radiation. Radiother Oncol 1990, **19**, 95–108.
2   Radford IR. The level of induced DNA double-strand breakage correlates with cell killing after X-irradiation. Int J Radiat Biol 1985, **48**, 45–54.
3   Frankenberg D, Frankenberg-Schwager M, Blöcher D, Harbich R. Evidence for DNA double-strand breaks as the critical lesions in yeast cells irradiated with sparsely or densely ionizing radiation under oxic or anoxic conditions. Radiat Res 1981, **88**, 524–532.
4   Hannawalt PC. Selective DNA repair in expressed genes in mammalian cells. In: Mutation and the Environment, Part A: Basic Mechanisms. ML Mendelsohn and RJ Albertini eds. Wiley–Liss, USA, 1990, pp 213–222.
5   Epstein RJ. Drug-induced DNA damage and tumour chemo-sensitivity. J Clin Oncol 1990, **8**, 2062–2084.
6   McMillan TJ. Residual DNA damage: what is left over and how does this determine cell fate? Eur J Cancer 1992, **28**, 267–269.
7   Whitaker SJ, McMillan TJ. Oxygen effect for DNA double-strand break induction determined by pulsed-field gel electrophoresis. Int J Radiat Biol 1992, **6**, 29–41.
8   Whitaker SJ, Powell SN, McMillan TJ. Molecular assays of radiation-induced DNA damage. Eur J Cancer 1991, **27**, 922–928.
9   Waldren C, Jones C, Pucj TT. Measurement of mutagenesis in mammalian cells. Proc Natl Acad Sci USA 1979, **76**, 1358–1362.

10 Yandell DW, Dryja TP, Little JB. Molecular genetic analysis of recessive mutations at a heterozygous autosomal locus in human cells. Mutat Res 1990, **229**, 89–102.

11 Kelland LR, Edwards SM, Steel GG. Induction and rejoining of DNA double-strand breaks in human cervix carcinoma cell lines of differing radiosensitivity. Radiat Res 1988, **116**, 526–538.

12 McMillan TJ, Cassoni AM, Edwards S, Holmes A, Peacock JH. The relationship of DNA double-strand break induction to radiosensitivity in human tumour cell lines. Int J Radiat Biol 1990, **58**, 427–438.

13 Chiu S-M, Oleinick NL. The sensitivity of active and inactive chromatin to ionizing radiation-induced DNA strand-breakage. Int J Radiat Biol 1982, **41**, 71–77.

14 Oleinick NL, Chiu S-M, Friedman LR. Gamma radiation as a probe of chromatin structure: damage to and repair of active chromatin in the metaphase chromosome. Radiat Res 1984, **98**, 629–641.

15 Whitaker SJ. DNA damage by drugs and radiation: What is important and how is it measured? Eur J Cancer 1992, **28**, 273–276.

16 Mattes WB, Hartley JA, Kohn KW, Matheson DW. GC rich regions in genomes as targets for DNA alkylation. Carcinogen 1988, **9**, 2065–2072.

17 Schwartz JL, Rotmensch J, Giovanazzi S, Cohen MB, Weichselbaum RR. Faster repair of DNA double-strand breaks in radioresistant human tumour cells. Int J Radiat Oncol Biol Phys 1988, **15**, 907–912.

18 Bohr VA, Evans MK, Fornace AJ. Biology of disease, DNA repair and its pathogenetic implications. Lab Invest 1989, **61**, 143–161.

19 Jackson DA, Cook PR. Transcription occurs at a nucleo-skeleton. EMBO J 1985, **4**, 919–925.

20 Mullenders LHF, Venema J, van Hoffen A, Mayne LV, Natarajan AT, van Zeeland AA. The role of the nuclear matrix in DNA repair. In: Mutation and the Environment, Part A: Basic Mechanisms. ML Mendelsohn, RJ Albertini eds. Wiley–Liss, USA, 1990, pp 223–232.

21 Mellon I, Spivak G, Hannawalt PC. Selective removal of transcription-blocking DNA damage from the transcribed strand of the mammalian DHFR gene. Cell 1987, **51**, 241–249.

22 Venema J, Mullenders LHF, Natarajan AT, van Zeeland AA, Mayne LV. The genetic defect in Cockayne syndrome is associated with a defect in repair of UV-induced damage in transcriptionally active DNA. Proc Natl Acad Sci USA 1990, **87**, 4707–4711.

Molecular Biology for Oncologists
Edited by John Yarnold, Michael Stratton, Trevor McMillan
© 1993, Elsevier Science Publishers B.V. All rights reserved

CHAPTER 16

How do cells repair DNA damage?

Gillian Ross

Radiotherapy Research Unit, The Institute of Cancer Research,
15 Cotswold Road, Sutton, Surrey, SM2 5NG, UK

What is DNA repair and why is it important in oncology?

There is a fundamental requirement for the accurate maintenance of genetic information, even though mutations are occasionally biologically advantageous. This requires an extremely accurate process for replicating DNA before cell division and mechanisms for the detection and elimination of accidental lesions. The frequent spontaneous changes occurring in DNA are largely transient by virtue of efficient DNA repair processes. Fewer than one in a thousand accidental base changes result in permanent change in information, i.e. mutation. DNA repair can be defined as those cellular processes associated with the preservation or restoration of DNA structure and function.

Clinical oncologists are increasingly intrigued by the prospect of identifying DNA repair processes relevant to both predisposition to cancer and response to cytotoxic therapy. The evidence for significant individual variability in DNA repair proficiency is currently limited to our knowledge of rare inherited human syndromes (see Chapter 9). In several of these syndromes, defects in DNA repair mechanisms are associated with hypersensitivity to DNA damaging agents, including ultraviolet light (UV), ionising radiation and cytotoxic drugs. There is also a predisposition to either lymphoid or epithelial neoplasia. These human genetic syndromes clearly provide very valuable models for the study of DNA repair in higher eukaryotes.

Although the syndromes are rare, the possibility of partial DNA repair defects in heterozygous cases raises critical questions in clinical oncology [1]. In particular, the concept of normal tissue tolerance in radiation oncology traditionally assumes a normal distribution of radiation sensitivity in the cancer population; clinical dose-fractionation regimens are chosen to maintain acceptable treatment complication rates based on a normal distribution of individual radiosensitivity. If, however, a proportion of patients demonstrating high complication levels represent a distinct subset of repair-deficient individuals, then they should be considered separately when determining radiation tolerance.

DNA repair is complex even in simple organisms

DNA repair pathways have been conserved through evolution, and much of our knowledge of DNA repair processes has been obtained by the detailed study of bacteria and yeast. Many genes and their protein products involved in repair have been identified (see Chapter 17). Many of these genes have been isolated because of a wealth of mutants defective in DNA repair pathways. This is partly a reflection of the simplicity of the unicellular state. The complexity of the multicellular eukaryotic state selects against such mutants. By comparison with bacteria, there are relatively few "spontaneous" rodent or human systems where a clear-cut defect in DNA repair is compatible with cell survival [2,3]. Given the complexity of DNA repair in simple organisms, even more genes are probably involved in the regulation of DNA repair in higher eukaryotes.

Excision repair pathways have been well characterised

The best characterised of the DNA repair processes are the excision repair pathways. Damaged or inappropriate bases are excised from DNA and the normal nucleotide sequence is restored. Defects in the excision repair pathway are implicated in the human syndromes associated with hypersensitivity to UV radiation, namely, xeroderma pigmentosum, Bloom's syndrome and Cockayne's syndrome, and a number of human genes involved in this process have been recognised. These are discussed further in Chapter 17.

Excision repair is involved in the removal of many classes of DNA lesion, including UV-induced pyrimidine dimers, bulky chemical adducts, alkylation and base damage [4]. As base damage, strand cross-links and DNA-protein links are part of the spectrum of damage induced by ionising radiation, it is possible that an excision repair pathway may play a role in the repair of some classes of ionising radiation-induced damage.

Several closely related enzymic actions can participate in the removal of altered nucleotides (see Figure 1). DNA glycosylases are enzymes which catalyse the

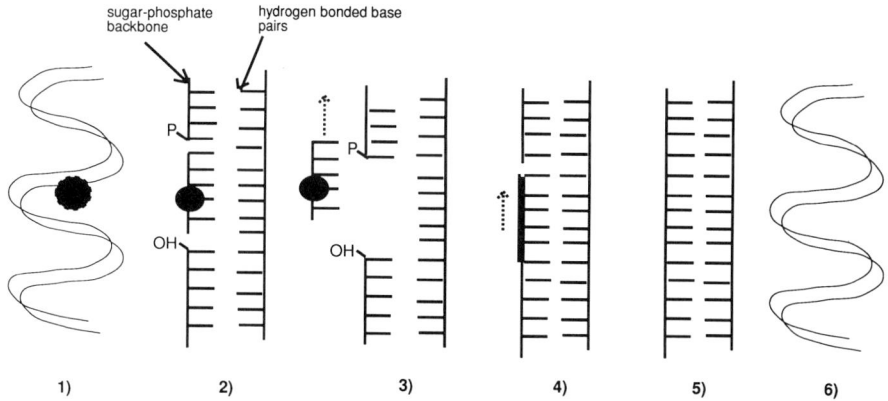

Figure 1. Excision repair pathway. (1) DNA molecule with base damage or bulky double-helix distorting lesion, e.g. UV photoproducts or cisplatin adducts. (2) Damage recognition and incision adjacent to site by DNA endonucleases. (3) Excision of damaged base and adjacent sequence by endonucleases. (4) Polymerisation or "fill-in" of missing bases, using undamaged DNA strand as template, mediated by DNA polymerases. (5) DNA ligases perform ligation of strands. (6) Restoration of intact DNA molecule with normal stereochemistry.

hydrolysis of the N-glycosylic bonds linking bases to the deoxyribose-phosphate backbone. Following base removal, the resulting "apurinic/apyrimidinic" (AP) site in the sugar-phosphate backbone must be excised by the sequential action of endo- and exonucleases. The integrity of the genetic information is restored by replacement of missing nucleotides by the action of DNA polymerases, and finally the rejoining of strands by DNA ligases.

An integral part of the enzyme activity of DNA glycosylases and AP endo-nucleases is the recognition of damaged DNA bases. This is in addition to their catalytic activities. There are important reasons to separate such functions mechanistically. Firstly, the precise biochemistry of such fundamental processes is important. Secondly, each discrete part of the repair process is potentially a candidate for therapeutic modulation. An area of increasing research interest is the identification of proteins which participate in the repair process at the damage recognition stage. Proteins have been detected in extracts of human cells which are capable of recognising a wide spectrum of damage to DNA, including that induced by UV, ionising radiation and cytotoxic drugs. It is postulated that overexpression of damage-recognition proteins might contribute to evolution of cytotoxic drug resistance [5]. No convincing evidence exists for such a mechanism operating in the repair of ionising radiation damage.

Accurate DNA repair in higher organisms depends on DNA repair enzymes working in a range of different chromatin conformations. Some cell lines estab-

G. Ross

lished from XP patients are capable of repairing pyrimidine dimers in purified or
"naked" DNA, but not in their native chromatin [6]. Thus, chromatin conformation
in XP cells may contribute to their defective DNA repair.

Does genetic recombination participate in DNA strand break rejoining?

Genetic recombination involves the exchange of DNA sequence between DNA
strands (Figure 2). Strand exchange and the homologous pairing of DNA strands
require a series of enzymic events. Recombination can occur both between and
within chromosomes, and different classes of DNA rearrangements are the end-
products of such events [7]. These processes are ubiquitous in nature, and can be
demonstrated in both the germline and somatic cells. Much of our understanding of
these events has been derived from the study of *E. coli* and yeasts. In *E. coli*, for
example, a gene (*Rec*A) has been identified which catalyses DNA strand exchange.
Strains of *E. coli* lacking the *Rec*A gene can be shown to be partially defective in
recombination.

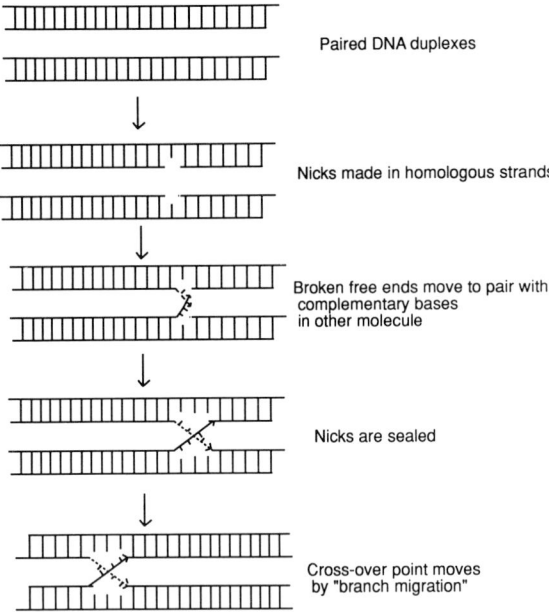

Paired DNA duplexes

Nicks made in homologous strands

Broken free ends move to pair with
complementary bases
in other molecule

Nicks are sealed

Cross-over point moves
by "branch migration"

Figure 2. A simplified scheme of genetic recombination.

Homologous recombination has been proposed as a mechanism for the repair of both DNA cytotoxic-induced adducts and radiation-induced strand breaks [8]. The most compelling evidence for the role of recombination in resolution of ionising radiation-induced double-strand breaks comes from experiments on SCID mice. These animals exhibit the Severe Combined Immune Deficiency phenotype attributed to a defect in immunoglobulin gene rearrangement. A high frequency of aberrant chromosomal rearrangements in the immunoglobulin and T-cell receptor genes is seen. Both lymphoid and fibroblast cell lines exhibit increased chromosomal damage and cell kill following ionising radiation exposure, and this is associated with a decrease in the rejoining of DNA double-strand breaks [9].

Does the repair of DNA double-strand breaks determine cellular radiosensitivity?

The nature of ionising radiation-induced lesions in genomic DNA has been reviewed in Chapter 15. The diversity of induced damage makes the identification of cytotoxic DNA lesions very difficult, and the extent to which specific DNA repair pathways participate in the repair of ionising radiation-induced DNA damage is not known. This makes it difficult to study whether variations in DNA repair proficiency are important in determining cellular response to ionising radiation. Although there is evidence associating proficiency of DNA double-strand break repair with cell lethality, it is clear that this represents a mechanistic oversimplification [10]. Mammalian cells repair the majority of double-strand breaks, and it seems likely that failure to repair subclasses of double-strand breaks, possibly in critical genomic sites, results in cell death following irradiation.

DNA repair can be observed in the test tube

A DNA repair pathway can be reconstituted from its molecular components in the laboratory, using "in vitro" repair systems [11]. The aim of a truly "in vitro" approach is to reconstruct specific DNA repair processes in the test tube. These techniques study the fate of lesions in simple DNA molecules, either oligonucleotides or recombinant plasmid DNA. Repair is inferred by the removal of damage from these molecules, and is mediated by the addition of active enzymes extracted from whole cells [12].

The "repair synthesis assay" permits the study and the purification of repair proteins involved in the processing of UV and cytotoxic damage via the excision repair pathway [13]. A simple plasmid DNA molecule containing either UV- or cisplatin-induced damage is incubated with proteins extracted from mammalian cells. The ability of the repair proteins to identify damaged bases, excise and restore normal sequence is monitored by the incorporation of radiolabelled nucleotides (Figure 3). The assay has provided a functional test of the ability of putative repair proteins to restore excision repair activity in extracts derived from repair-defective cell lines from patients with xeroderma pigmentosum.

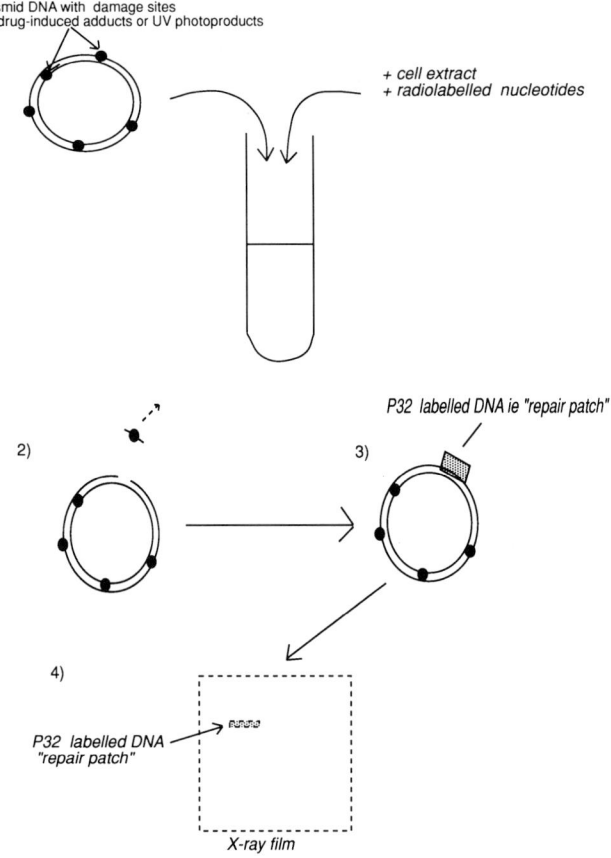

Figure 3. Excision repair synthesis assay. (1) Substrate plasmid DNA for reaction is pretreated with DNA damaging agent, e.g. UV light or cisplatin. Extract prepared from a cell line of interest is added to an in vitro reaction containing plasmid substrate molecules. (2) Active proteins within cell extract recognise DNA damage and initiate incision at damage site. (3) Cell extract contains DNA polymerases which incorporate radiolabelled nucleotides at site of "repair synthesis" of plasmid DNA molecule. (4) Plasmid DNA is subjected to gel electrophoresis and transferred to a nitrocellulose filter by Southern blotting. Incorporation of radiolabelled nucleotides in repair patch detected by autoradiography.

In vitro studies have been applied to the study of DNA strand break rejoining

A novel approach to the study of DNA double-strand breaks involves the use of bacterial restriction enzymes, which generate differing biochemical classes of DNA double-strand breaks similar to those produced by ionising radiation in genomic DNA. Whilst these lesions represent a simplification of the DNA strand breaks induced by ionising radiation, they have the particular advantage of allowing study of the fate of subclasses of strand break in an analytical manner.

Rejoining of restriction-induced DNA strand breaks in plasmid DNA can be achieved by proteins derived from human cell lines. Using this system, a defect has been proposed in the accuracy of DNA double-strand break rejoining in cell extracts prepared from ataxia telangiectasia. The importance of this study is two-fold. The observation of reduced fidelity of DNA double-strand break rejoining in ataxia telangiectasia has provided a new clue to the mechanistic basis of this rare disorder. Secondly, it has added impetus to the notion that in vitro systems might make a real contribution to our step-wise analysis the repair of ionising radiation-induced DNA damage [14].

The analytical power of in vitro assays can be adapted to study repair in the cellular environment

Modelled double-strand breaks can also be used to study repair in the cellular environment. Plasmid DNA bearing mammalian selectable genes (*gpt* and *neo*) is digested by restriction enzymes at a site within one of the selectable genes, rendering it inactive. This cut plasmid molecule can be successfully transfected into rodent or human cells. Accurate rejoining of the plasmid ends restitutes the gene sequence, resulting in gene expression within the host cell [12, 15].

Studies of DNA organisation within the nucleus have recently emphasised the importance of higher order structure on replication, RNA transcription and possibly repair (see Chapter 17). The intact higher order structure of DNA and chromatin is maintained in human cells embedded in agarose and lysed isotonically. The resulting structures can be shown to preserve critical molecular functions such as DNA replication and transcription [16]. Repair can therefore be analysed and manipulated in a similar manner to those described in the previous section but the DNA structure is more relevant. Whole cells can be permeabilised to permit introduction of molecules. These might be restriction enzymes which introduce DNA damage or purified proteins felt to be participants in a repair pathway under study. Such a system has also been used to assess the activity of DNA repair inhibitors [17]. The action of molecules such as enzymes and antibodies can also be studied by microinjection directly into cells [18,19].

It appears that the analytical power of in vitro methodogy can be merged with viable cell systems to provide powerful new tools to investigate DNA repair in the mammalian genome. It is likely that such techniques will continue to contribute to our knowledge of both fundamental DNA repair processes, and the fate of drug and radiation-induced DNA damage.

Can DNA repair be induced following DNA damage?

It has long been recognised that bacteria respond to DNA damage by the induction of several gene families. The term "SOS repair" was introduced to describe such a response. Included in the phenomena of the SOS response are cell division delay, inhibition of DNA degradation, inaccurate initiation of DNA replication, long-patch nucleotide repair and interruption of cellular respiration [20].

Given the existence of such damage-inducible pathways in bacteria, it is tempting to postulate that more complex mammalian cells might have similar or even more efficient responses to genotoxic insults. The main evidence for this comes from the observation that lymphocytes exposed to a low dose of ionising radiation are rendered more resistant to a second treatment. This has been termed an adaptive response. Molecular analyses have now demonstrated that a large number of genes are induced by DNA damage (see Chapter 18). UV irradiation increases synthesis of protein products from genes as diverse as c-*fos*, HIV 1, collagenase and human metallothionine II-A. Ionising radiation can also induce several classes of so-called "early response" genes, including *fos*, *jun* and the *Egr*-1 families, which themselves code for transcription factors participating in the further regulation of cell responses. It is possible that some of these genes might be involved in regulating cell cycle progression such that DNA synthesis is inhibited following irradiation. This is believed to allow a long time for repair to take place. Such genes have been designated Growth Arrest/Damage Inducible or *gadd*. In ataxia telangiectasia cells, the radiation hypersensitive phenotype is associated with a deficiency in cell cycle arrest, and a reduced induction of the gadd 45 gene has recently been observed.

Whilst defects in specific DNA repair genes might lead to expression of a radiation hypersensitive phenotype, it is also feasible that abnormalities of repair induction might lead to abnormal sensitivity. The extent to which incomplete expression of such phenotypes might account for any inherent variation in individual response to cancer therapy is intriguing but speculative.

Conclusions

Advances in molecular biology and somatic cell genetics have made a major contribution to our understanding of DNA repair pathways in simple organisms. It is hoped that the application of similar technology to human cells will make a similar impact. New methodologies now permit the modelling of repair processes relevant to clinical DNA damaging modalities, including ionising radiation and cytotoxic drugs.

The isolation of genes and their protein products involved in human DNA repair represents a major biological advance. It is hoped that characterisation of fundamental mechanisms might facilitate the design of specific repair inhibitors. At the clinical level, the relevance of variability of DNA repair proficiency to determining an individual's tumour or normal tissue response to cancer therapy remains to be elucidated.

References

1 Swift M, Reitnauer P, Morrell D. Chase C. Breast and other cancers in families with ataxia telangiectasia. N Engl J Med 1987, **316**, 1289–1294.

2 Hickson ID, Davies S. Robson C. DNA repair in radiation sensitive mutants of mammalian cells: possible involvement of DNA topoisomerases. Int J Radiat Biol 1990, **58**, 561–568.

3 Collins A, Johnson R. DNA repair mutants in higher eukaryotes. J Cell Sci Suppl 1987, **6**, 61–82.

4 Epstein R. Drug-induced DNA damage and tumour chemosensitivity. J Clin Oncol 1990, **8**, 2062–2084.

5 Chu G, Chang E. Cisplatin-resistant cells express increased levels of a factor that recognizes damaged DNA. Proc Natl Acad Sci USA 1990, **87**, 3324–3328.

6 Fujiwara Y, Kano Y. Cellular responses to DNA damage. UCLA Symp Mol Cell Biol 1983, **2**, 215–224.

7 Bollag R, Waldman A, Liskay R. Homologous recombination in mammalian cells. Annu Rev Genet 1989, **23**, 199–225.

8 Lindahl T, Wood RD. DNA Repair and recombination (27 refs.). Curr Opin Cell Biol 1989, **1**, 475–480.

9 Fulop GM, Phillips RA. The scid mutation in mice causes a general defect in DNA repair. Nature 1990, **347**, 479–482.

10 Bryant P. Enzymatic restriction of mammalian cellular DNA: evidence for double strand breaks as potentially lethal lesions. Int J Radiat Biol 1985, **48**, 55–60.

11 Hoeijmakers J, Eker A, Wood R, Robins P. Use of in vivo and in vitro assays for the characterisation of mammalian excision repair and isolation of repair proteins. Mutat Res 1990, **236**, 223–238.

12 Thacker J. The use of integrating DNA vectors to analyse the molecular defects in

ionising radiation-sensitive mutants of mammalian cells including ataxia telangectasia. Mutat Res 1989, **220**, 187–204.

13   Wood R. Repair of pyrimidine dimer ultraviolet photoproducts by human cell extracts. Biochemistry 1989, **28**, 8287–8292.

14   North P, Ganesh A, Thacker J. The rejoining of double strand breaks in DNA by human cell extracts. Nucleic Acids Res 1990, **18**, 6205–6210.

15   Powell SN, McMillan TJ. Clonal variation of DNA repair in a human glioma cell line. Radiother Oncol 1991, **21**, 225–232.

16   Jackson D, Cook P. Replication occurs at a nucleoskeleton. EMBO J 1986, **5**, 1403–1410.

17   Gedik CM, Collins AR. Comparison of effects of fostriecin, novobiocin, and camptothecin, inhibitors of DNA topoisomerases, on DNA replication and repair in human cells. Nucleic Acids Res 1990, **18**, 1007–1013.

18   Hoeijmakers J, Eker A, Wood R, Robins P. Use of in vivo and in vitro assays for the characterisation of mammalian excision repair and isolation of repair proteins. Mutat Res 1990, **236**, 223–238.

19   Lehmann A, Hoeijmakers J, van Zeeland A, Backendorf C, Bridges B, Collins A. Workshop on DNA repair. Mutat Res 1992, **273**, 1–28.

20   Friedberg E. DNA Repair. WH Freeman, San Francisco, 1985.

Molecular Biology for Oncologists
Edited by John Yarnold, Michael Stratton, Trevor McMillan
© *1993, Elsevier Science Publishers B.V. All rights reserved*

CHAPTER 17

The isolation of DNA repair genes

Richard Cartwright and Trevor J. McMillan

Radiotherapy Research Unit, The Institute of Cancer Research,
15 Cotswold Road, Sutton, Surrey, SM2 5NG, UK

The genome is vulnerable to spontaneous and induced damage and processes must exist for recognising and repairing this

The fidelity of DNA of eukaryotic cells is under constant threat from errors introduced during cell replication and from damage induced by products of cell metabolism. In addition, external physical and chemical agents can introduce DNA damage that is potentially mutagenic or lethal to the cell. Thus, the repair of DNA damage is fundamental to cell survival and complex processes have evolved to fulfil this demand. As outlined in Chapters 15, 16 and 19 the interest in DNA repair for oncologists lies in the role of DNA damage in carcinogenesis and in its significant impact on the effectiveness of cytotoxic chemo- and radiotherapy. The isolation of DNA repair genes is a useful approach to the study of the functions of DNA repair. The genes and gene products may also be new targets which can be modified in attempts to alter the sensitivity of cells to various agents.

DNA repair in all organisms is controlled by the complex action of many interacting genes with different functions. We know a great deal about the repair processes in prokaryotic systems, especially *E. coli*, and many repair genes have been cloned in these systems. However, our knowledge even of the simplest eukaryotic systems, such as yeast, is far less advanced. Only in the last 3–4 years have the first human repair genes been isolated.

Repair deficient cells are important resources for repair gene cloning

A cell that is repair-deficient usually has a high sensitivity to a DNA damaging agent. At its extreme, however, a repair deficiency may also make a cell non-viable due to the fundamental role of DNA repair in cell metabolism. Nevertheless, repair deficient cells do occur naturally in the form of human repair deficient syndromes and a number of laboratory-derived DNA-damage-sensitive mutants have been isolated. It will be seen later that the reversal of the sensitivity of such mutants by DNA transfer techniques is an important approach to the isolation of repair genes.

Some important human syndromes are caused by defects in repair genes

It is possible that minor repair defects occur very commonly in the human population and that these may be sufficient to cause small but significant increases in the sensitivity of individuals to DNA damaging agents. It is the extremes, however, that are the most useful for the study of DNA repair. Xeroderma pigmentosum, ataxia telangiectasia and Fanconi's anaemia are three of the most studied members of a group of human genetic disorders characterised by hypersensitivity to DNA damaging agents. They have been classified as DNA repair disorders although the basis of this deficiency is not always clear. Cell lines obtained by establishing cells from these individuals in tissue culture in the laboratory can be used to give information on the DNA repair processes and allow the isolation of the defective genes involved. Several of these repair deficient disease cell lines have been separated into numerous complementation groups (see Table 1), suggesting that the diseases are due to defects in several genes. The products of these genes are presumed to be part of the same biochemical pathway so that the features of the cells and the clinical features of the patients are similar in all the complementation groups.

Xeroderma pigmentosum is probably the most studied of the repair deficient diseases. It is characterised by extreme photosensitivity leading to the occurrence of multiple skin carcinomas. The extreme sensitivity of cells taken from individuals with this syndrome to ultraviolet (UV) light is well established (Figure 1b) and it is believed to be due to a defect in the genes responsible for the early stages of the excision repair pathway. Mutations in up to seven different genes can each produce the characteristics of xeroderma pigmentosum and some of these genes have been isolated (see later).

Patients with ataxia telangiectasia are characteristically highly sensitive to ionising radiation when given as therapy. In addition, they have an increased susceptibility to various types of cancer. An increased cancer risk has even been postulated in those individuals heterozygous for a defect in genes leading to this syn-

Table 1. Complementation group assignment of human DNA repair deficient diseases.

Disease	Deficiency	Agent	Number of comple-mentation groups
Xeroderma pigmentosum	Excision repair	UV light	7
Cockayne's syndrome	Recovery of DNA/RNA synthesis	UV light	3
Trichothiodystrophy	?	UV light	1
Ataxia telangiectasia	Chromosome repair	X-Rays	4
Nijmegen breakage syndrome	DNA breakage	X-Rays	2
Fanconi's anaemia	DNA/DNA cross-linkage repair	MMC	4
Bloom's syndrome	Ligase function	–	1

drome. It is believed that there are four genes that can be altered to produce this syndrome. None have yet been isolated but they have been localised to the chromosomal region 11q22/23.

Bloom's syndrome is a chromosome instability syndrome that is defective in some aspects of its DNA repair leading to spontaneous chromosomal damage, high frequencies of sister chromatid exchanges and abnormal responses to DNA damage. Only a single gene locus appears to be involved in Bloom's syndrome. Biochemically, cells from Bloom's syndrome patients have an altered DNA ligase

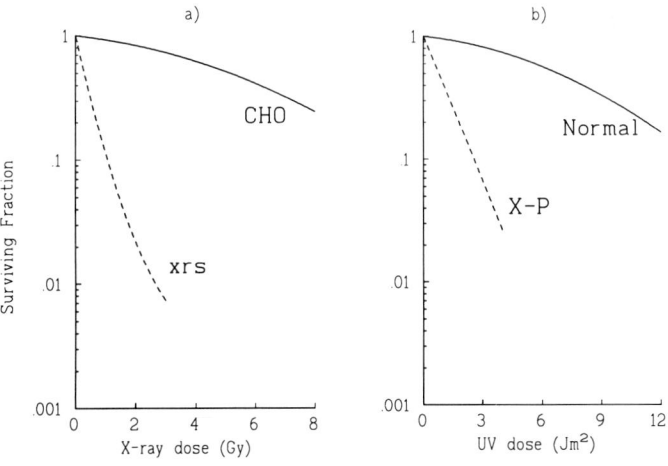

Figure 1. Examples of the extreme sensitivity of repair deficient cells. (a) Repair-deficient mutant (xrs) derived from Chinese hamster ovary cells. Parent CHO cells and xrs cells treated with ionising radiation. (b) Fibroblasts from normal or xeroderma pigmentosum patients treated with UV light.

function (i.e. they have a reduced ability to join together breaks in one of the strands in the DNA molecule), but mutation in the most abundant DNA ligase in human cells, DNA ligase 1, has not been found.

Fanconi's anaemia cells have a high frequency of spontaneous chromosomal aberrations compared with cells of unaffected individuals. There are two theories to explain the features of this disease. The first is that cells cannot repair damaged DNA because the defective protein is directly involved in the recognition, modification or repair of DNA cross-links. Alternatively, the cells may be unable to respond to remove free radicals caused by DNA cross-linking agents. So far there are four genes believed to be involved in this syndrome and one of these has been isolated [1].

Many repair deficient cell lines have been isolated in the laboratory

There are now available a large number of mutant cell lines that can be used for the identification and cloning of possible repair genes. With both UV and ionising radiation, the emphasis has been on sensitive mutants; with drugs, more emphasis has been placed on resistant variants. During the last decade, a large number of UV sensitive Chinese hamster ovary (CHO) cell lines have been isolated that are deficient in DNA excision repair. Genetic characterisation performed by fusion of cells from different mutant cell lines has revealed the presence of at least eight complementation groups [2]. Mutant rodent cell lines that are sensitive to ionising radiation are less abundant but at least eight or nine complementation groups have been identified [3,4]; and three of these groups have been characterised as having a reduced ability to rejoin DNA double-strand breaks (Figure 1a).

The advantage of mutants from rodent systems is that they are more amenable to DNA transfer (see below) and that they can be the recipients for human DNA which can then be easily identified. The disadvantage is that some repair functions differ between rodent and human cells so some repair functions important in humans may be missed. An important target is therefore to isolate and characterise mutant human cell lines, such as one recently derived from a human bladder carcinoma cell line which is sensitive to ionising radiation and demonstrates a reduced fidelity of repair [5].

There are several approaches to identifying and isolating the genes involved in DNA repair

Repair ability is restored when two repair deficient cells containing different mutations are fused

Once mutant cell lines have been isolated, the next step is to determine the complexity of control of the repair process and how many genes are likely to be involved. When two sensitive mutant cells are fused together so that the genetic material of both cells are represented in a single cell, one can conclude that they are defective in the same gene if the fused cell is still sensitive. However, if cell fusion between two mutants leads to a resistant cell then the implication is that the gene that was defective in one cell is still intact in the other and vice versa. One concludes that they are defective in different genes and that at least two genes are involved in that sensitivity (Figure 2). The two mutants are said to belong to different complementation groups. By performing the fusions between the full permutations of pairs of mutants that are available, one can calculate the total number of complementation groups and this can be equated to the minimum number of genes involved in the repair process.

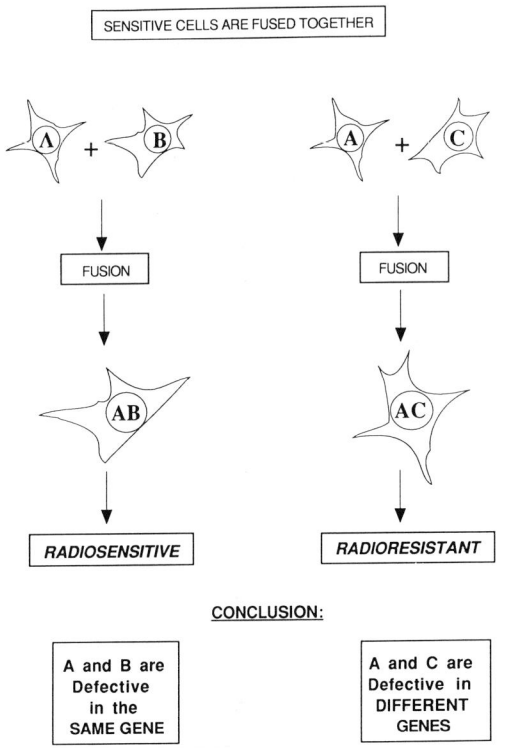

SENSITIVE CELLS ARE FUSED TOGETHER

A + B   A + C

FUSION   FUSION

AB   AC

RADIOSENSITIVE   *RADIORESISTANT*

CONCLUSION:

A and B are Defective in the SAME GENE   A and C are Defective in DIFFERENT GENES

Figure 2. Complementation analysis is used to give an indication of the number of genes which can determine a specific phenotype. Two sensitive cells are fused and if the resulting hybrid is still sensitive the implication is that the two cells are defective in the same gene. If the hybrid is resistant then this suggests that the cells are defective in different genes.

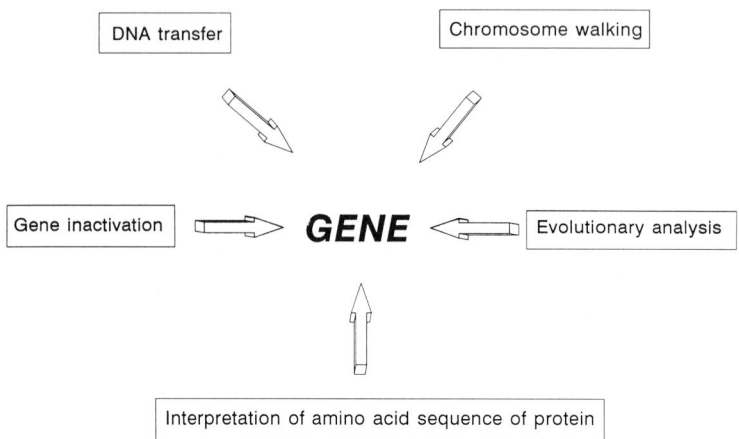

Figure 3. A number of different approaches have been used to identify DNA repair genes (see text for details).

Repair ability can also be restored by introducing fragments of DNA into repair deficient cells

A variety of strategies have been adopted to isolate repair genes (Figure 3). The most widely used and the most straightforward approach is based on correction of the repair deficient mutants by the transfer of normal genes from the genome of repair proficient cells followed by identification and recovery of the correcting DNA sequence. Whole chromosomes or random small fragments of genomic DNA can be transferred into the mutant cells. The chromosomes introduced may remain independent or may attach onto the existing chromosomes. Small fragments of DNA must be integrated into the cells' chromosomes in order to be retained. The result of this transfer is a large number of cells each containing a small fragment of foreign DNA. Of these, only a very small proportion will contain repair genes. This small fraction is usually selected by exposure to the relevant cytotoxic agent, which kills the sensitive cells. The cells containing the correct repair genes survive because these genes have made them resistant to treatment.

Rather than introducing random short lengths of DNA taken directly from donor cells, one may use DNA that has been integrated into various DNA vectors. Collections of vectors containing a fully representative sample of a cell's whole DNA or cDNA are known as libraries and these are maintained by growing individual vectors within appropriate bacterial hosts. The use of a cDNA library in this context has its advantages because it reflects only those genes being expressed in the repair proficient cells and each gene has a reduced size due to the lack of introns in the cDNA. The fate of the DNA once introduced in the cells is dictated by the nature of the vector. Some vectors (e.g. those based on the SV40 virus) get inserted into the chromosomes of the recipient cell genome, while others (e.g. those based on the Epstein-Barr virus) remain episomal in the recipient cell. The latter vectors are only recently available but are already proving to be very useful because they can be re-isolated from the cells by a very simple cell lysis and extraction procedure [6]. This is in contrast to systems in which the DNA is integrated where more complex recovery methods need to be used.

An important tool in the recognition and subsequent retrieval of transferred sequences that contain repair genes revolves around the existence of novel interspersed repetitive sequences (irs) in DNA. These sequences are, as their name suggests, short sequences which are repeated very frequently within the genome and they are specific to a particular species. If human DNA is used as the donor into sensitive rodent cells, the human DNA can be identified using a human irs. The most extensively studied human irs is the Alu repeat sequence which is about 300 bases long and is present in almost 2 million copies in a diploid human cell [7]. The identification of these sequences can be at the level of visualisation on chromosomes by fluorescent in situ hybridisation on chromosome preparations using a gene probe of the alu repeat or on a Southern blot using the same probe. Alternatively, the human sequences can be amplified in the polymerase chain reaction using parts of the alu sequence as primers.

There are other ways of identifying and isolating repair genes

The isolation of repair genes by DNA transfer relies to a great degree on the existence of a wide range of mutant cells which are easy to work with in the laboratory. This approach is therefore limited by the increased difficulties in using human cells in the types of experiments that we have just described and in the possibility that repair deficiencies may be catastrophic to a cell such that only more minor repair deficiencies can be represented in the bank of mutant cell lines. A number of alternative approaches have therefore been used.

DNA repair is such a fundamental process for the maintenance of life that the genes involved in repair tend to show a large degree of homology between different species. This means that the types of experiments described above can be performed in lower organisms such as bacteria and yeast which can be handled very easily in the laboratory and repair genes can be isolated. These genes can then be used to probe genomic DNA or cDNA libraries of higher species in order to identify the corresponding genes. Mutations that are lethal in mammalian cells may not be so in other organisms. This offers scope for isolating a wider spectrum of DNA repair genes.

In some situations (e.g. with human DNA ligase I) it has been easier to identify a repair protein rather than directly isolate a repair gene. This has especially been the case when a specific repair function rather than a specific repair deficiency is being characterised. Once a protein has been isolated, there are two routes to obtaining the corresponding gene. First, an antibody may be raised against the protein and this can be used to screen a cDNA library that has been specifically designed for the integrated genes to be expressed within the bacteria in which the vectors reside. Thus, the antibodies should detect bacteria containing the relevant gene by virtue of their expression of the repair protein. Alternatively, the amino acid sequence of the protein can be converted into a likely DNA sequence which can be synthesised and used to probe genomic DNA or cDNA libraries.

An approach that is specific to the human repair deficient diseases involves the evolving technology of molecular epidemiology. Genetic markers that are known to vary between individuals can be analysed in families containing individuals with the repair deficient disease in question. The aim is to determine which markers are always present in individuals with the disease. Since the position within the human genome of these markers will be known, it is then possible to locate a repair gene to a specific part of a chromosome. An important aim of the human genome project is to produce DNA libraries (either in bacteria or in yeast) with lengths of DNA that are continuous along whole human chromosomes. In the future, it will therefore be possible to translate a knowledge of a location of a gene on a chromosome fairly quickly into the isolation of a gene.

Some genes responsible for repair of UV and ionising radiation-induced damage have been isolated

Bacteria and yeast have proved to be good models for the isolation of DNA repair genes

The ability to manipulate the genetic material of bacteria and yeast is far more advanced than in mammalian cells. Consequently gene isolation from these systems is easier. In the bacterium *E. coli,* it has been shown that the process of excision repair involves at least 28 genes [8], either directly involved in the repair process or in the regulation of the activity of the repair genes. Thus, in perhaps the most well-characterized repair system, it is clear that DNA repair is a highly complex interaction of proteins and in higher eukaryotes the systems are likely to be even more complex.

A progression in evolutionary terms comes from the use of yeast. At least 50 DNA repair genes have been identified in *Saccharomyces cerevisiae* and 25 of these have been cloned [9]. These genes have been shown to be involved in various forms of repair including excision repair, post replication repair and recombinational strand break repair. Consequently, a sensitivity to a wide range of agents is seen in the large number of cell lines mutant in these various genes. Similar work is now being done on another yeast, *Schizosaccharomyces pombe*. This has the advantage that it is much higher up the evolutionary tree than *S. cerevisiae* and therefore has a higher likelihood of possessing genes that are very similar to those in man. To date, over 20 complementation groups have been shown to be involved in DNA repair in *S. pombe*.

Progress is being made in the isolation of human repair genes

Two approaches have been successful at identifying human repair genes. The first has been to isolate human DNA that can complement repair deficiencies in rodent cells. The second approach has involved the direct reversal of the sensitivity of cells from patients with repair-deficient syndromes. The majority of the human repair genes that have been cloned were isolated by their ability to complement UV-sensitive rodent cell lines. Because these rodent mutants are deficient in the excision repair process, the cloned genes have been designated as Excision Repair Cross Complementing (ERCC) genes. Table 2 shows the characteristics of some of these genes. It can be seen that most seem to have a DNA helicase function (i.e. it facilitates the unwinding of the DNA helix). This suggests that they are involved in the very early stages of the excision repair pathway in which the DNA is unwound in order to allow the other repair enzymes access to the damaged DNA.

The first cloned human repair gene (ERCC1) has been well characterized

The ERCC1 gene [10] was the first human repair gene to be cloned and as such is the most characterized to date. It therefore provides a good example of the isolation of a repair gene and the information which can be gained after cloning.

The ERCC1 gene was cloned by its ability to complement the excision repair defect in CHO mutant 43-3B. It was originally introduced into the mutant as a random genomic DNA fragment. Subsequent analysis has shown it to be approximately 15–17 kb in length, of which only 1.1 kb is represented as the mRNA. This in turn is translated into a protein of 297 amino acids.

Table 2. Human excision repair genes that have been identified.

Gene	Size (kb)	Product	Function	Location
ERCC1	15–17	297 amino acids	DNA helicase	Ch19q 13.2
ERCC2	?	760 amino acids	DNA helicase	Ch19q 13.2
ERCC3	45	782 amino acids	DNA helicase	Ch2q 21
ERCC5	32	?	?	Ch13q 14-34
ERCC6	100	1493 amino acids	DNA helicase	Ch10q 11

R. Cartwright and T.J. McMillan

ERCC1 has been shown to have interesting homology to proteins from lower organisms (Figure 4). It seems to be a gene which is derived from segments of several other genes. It includes a large region with appreciable homology to the RAD10 protein of *S. cerevisiae*. In addition, it has amino acid sequences which are highly homologous to two proteins involved in excision repair in *E. coli*, uvrA and uvrC. The ERCC1 gene has been demonstrated to hybridize to homologous DNA in all vertebrates tested and with *Drosophila*. The amino acid sequence of ERCC1 has regions which are similar to proteins that are known to be localized in the nucleus and bind to DNA. Thus, these two properties can be inferred for ERCC1.

Genes defective in xeroderma pigmentosum and Fanconi's anaemia have been isolated

Two of the genes isolated from the analysis of UV-sensitive rodent cells, ERCC2 and ERCC3, have been shown to correct defects in two of the complementation groups of xeroderma pigmentosum (XPD and XPB, respectively). In addition, cells of XPA and XPC have been made resistant by the direct introduction of DNA. In the first instance, this was done by the introduction of mouse genomic DNA into XPA cells [11]. In the case of XPC, a human cDNA library in an EBV-based vector was introduced into XPC cells and the appropriate complementing sequence isolated [12].

The cDNA approach has also been used to isolate the gene which is defective in Fanconi's anaemia group C cells [1]. This gene has been designated FACC

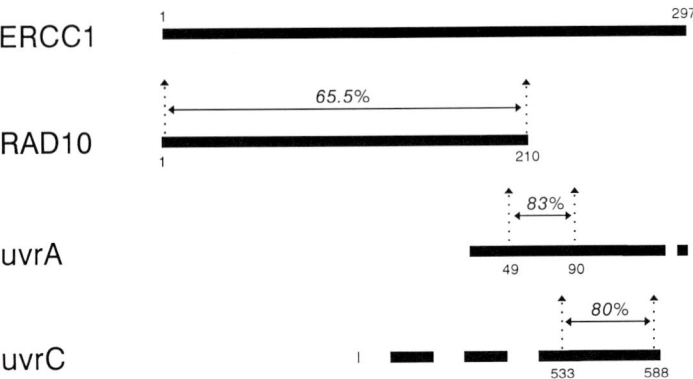

Figure 4. The protein encoded by the human ERCC1 gene has regions which are homologous to repair proteins from lower organisms. The first 210 amino acids show 65.5% homology to the RAD10 protein from yeast. The final part of the protein has regions that show a high degree of homology with two bacterial repair proteins, uvrA and uvrC. The percentage values are the extent of the homology. The other numbers are the amino acid numbers for each protein.

(Fanconi's anaemia group C complementing) and the cDNA is 4.6 kb long. Analysis done to date has not demonstrated any motifs common to other known repair proteins so it is highly likely to be a novel repair gene.

Genes involved in repair of ionising radiation-induced damage are proving to be more difficult to isolate
The gene defect responsible for ataxia telangiectasia has proved to be remarkably elusive, as have other genes involved in the repair of damage induced by ionising radiation. The only X-ray repair cross complementing (XRCC) gene isolated to date is the XRCC1 gene which corrects the sensitivity of the EM9 mutant of CHO cells [13]. XRCC1 increases the ability of EM9 cells to repair DNA single-strand breaks and reduces the frequency of sister chromatid exchanges in these cells. Other XRCC genes have been named and in some cases their positions within the genome have been identified [14] but no more have yet been cloned.

Drug-induced damage requires a distinct set of repair enzymes

Chemical agents produce a wide variety of types of DNA lesions, ranging from strand breaks to DNA cross-links and specific modifications of bases. Some of the enzymes involved in the removal of this damage are likely to be common to processes involved in damage induced by radiation. However, it is clear that some specific enzymes have evolved to remove some of the unique lesions in DNA. A good example of this is O^6-methyl guanine DNA methyltransferase. As its name suggests, this enzyme removes the methyl group from the O^6 position of guanine, a lesion which if unrepaired leads to mutation and cell death. The genes encoding methyltransferases were first isolated from bacteria and more recently the human equivalent has been isolated [15]. Depletion of these gene products in bacteria and mammalian cells leads to an increased sensitivity to methylating agents and introducing the gene into cells which normally have a defect in this function can increase resistance to these agents.

Abasic sites (i.e. sites of base loss from the DNA double helix) are examples of lesions that are a common product of treatment with many types of DNA damaging agents. They can either be induced by an agent or they are produced by the early stages of DNA repair. Endonucleases incise the DNA phosphodiester backbone and excise the sugar phosphate residue at the site of damage. Again the genes coding for endonucleases were first isolated from bacteria (the *nth* and *nfo* genes encode the major endonucleases Endo III and Endo IV, respectively). Equivalent genes have been isolated from yeast (*Apn*-1), *Drosophila melanogaster* (AP3) and mouse (APEX1) and recently the human AP-endonuclease gene, *APE*, has been cloned [16].

DNA repair genes will be useful in the laboratory and the clinic

A great deal of effort is going into the isolation of repair genes because DNA repair is a fundamental process in normal cell metabolism. From a more immediate applied viewpoint, the characterisation of repair genes will hopefully allow the more rapid identification of repair-deficient individuals. At its extreme this is important in syndromes such as XP or A-T where a repair deficiency has a devastating effect on an individual, either directly because of the physiological effects (e.g. immunodeficiency) or because of their sensitivity to carcinogens. The recognition of a sensitivity to external agents is also potentially important where such agents are used in therapy. Abnormal sensitivity to ionising radiation, for example, greatly limits the usefulness of radiotherapy in the treatment of cancer. The total dose that can be given is limited by the normal tissue tolerance of the most sensitive individuals. If these patients could be identified, the dose to the remaining patients could be increased with the hope of improved tumour control. Ultimately the aim will be not only to use repair genes to identify such patients but also to use developing molecular therapy techniques to manipulate the function of these genes in order to increase normal tissue tolerance or decrease tumour cell resistance.

References

1 Strathdee CA, Gavish H, Shannon WR, Buchwald M. Cloning of cDNAs for Fanconi's anaemia by functional complementation. Nature 1992, **356**, 763–767.
2 Thompson LH, Shiomi T, Salazar EP, Stewart SA. An eighth complementation group of rodent cells hypersensitive to ultraviolet radiation. Somat Cell Mol Genet 1988, **14**, 605–612.
3 Jeggo PA, Tesmer J, Chen DJ. Genetic analysis of ionizing radiation sensitive mutants of cultured mammalian cell lines. Mutat Res 1991, **254**, 125–133.
4 Thacker J, Wilkinson RE. The genetic basis of resistance to ionizing radiation damage in cultured mammalian cells. Mutat Res 1991, **254**, 135–142.
5 McMillan TJ, Holmes A. The isolation and partial characterisation of a radiation-sensitive clone of a bladder carcinoma cell line. Radiat Res 1991, **128**, 301–305.
6 Swirski RA, van Den Berg D, Murphy A et al. An Epstein-Barr based shuttle vector system for direct cloning in human tissue culture cells. Proc Natl Acad Sci USA 1992, in press.
7 Nelson DL. Interspersed repetitive sequence polymerase chain reaction (IRSPCR) for generation of human DNA fragments from complex sources. Methods: a companion to Methods in Enzymology 1991, **2**, 60–74.
8 Friedberg EC. DNA Repair. Freeman, New York, 1985, p 614.
9 Friedberg EC. Eukaryotic DNA repair: glimpses through the yeast *Saccharomyces cerevisiae*. Bioassays 1991, **13**, 295–301.
10 Duin van M, De Wit J, Odijk H et al. Molecular characterisation of the human excision repair gene ERCC-1: cDNA cloning and amino acid homology with the yeast DNA repair gene RAD10. Cell 1986, **44**, 913–923.

11 Tanaka K, Satokato I, Ogita et al. Molecular cloning of a mouse DNA repair gene that complements the defect of group A xeroderma pigmentosum. Proc Natl Acad Sci USA 1989, **86**, 5512–5516.

12 Legerski R, Peterson C. Expression cloning of a human DNA repair gene involved in *xeroderma pigmentosum* group C. Nature 1992, **359**, 70–73.

13 Thompson LH, Brookman KW, Jones NJ, Carrano AV. Molecular cloning of the human XRCC-1 gene which corrects defective DNA strand break repair and sister chromatid exchange. Mol Cell Biol 1990, **10**, 6160–6171.

14 Jeggo PA, Hafezparast M, Thompson AF et al. Localization of a DNA repair gene (XRCC5) involved in double strand break rejoining to human chromosome 2. Proc Natl Acad Sci 1992, **89**, 6423–6427.

15 Tano K, Shiota S, Collier J et al. Isolation and structural characterisation of a cDNA clone encoding the human DNA repair protein for O6-alkylguanine. Proc Natl Acad Sci USA 1990, **87**, 686–690.

16 Demple B, Herman T, Chen DS. Cloning and expression of APE, the cDNA encoding the major human apurinic endonuclease: definition of a family of DNA repair enzymes. Proc Natl Acad Sci USA 1991, **88**, 11450–11454.

Molecular Biology for Oncologists
Edited by John Yarnold, Michael Stratton, Trevor McMillan
© *1993, Elsevier Science Publishers B.V. All rights reserved*

CHAPTER 18

Radiation-induced cytokines and growth factors: cellular and molecular basis of the modification of radiation damage

Ralph Weichselbaum[a], Zvi Fuks[b], Dennis Hallahan[a],
Adrianne Haimovitz-Friedman[b] and Donald Kufe[c]

[a]*Department of Radiation and Cellular Oncology, University of Chicago, Chicago, IL, USA*
[b]*Department of Radiation Oncology, Memorial Sloan-Kettering Cancer Center, New York, NY, USA*
[c]*Department of Pharmacology, Dana-Farber Cancer Center, Boston, MA, USA*

Introduction

Advances in molecular genetics and cell biology are providing insights into how cells mediate their responses to many forms of environmental stress, including radiation injury (see Chapters 14 and 17). Recent studies have demonstrated that growth factors and cytokines are induced by ionising radiation in mammalian cells. Radiation-induced growth factors and cytokines include TNF-α, IL-1, FGF, PDGF-α and TGF-β. In this chapter, the mechanism of induction of growth factors and cytokines is described, as well as their potential role in the cellular and tissue responses to ionising radiation [1,2].

TNF-α is induced by ionising radiation

Tumour necrosis factor (TNF) is a mediator of the inflammatory response produced by monocytes and other cell types. TNF causes vascular endothelial obliteration and necrosis in experimental tumours, and is directly cytotoxic to

213

certain human tumour cells in vitro [3]. It has been suggested that cell killing by TNF is mediated by free radicals [4,5]. TNF-α mRNA and protein are induced in some human tumour cell lines following X-irradiation [6]. Moreover, the addition of TNF to the culture media of TNF-producing and non-producing tumours enhanced radiation killing of human sarcoma and epithelial tumour lines [7]. The mechanism of interactive killing between TNF and radiation is unknown but it is thought to be associated with free radical production and enhanced production of DNA strand breaks (see Figure 1).

Paradoxically, TNF administered to mice prior to irradiation has been shown to protect bone marrow progenitors from the cytotoxic effects of ionising radiation [8–10]. Bacterial lipopolysaccharide (LPS) induces TNF and protects mice from radiation lethality. The link between these two effects is shown by pretreatment of animals with anti-TNF antibodies which reverses the radioprotective effects of LPS.

The different effects of TNF on irradiated transformed and normal cells may be explained by the fact that TNF interacts with two cell surface receptors, the p70 and the p55 receptors. The p70 receptor is thought to mediate cell killing, while the p55 receptor mediates induction of early response genes. It is possible that the diverse effects of TNF on the radiation response in different cell types is associated with the relative abundance of each receptor type.

One of the genes induced by TNF is manganese superoxide dismutase (MnSOD) which has the ability to scavenge radiation-induced free radicals. Human T-cell lines infected with the human immunodeficiency virus (HIV) have reduced levels of the dismutase and are more sensitive to ionising radiation than control cells. Furthermore, sarcoma cells transfected with a sense c-DNA for MnSOD are more radioresistant than untransfected cells, while cells transfected with anti-sense cDNA for MnSOD are more radiosensitive [10]. These data suggest that TNF may

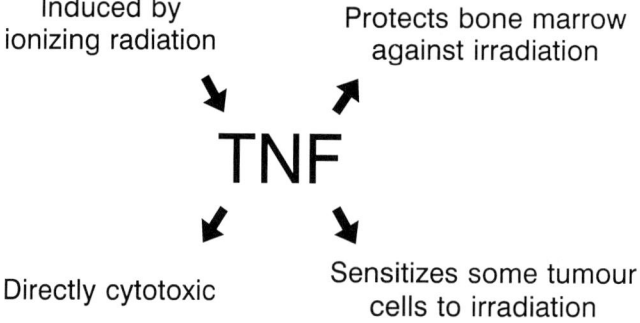

Figure 1. Tumour necrosis factor is induced by ionising radiation and has a diversity of effects on mammalian cells.

contribute to radiation sensitivity or protection by regulating levels of radical oxygen intermediates through expression of MnSOD or other free radical scavenging proteins. Thus, tissue responses to radiation may depend upon responses to TNF mediated via cellular p55 and p70 receptors.

Induction of IL-1 protects the hematopoietic compartment from X-ray damage

Interleukin-1 (IL-1) protects bone marrow cells irradiated in vitro. Pretreatment of mice with IL-1 receptor antibody before irradiation results in reduced survival rates [11]. Thus, endogenous IL-1 as well as TNF can contribute to tissue radio-protection. This concept is supported by reports of the induction of IL-1 mRNA following irradiation [12]. TNF and IL-1 seem to mediate their radioprotective effects on early bone marrow progenitor cells in different ways. IL-1 promotes the growth and proliferation of early progenitor cells expressed as unique "high-proliferative-potential" colonies. Conversely, TNF inhibits the development of early progenitor cells. Whereas IL-1 increases cycling of hematopoietic progenitor cells, TNF suppresses the cycling of such cells. IL-1 and TNF exhibit synergistic radioprotection in vivo, and this also suggests that the two cytokines act through different pathways. In addition, TNF and IL-1 stimulate stromal cells of the bone marrow to release other growth factors, and both IL-1 and TNF have been reported to stimulate the production of oxygen radical scavengers. Although the cellular and biochemical mechanisms remain unclear, it appears that induction of IL-1 by ionising radiation may influence cellular radiosensitivity and the tissue response to radiation [10,13].

Induction of bFGF may play a role in cellular repair and late normal tissue effects of radiotherapy

PDGF-α and basic fibroblast growth factor (bFGF) are released from vascular endothelial cells after radiation exposure [14]. They may therefore be responsible for the proliferation of smooth muscle and endothelial cells observed in smaller arterioles following irradiation in vivo. When cells are held in a non-proliferative state for a period up to 24 h following irradiation, they suffer a lower degree of cell kill than if they were allowed to proliferate immediately after treatment. This is believed to be due to repair of radiation-induced damage and is termed potentially lethal damage repair (PLDR). It has been demonstrated that bFGF is expressed following irradiation and this induces PLDR. This effect can be inhibited by a neutralizing monoclonal antibody against bFGF [15].

Pathological and ultrastructural analysis reveal that damage to the endothelial component of the microvasculature is a dominant cause of radiation injury to the skin, heart, liver and the central nervous system. The endothelium of large vessels is more resistant to the effects of ionising radiation in vivo. It has been demonstrated that bFGF is detected in the basement membrane of all blood vessels but its level varies with vessel size and with the presence of branching points from vessels [16]. This variation in bFGF in the basement membrane parallels the sensitivity of different portions of the vasculature to irradiation. The presence of bFGF is proposed to enhance the repair of X-ray induced damage to capillary endothelium and may be one factor in the relative radioresistance of large vessels.

TGF-β: a possible role in radiation fibrosis

Transforming growth factor-β (TGF-β) plays an important role in regulating proliferation in a variety of cell types. For example, TGF-β is reported to inhibit epithelial cell growth but to stimulate fibroblast proliferation and collagen production. Fibrosis is an important late event in radiation injury in many tissues. TGF-β has been shown to be associated with pathological changes of late radiation damage in normal hepatocytes as compared to unirradiated hepatocytes [17,18]. Irradiated hepatic tissue shows high levels of TGF-β, which correlates with the extent of hepatic fibrosis. To support this correlation, TFG-β levels are higher in human hepatocytes than in rat hepatocytes, and hepatocytes in children have higher concentrations of TGF-β than adults. Thus, it has been proposed that differential concentrations of TGF-β may explain why rat livers are less prone than human livers to fibrosis after irradiation, and why children appear more sensitive to liver irradiation than adults.

TGF-β was also found to sensitize mice to whole body irradiation, possibly by down-regulating the IL-1 receptor, and by reducing IL-1 and TNF production [8,10]. Moreover, it has been demonstrated that TGF-β secretion is increased from monocytes in irradiated lung tissue. Thus, increased tissue concentrations of TGF-β may be associated with the pathogenesis of the late effects of radiation in pulmonary tissue [19]. Thus, radiation-induced TGF-β production as well as other cytokines may be associated with a number of late radiation sequelae [20].

Early response genes c-*jun* and *egr*-1 are induced by exposure to ionising radiation

Genes induced by radiation without the need for new protein synthesis are called immediate early or primary response genes [21,22]. Some primary genes encode transcription factors that initiate a cascade of responses in other genes. These protein-gene interactions lead to biological responses depending on the state of the stimulated cell and the type of stimulus. Examples of early response genes encoding transcription factors induced by ionising radiation include the *Jun/Fos*, *Egr*-1, and NF-κB/rel families.

The c-*jun* gene encodes a 40 kDa protein that binds to a highly conserved promoter/enhancer sequence (5'TGAC/GTCA3'), called the AP-1 domain, bound by a number of different transcription factors, including jun-B, jun-d and c-fos [21,22]. Jun and Fos proteins frequently form a heterodimer that binds to the AP-1 domain, stimulating the transcription of genes induced by some growth factors and phorbol esters. Induction of c-*jun* is associated with cellular proliferation and the transition from G_0 to G_1 phases of the cell cycle. Transcriptional activation of the c-*jun* and c-*fos* genes is seen after exposure of human HL-60 cells to X-rays (see below).

Egr-1 encodes a nuclear phosphoprotein with a zinc finger motif that has a partial homology to the Wilm's tumour suppressor gene [22]. Like the c-*jun* gene product, *Egr*-1 may also participate in the transition of quiescent cells from G_0 to G_1. *Egr*-1 is expressed during differentiation of hematopoietic and neuronal cells and may be involved in negative regulation of cell growth.

NF-κB is a protein which, through its binding to gene enhancers, is intimately involved in the regulation of several genes. It has been shown that ionising radiation regulates NF-κB expression at both the mRNA and protein levels. Of particular interest is the finding that NF-κB protein binds to DNA at doses as low as 2 Gy in the absence of new protein synthesis. NF-κB exists as a cytoplasmic protein prior to activation. The data suggest that ionising radiation activates signalling pathways that include cytoplasmic events, and that NF-κB dissociates from the inhibitor IκB before nuclear localization and binding [24]. Thus, dissociation from an inhibitor prior to nuclear binding may be one mechanism by which radiation activates transcription factors and induces later response genes.

Transcription and signal transduction following irradiation

Radiation-induced transcription of the c-*jun* proto-oncogene in human HL-60 promyelocytic leukemia cells has been described [23]. The increase in c-*jun* mRNA production is time and dose dependent, and inversely proportional to dose rate. Transcriptional run-on analyses reveals that radiation stimulates the rate of c-*jun* transcription. Radiation treatment is also associated with prolongation of the half-life of c-*jun* mRNA which is possibly due to the inhibition of a labile protein that degrades c-*jun* mRNA. Inhibition of protein synthesis with cycloheximide prior to radiation does not inhibit X-ray-induced c-*jun* expression, suggesting that

de novo protein synthesis is not necessary for c-*jun* mRNA induction following exposure to ionising radiation. It appears that radiation exerts its influence on gene
induction at the level of transcription since TNF and bFGF are also controlled in
this way. The number of genes studied in this way, however, remains small.

Signalling events prior to transcriptional induction by X-irradiation include the
activation of protein kinase C (PKC) [25]. Protein kinase C activity extracted 15 s
following gamma-irradiation is 4.5-fold higher than control. Peak phosphotransferase (PKC) activity is observed after 30 s, and returns to baseline within 60 s.
This suggests that PKC activation following irradiation is one way of activating
transcription factors to induce gene expression. There are likely to be other kinases
in addition to PKC which are activated by irradiation of specific cells and tissues.
This conclusion is supported by the finding that the induction of c-*jun* and TNF by
radiation can be abrogated following depletion of PKC by the phorbol ester, TPA,
and by treatment of cells with a kinase inhibitor, H7.

Haimovitz-Friedman (personal communication) have studied the role of PKC
following bFGF stimulation. Stimulation of unirradiated bovine aortic endothelial
cells with bFGF results in translocation of cytoplasmic PKC-α into the membrane
within 30 s of stimulation. This translocation is associated with an enzymatic activation of PKC within 30 s of stimulation that was down-regulated to normal levels
by 120 s. Radiation exposure of bovine aortic endothelial cells in the presence of
non-toxic concentrations of H7, or after complete cellular depletion of PKC by
prolonged exposure to TPA, eliminates bFGF-induced radiation damage repair.
These data demonstrate an important role for PKC in modifying the cellular response to radiation in addition to its role in the activation of early response genes.

Conclusions and clinical implications

Growth factors and cytokines preserve the homeostatic function of the organism by
responding to stress such as infection or tissue injury. IL-1, bFGF, PDGF and TGF-
β are secreted by a wide variety of cell types following a number of stimuli. In this
context, the induction of cytokines in mammalian cells following X-irradiation
appears to be a general response to stress. It is important to note, however, that
these growth factors and cytokines have evolved relatively specialized functions in
cells and tissues of the adult and embryo which give rise to a variety of phenotypic
effects [26]. One common property of TNF, IL-1, bFGF, PDGF and TGF-β is that
they promote recruitment of quiescent target cells into the cell cycle and thus
promote overall cellular growth in several systems. G_1 is a radiosensitive phase of
the cell cycle and S is generally the most radioresistant phase. We propose that
there may be a selective survival advantage for irradiated cells to traverse G_1 into

S. Secretion of growth factors and cytokines following irradiation may also be important in normal tissue and tumour re-population. Induction of radical-scavenging proteins such as MnSOD following TNF and IL-1 exposure may also provide protection against induction of lethality following ionising radiation. The cellular specificity of cytokine response is likely to be linked to ligand secretion and concentration, receptor type and factors such as the extracellular matrix.

The induction of the early response genes by DNA damaging agents has been proposed to be analogous to the induction of the SOS response in *E. coli* because the sequence of protein modifications, DNA binding and transcriptional induction of downstream genes have resemblance to the SOS system [27] . A potential repair function as a result of the early gene response cascade is, however, speculative. A wide variety of genes have DNA binding sequences for *Jun*, *Egr*-1, and NF-κB, and further investigation may reveal that combinations of transcription factors induce gene encoding proteins which are important in the cellular responses to radiation.

The above investigations provide insights into potential uses for cytokines in radiotherapy. We have initiated a phase I trial combining TNF with localized radiation with the goal of increasing tumour radiosensitization. The work of Neta and Oppenheim suggests that TNF and IL-1 may be used together to protect against the myelosuppressive effects of wide-field radiotherapy. The use of intravenous bFGF as a radioprotector is intriguing and is proposed by Fuks and colleagues. The potential radioprotection produced by bFGF must be balanced against potential tumour radioprotection or stimulation of proliferation of endothelial cells causing microvascular obliteration. Inhibition of TGF-β might decrease late effects of therapeutic radiation that results from fibrosis. We have recently proposed that DNA elements cloned from the promoter regions of radiation-inducible factors that activate these genes might be linked to toxins or radioprotectors and targeted to tumours or normal tissues [28]. Knowledge of the signalling events following the cellular response to X-rays has opened new avenues for investigations.

Acknowledgements

This paper is in support of grants: Ralph Weichselbaum, CA41068, CA42596; Zvi Fuks, CA52462; Donald Kufe, CA55241; Dennis Hallahan, CA58508.

References

1   Hill RP. Cellular basis of radiotherapy. In: Basic Science of Oncology, Tannock IF, Hill RP eds. Pergamon Press, Toronto, 1987, pp 237–255.

2   Fuks Z, Weichselbaum RR. Molecular basis of cancer therapy: radiation therapy. In: Molecular Basis of Cancer, Mendelsohn J ed., 1993, in press.
3   Sugarman BJ, Aggarwal BB, Hass PE et al. Recombinant tumor necrosis factor alpha: effects of proliferation on normal and transformed. Science 1985, **230**, 943–945.
4   Zimmerman RJ, Chan A, Leadon SA. Oxidative damage in murine tumor cells treated in vitro by recombinant human tumor necrosis factor. Cancer Res 1989, **49**, 1644–1648.
5   Yamauchi N, Kuriyama H, Watanabe N et al. Intracellular hydroxyl radical production induced by recombinant human tumor necrosis factor and its implication in the killing of tumor cells in vitro. Cancer Res 1989, **49**, 1671–1675.
6   Hallahan DE, Sriggs DR, Beckett MA et al. Increased tumor necrosis factor alpha mRNA after cellular exposure to ionizing radiation. Proc Natl Acad Sci 1989, **88**, 10104–10107.
7   Hallahan DE, Beckett MA, Kufe DW, Weichselbaum RR. Interaction between recombinant tumor necrosis factor and radiation in 13 human tumor cell lines. Int J Radiat Oncol Biol Phys 1990, **19**, 69–74.
8   Neta R, Oppenheim JJ, Schreiber R et al. Role of cytokines (interleukin-1, tumor necrosis factor, and transforming growth factor α) in natural and lipopolysaccharide enhanced radioresistance. J Exp Med 1981, **173**, 1177–1182.
9   Neta R, Oppenheim JJ, Douches SD. Interdependence of the radioprotective effects of human recombinant IL-1 TNF-α G-CSF in murine recombinant G-CFS. J Immunol 1988, **140**, 108–111.
10 Neta R, Oppenheim J. Radioprotection with cytokines – learning from nature to cope with radiation damage. Cancer Cells 1991, **3**, 391–396.
11 Neta R, Vogel SN, Plocinski JM et al. *In vivo* modulation with anti-interleukin-1 IL-1 receptor (p80) antibody 35F5 of the response to IL-1. The relationship of radio-protection colonies stimulating factor and IL-6. Blood 1990, **76**, 57–62.
12 Woloschak GE, Chang-Liu CM, Jones PS, Jones CA. Modulation of gene expression to serine and hamster embryo cells following ionizing radiation. Cancer Res 1990, **53**, 339–344.
13 Weichselbaum RR, Hallahan DE, Sukhatme V et al. Biological consequences of gene regulation after ionizing radiation. J Natl Cancer Inst 1991, **83**, 480–484.
14 Witte L, Fuks Z, Haimovitz-Friedman A et al. Effects of irradiation on the release of growth factors from cultured bovine porcine and human endothelial cells. Cancer Res 1989, **49**, 5066–5072.
15 Haimovitz-Friedman A, Vlodavsky I, Chaudhuri A et al. Autocrine effects of fibroblast growth factor and radiation damage repair in endothelial cells. Cancer Res 1991, **51**, 2552–2558.
16 Cordon-Cardo C, Vlodavsky I, Haimovitz-Friedman A et al. Expression of basic fibroblast growth factor in normal human tissues. Lab Invest 1990, **63**, 832–840.
17 Anscher M, Crocker I, Jirtle R. Transforming growth factor beta-1: expression in the radiated liver. Radiat Res 1990, **122**, 77–85.
18 Jirtle RL, Anscher MS. The role of TGFb-1 and the pathogenesis of radiation induced hepatic fibrosis. Radiation Research: A 20th Century Prospective. Congress Proc 1991, **2**, 819–823.
19 Rubin P, Finkelstein J, Shapiro D. Molecular biology mechanisms in the radiation induction of pulmonary injury syndromes: interrelationship between the alveolar macrophage and the septal fibroblasts. Int J Radiat Oncol Biol Phys 1992, **24**, 93–101.
20 Fuks Z, Weichselbaum RR. Radiation tolerance and the new biology: growth factor

involvement in radiation injury to the lung. Int J Radiat Oncol Biol Phys 1992, **24**, 183–184.

21 Ziff EP. Transcription factors: a new family gathers at the cyclic amp response site. Trends Genet 1990, **6**, 69–72.

22 Sukhatme VP. Early transcriptional events in cell growth: the EGR family. J Am Soc Nephrol 1990, **1**, 859–866.

23 Sherman ML, Datta R, Hallahan DE et al. Ionizing radiation regulates the expression of the c-jun proto-oncogene. Proc Natl Acad Sci 1990, **87**, 5663–5666.

24 Brach MA, Sherman ML, Gunji H et al. Ionizing radiation induces expression and binding activity of nuclear factor κB. J Clin Invest 1991, **88**, 691–695.

25 Hallahan DE, Virudachalam S, Sherman ML et al. Tumor necrosis factor gene expression is mediated by protein kinase-C following activation by ionizing radiation. Cancer Res 1991, **51**, 4565–4569.

26 Green JBA, Smith JC. Growth factors as morphogens. Trends Genet 1991, **7**, 245–250.

27 DeVary Y, Gottlieb R, Lau LF, Karin M. Rapid and preferential activation of the c-jun gene during the mammalian UV-response. Mol Cell Biol 1991, **11**, 2804–2811.

28 Weichselbaum RR, Hallahan DE, Sukhatme V, Kufe DW. Gene therapy targeting by ionizing radiation. Int J Radiat Oncol Biol Phys 1993, in press.

Molecular Biology for Oncologists
Edited by John Yarnold, Michael Stratton, Trevor McMillan
© *1993, Elsevier Science Publishers B.V. All rights reserved*

CHAPTER 19

Chemotherapeutic drug-induced DNA damage and repair

Robert Brown and Donald Bissett

CRC Department of Medical Oncology, Beatson Laboratories,
Garscube Estate, Glasgow, UK

DNA is an important target for chemotherapeutic drugs

The basis of chemotherapy is that drugs are selectively more toxic to the tumour than to the host. As our understanding of the biochemical differences between normal and tumour cells increases, the possibility of rationally designing drugs targeted to tumour specific biochemical pathways becomes feasible [1]. However, most existing clinically active drugs have been discovered by a combination of random screening and serendipity. In the National Cancer Institute (USA) screening programme for new anti-tumour drugs, it was estimated that up to 1984 more than 370 000 compounds had been screened but only 8 drugs had proved of clinical benefit [2]. This leads to the question – what are the cellular targets of the successful drugs that make them effective anti-tumour agents, and can this knowledge be used to aid the design of novel drugs or to improve therapeutic strategies with existing agents?

Many effective chemotherapeutic drugs act indirectly or directly on DNA and DNA metabolising enzymes [3]. Drugs that cause direct damage to DNA include:

(i) platinum coordination complexes, such as *cis*-diamminedichlorplatinum(II) (CDDP). These compounds form bidentate adducts, with the platinum moiety bridging adjacent bases on the same DNA strand (intrastrand) or bases on opposite strands (interstrand) (Figure 1).

(ii) alkylating agents, such as nitrosoureas, mitomycin C, cyclophosphamide, melphalan, chlorambucil and busulfan. These compounds form monofunctional or bifunctional covalent bonds between the carbon of an alkyl moiety and a nucleophilic base of DNA. Bifunctional agents can give rise to intrastrand and interstrand crosslinks.

Indirect damage to DNA can occur by:

 (i) blockage of DNA synthesis by nucleotide analogues such as 5-fluorouracil or cytosine arabinoside, or inhibition of dihydrofolate reductase by methotrexate;
(ii) generation of free radicals which cause base damage and DNA strand breaks, for instance from bleomycin and quinones such as mitomycin C or doxorubicin;
(iii) interference with proteins involved in DNA function, such as anthracycline-induced stabilisation of the DNA topoisomerase II cleavage complex.

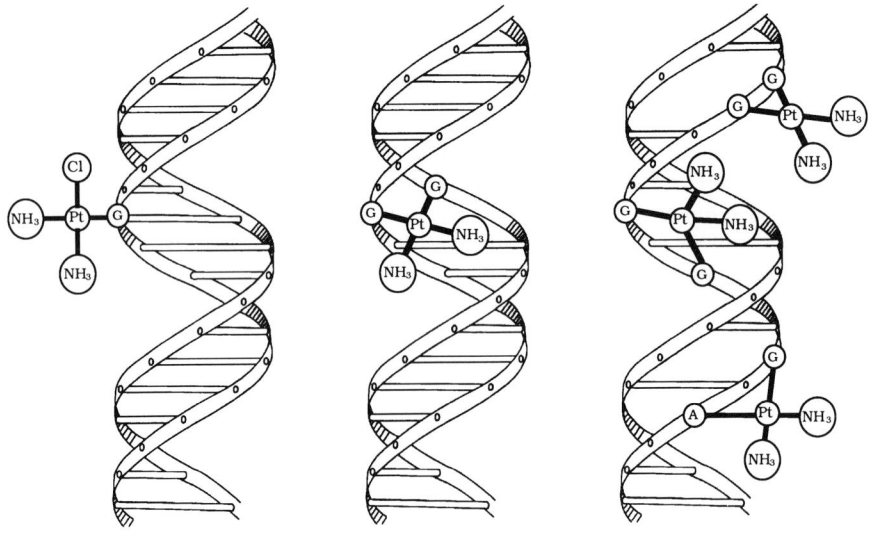

Monofunctionally Bound   Interstrand Crosslink   Intrastrand Crosslinks

Figure 1. Schematic diagram of CDDP–DNA adducts.

Although chemotherapeutic drugs also exert effects on other cellular components, many studies have correlated DNA damage induced by chemotherapeutic drugs with cellular toxicity [4]. Tumour cells may be more susceptible than normal cells to the effects of these agents because of increased initial levels of DNA damage, decreased removal of the drug-DNA lesions, or decreased tolerance of DNA damage. The remainder of this chapter describes approaches to the measurement of drug-induced DNA damage and repair in tumour cells, and their implications for future cancer chemotherapy.

Measurement of drug-induced DNA damage

Measurement of DNA damage at a cellular level is complicated by:

  (i)   the variety of lesions that result from exposure to any one agent;
 (ii)   the heterogeneity of DNA damage throughout a cell population;
(iii)   the heterogeneity of damage throughout the genome.

Cytogenetic studies provide a crude guide to the level of DNA damage induced by a drug; following exposure to a platinum coordination complex, one chromosomal aberration per cell approximates to a 37% survival rate [5]. In addition, useful information can be obtained by measurement of the amount of drug bound to DNA following drug exposure. Assays exist for classical alkylating agents (gas chromatography), platinum coordination complexes (atomic absorption spectrometry, AAS), and intercalating agents (high performance liquid chromatography, HPLC). Although the levels of DNA damage measured in this way correlate with cytotoxicity, they provide no information about the specific DNA lesions that cause cell death.

Many cytotoxics induce DNA strand breaks

DNA filter elution methods have proved extremely useful in the identification and quantification of specific DNA lesions, including single- and double-strand breaks, DNA–protein crosslinks and interstrand DNA crosslinks [6]. In this technique, cells are lysed on a filter, most RNA and protein is removed by washing, and DNA is eluted by pumping buffer through the filter. The basic principle is that the rate of elution depends on the length of the DNA fragments and the nature of the lesion detected is determined by the characteristics of the lysis and elution buffers. DNA double-strand breaks are measured when the elution buffer has a neutral pH because the double-strandedness of DNA is maintained. Alkaline elution buffer denatures the DNA so that single-strand breaks are assessed.

An alternative method for the detection of double-strand DNA breaks is pulsed field gel electrophoresis [7]. This technique was developed for the separation of large DNA molecules of differing size and is therefore sensitive to changes in DNA fragment lengths which result from drug-induced double-strand DNA breaks.

Measurement of specific drug-DNA lesions is sometimes possible

DNA damage induced by the platinum coordination complex CDDP has been well characterised, and methods for the detection of specific CDDP–DNA adducts have been developed. Drug-damaged DNA is enzymically digested to produce a mixture of mononucleotides and CDDP-dinucleotide adducts which can be separated by HPLC [8]. More recently, monoclonal antibodies have been raised against CDDP–DNA adducts. These have been used to quantify DNA damage in cell lines and in tumour biopsies, and can also be used to detect adducts at the cellular level by immunocytochemistry [9]. This offers a simple means of examining both the frequency and the distribution of DNA lesions throughout the malignant and non-malignant cells in a solid tumour after treatment with CDDP in vivo. Because of the practical difficulty of obtaining serial tumour biopsies in patients, CDDP–DNA adducts have been measured in more accessible tissue samples (peripheral leukocytes and buccal smears) during and after chemotherapy. Although correlations have been made between the number of adducts per normal cell and tumour response, this is probably a rather complex method of measuring drug exposure and does not assess the DNA repair capacity of the tumour.

Similar approaches to the detection of drug-induced DNA damage are being pursued for other drugs. For example, monoclonal antibodies have now been raised against melphalan–DNA adducts. Such approaches can detect down to femtomole levels of DNA damage and they find immediate application in the study of drug–DNA interactions in tumour cell lines and murine tumours, but their role, if any, in clinical management remains to be defined. Two important questions remain unanswered: do normal and malignant cells differ in their ability to repair drug–DNA adducts, and are drug-resistant tumours more efficient at adduct removal?

Cells vary in their response to drug-induced DNA damage

It has been postulated that the drug sensitivity of certain tumour types may be due to deficiencies in DNA repair in these cells, acquired during development of the tumour. Evidence for the importance of DNA repair pathways in carcinogenesis comes most directly from inherited tumour-prone human disorders, in which cellular hypersensitivity to radiation and DNA damaging drugs is associated with defects in DNA repair. Although these syndromes are themselves rare in the general population, heterozygous carriers have been estimated at frequencies of 0.1–1% and an increased risk of cancer has been suggested in certain heterozygote carriers [10]. Further evidence for a possible involvement of DNA repair in carcinogenesis comes from chemical carcinogenesis models in animals. It has been suggested that lack of repair of adducts allows a premutagenic lesion to persist long enough to be fixed upon DNA replication [11]. If defective DNA repair processes are involved in tumour development, identification of the particular defect in tumours and hence the DNA damage sensitivity could have important implications for the choice of chemotherapy treatment regimes.

As techniques for separating genomic DNA into functionally distinct subfractions have been developed, it has become apparent that DNA damage and repair are not homogeneous throughout the genome [12]. Active chromatin has certain structural features, such as a lower level of DNA condensation, which appear to render it more susceptible to cytotoxic damage. It has also recently been shown that repair of DNA damage is inhomogeneous; for example transcriptionally active genes, such as the DHFR gene, are repaired more efficiently than the genome as a whole [12]. The importance of repair of transcriptionally active genes is reflected in Cockayne's syndrome, where failure of preferential repair leads to neurodegeneration and photosensitivity, but not, interestingly, to an increase frequency of malignancy. As yet it is unknown whether preferential DNA repair plays any role in carcinogenesis or drug resistance.

Analysis of DNA repair in tumours is not straightforward. Many of the assays used to examine repair and cellular sensitivity of cells grown in tissue culture are not applicable to tumours. Even when cell lines from tumours are available, it is impossible to compare the tumour with its normal cell of origin. Therefore much of the work on DNA repair and response of tumours to chemotherapy is based on the analysis of alterations in cell lines which have been selected for acquired resistance to chemotherapeutic drugs (Table 1).

Drug resistance may be due to enhanced DNA repair

Evidence for increased DNA repair as a mechanism of chemotherapeutic drug resistance has been shown in CDDP-resistant cells [13]. Resistant cells selected by chronic exposure to CDDP have been shown to have increased unscheduled DNA synthesis (i.e. DNA synthesis which is part of patch-filling during repair, see Chapter 16) after exposure to the drug. Some resistant cells also remove the major CDDP-induced lesion (the Pt-GG intrastrand crosslink) more rapidly than do sensitive cells. Similarly, CDDP-sensitive testicular tumour cells appear to be less

Table 1. The role of DNA repair in drug resistance.

Human tumour	Cell line	Drug (fold resistance)	Changes in DNA repair[a]
Ovarian	A2780 PEO4	CDDP (3)	Increased repair patches Increased UDS[b]
Bladder and testicular	RT112 833K SUSA	CDDP (5)	Increased interstrand crosslink repair
Colon	HCT8	CDDP (4)	Increased DNA pol α and β
Cervical	HeLa	CDDP (20)	Increased CAT reactivation
	HeLa	Bleomycin (20)	Increased repair of single-strand breaks
Head and neck	A253	Bleomycin (4–21)	Increased repair of single-strand breaks
Melanoma	MM253	DTIC (5–10)	Increased UV repair synthesis
CLL patients	Lymphocytes from 14 patients	Melphalan	Increased removal of interstrand crosslinks
Myeloid leukaemia	Myelocytes from 10 patients	BCNU (6)	Increase in alkyltransferase
Breast	MCF-7	Melphalan (3)	Increased removal of interstrand crosslinks

[a]See reference 12 and references therein.
[b]UDS, unscheduled DNA synthesis

proficient in the removal of adducts than resistant bladder tumour cells. One disadvantage of these types of repair assays which treat cells with CDDP is the unknown effects of the drug on cellular components, other than DNA, which could affect the cellular repair machinery. An assay system that provides information concerning the cellular response to damaged DNA, without the confounding effects of other damaged cellular molecules, is the plasmid reactivation assay (Figure 2). In this assay plasmid DNA is treated in vitro with CDDP and then introduced into the cells of interest. The plasmid DNA contains a "reporter" gene whose expression can easily be quantified. Platination of the DNA inhibits expression of this reporter gene, but after transfection into repair proficient cells, removal of the platinum adducts increases reporter gene expression. Increased reactivation of a CDDP-treated reporter gene has been observed in CDDP-resistant cells, again supporting an involvement of enhanced DNA repair in CDDP resistance [14].

While the types of assay described above can be used to show increased removal of damaged DNA in cell lines, they cannot readily be used to examine repair levels in vivo in tumour biopsies. The relevance of DNA repair mechanisms to clinical chemotherapy treatment failure can only be assessed when suitable means of assaying repair in tumour biopsies become available. Possible approaches are the identification of repair genes and measurement of their expression in tumours, or the measurement of specific biochemical activities in tumour cell extracts.

Some proteins involved in the response to DNA damage have been identified

DNA repair synthesis assay

DNA excision repair by cell extracts can be measured in vitro by the incorporation of radiolabelled nucleotides into damaged DNA (Figure 3) [15]. This assay has been used to measure repair of UV-light-, acetylaminofluorine-, and CDDP-induced DNA damage by human cell extracts. Extracts from xeroderma pigmentosum (XP) cells, known to be defective in this type of DNA repair, have been shown

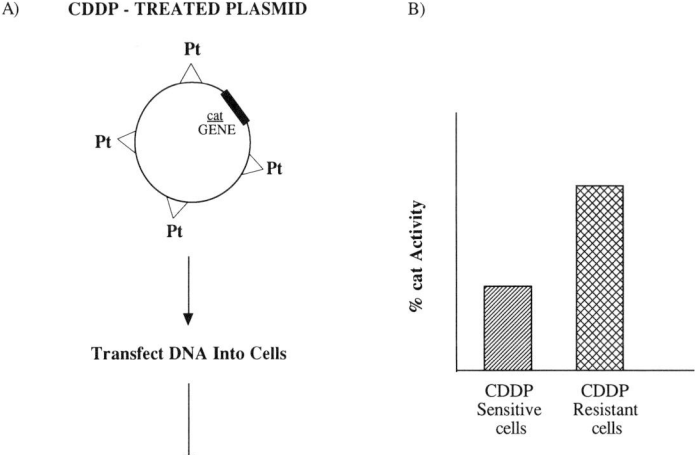

Figure 2. The plasmid reactivation assay. (A) A plasmid containing a reporter gene such as the *cat* gene, which encodes the enzyme chloramphenicol acetyl transferase (CAT), is treated with CDDP. After transfection of undamaged or CDDP-treated plasmid into cells, the cells are grown to allow the expression and repair of the *cat* gene. (B) CAT activity can then be measured in extracts from the cells and can be compared between CDDP-sensitive and resistant lines.

to be deficient in repair synthesis activity. The assay has been used to identify and purify some of the proteins required for excision repair, namely the XP-A protein, human single-stranded DNA-binding protein (hSSB) and proliferating cell nuclear antigen (PCNA) [16]. This information may be useful for designing

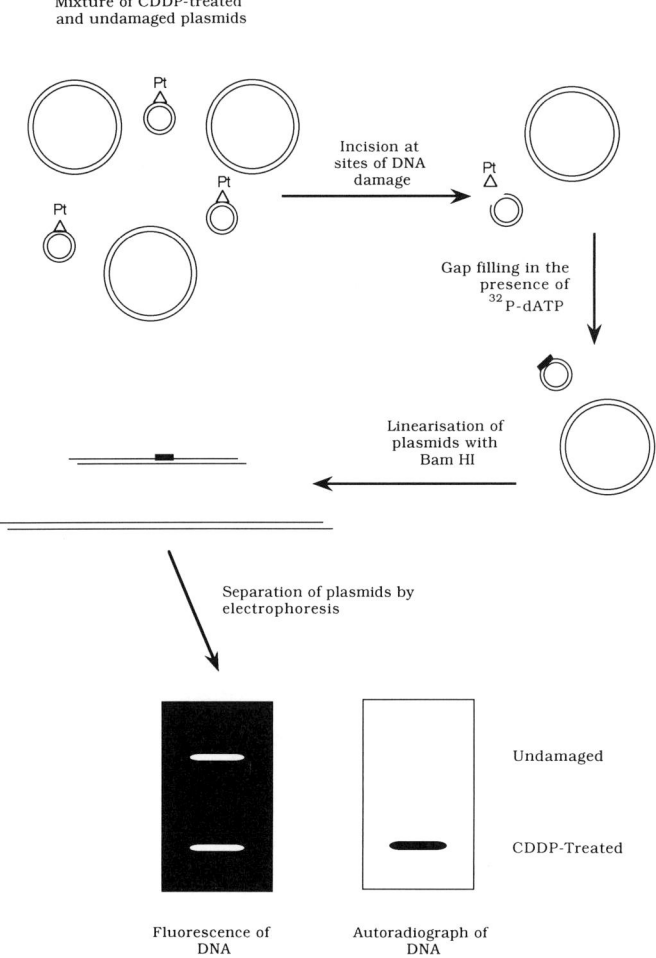

Figure 3. The repair synthesis assay. Undamaged and CDDP-treated plasmids of different sizes are incubated with cell extract in the presence of radiolabelled dATP. Following the repair reaction, plasmids are separated by agarose gel electrophoresis and monitored for incorporation of labelled dATP to detect damage-dependent repair synthesis.

drugs that can target specific proteins and inhibit DNA excision repair and there is obvious potential to use this assay to analyse the repair activity of resistant tumour cell lines.

Damage recognition proteins

Protein–DNA complexes have a decreased mobility compared with either DNA or protein alone under appropriate electrophoresis conditions. Damaged oligonucleotides are retarded in a gel when proteins are attached (Figure 4A). Western blots of proteins can be probed with damaged oligonucleotides to allow the formation of DNA–protein complexes (Figure 4B). Damage recognition proteins (DRPs) binding to UV- and CDDP-damaged DNA have been identified in cell extracts. Certain xeroderma pigmentosum (XP) cell lines of complementation group E have been shown to be deficient in UV-damage recognition proteins. Further, a CDDP-resistant cell line, that has increased levels of damaged plasmid DNA reactivation, has overexpression of a UV-DRP [17]. Although these observations suggest that UV-DRP may be involved in the damage recognition step of human excision repair, it should be noted that not all XP-E cells have a detectable deficiency in a UV-DRP. The CDDP-DRP have been shown to bind only to certain types of platinum adducts in DNA [18]; they bind to the Pt–GG intrastrand crosslink, but not to interstrand crosslinks or monofunctional forms characteristic of the inactive *trans*-DDP. This specificity of interaction may implicate CDDP-DRP in the anti-tumour activity of the platinum coordination complexes.

The precise role of these DRPs, in response to DNA damage and in drug resistance, remains uncertain. Possible functions for DRP include:

(i)   a recognition step in a DNA repair pathway which allows access of other repair proteins to the site of damage;
(ii) structural proteins in chromatin, binding to conformational changes in DNA, and perhaps actually impeding DNA repair by binding to the intrastrand adducts of CDDP;
(iii) the DRPs may have another role in DNA metabolism, for instance DNA replication or gene transcription;
(iv) regulation of the cell cycle in response to DNA damage.

Alkyltransferases

Alkylation of DNA by chemotherapeutic drugs such as CCNU and BCNU can be directly reversed by the transfer of the alkyl moiety via an alkyltransferase from the alkylated base to a cysteine residue in the active site of the enzyme [19]. Alkyltransferases are thus suicide proteins, irreversibly inactivated after binding an alkyl group. Cells which are deficient in O^6-alkylguanine alkyltransferase are unable to repair methylation of the O^6 position of guanine. These are termed mer– (methyl excision repair deficient), while cells that are methyl excision repair proficient are termed mer+. Mer– cells are hypersensitive to simple methylating, ethylating and chloroethylating agents, in comparison to mer+ cells. However, human tumours have generally been found to express the mer+ phenotype, and

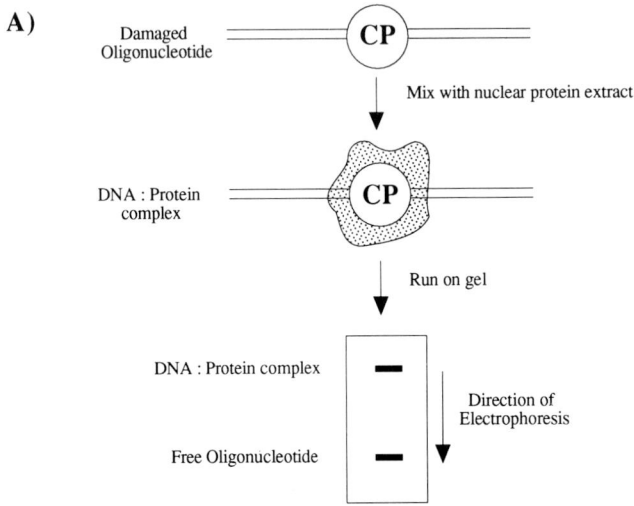

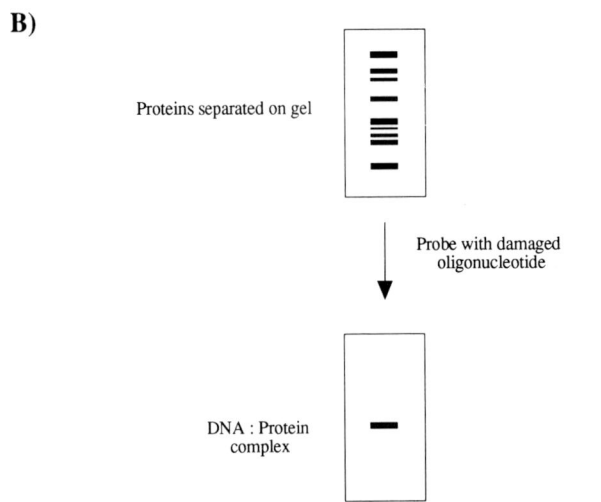

evidence correlating enhanced alkylation repair in human tumours and alkylating agent treatment failure is scanty. It is considered likely that resistance to the methylating agent dacarbazine and the chloroethylnitrosoureas (BCNU, CCNU and methyl-CCNU) is due to enhanced repair by alkyltransferase.

DNA polymerases

There are at least five mammalian DNA polymerases [20]. Since polymerase-β (pol-β) can be induced by some DNA damaging agents and its activity is independent of the replicative state, it has been assigned a DNA repair function. Pol-β activity has been shown to be elevated in CDDP-resistant P388 leukaemia cells that have elevated levels of excision repair; and Pol-α and -β levels are elevated in certain CDDP-resistant colonic and ovarian carcinoma cell lines.

DNA topoisomerases

An important target for the cytotoxicity of chemotherapeutic drugs are DNA topoisomerases [21]. These are enzymes that change the topological state of DNA by breaking and rejoining its backbone. Topoisomerases are involved in a number of DNA processes in the cell, including replication, transcription and recombination, and are generally viewed as being important in unwinding DNA to allow access of enzymes involved in these processes. DNA topoisomerases are classified into two types. Type I DNA topoisomerases change the topological state of DNA by transiently breaking one strand of the DNA double helix. Type II topoisomerases catalyse transient, enzyme-bridged, double-strand breaks and allow other strands of DNA to pass through. Many anti-tumour drugs have been shown to interfere with the breakage-rejoining activity of topoisomerase II, by trapping an important reaction intermediate, termed the cleavable complex. Drug-stabilised cleavable com-

---

Figure 4. Assays of DNA damage recognition proteins. (A) Mobility shift assay for CDDP–DNA binding proteins. Oligonucleotide DNA is treated in vitro with CDDP to produce CDDP–DNA adducts. After labelling with ^{32}P, the oligonucleotide is incubated with protein extract prepared from the nuclei of the cells of interest. The ability of proteins to bind to the damaged DNA can be assessed by gel electrophoresis and autoradiography. Proteins binding to the DNA are observed as a slower migrating retardation complex, and if they are damage specific, they bind to the damaged DNA but not to undamaged DNA. (B) Southwestern assay of cell extract. Cell extracts are separated by polyacrylamide gel electrophoresis and transferred to nitrocellulose. The nitrocellulose blot is incubated with labelled oligonucleotide, either undamaged or containing CDDP–DNA adducts. Proteins binding to the DNA can be assessed by gel electrophoresis and autoradiography. Damage recognition proteins should only be observed on the blot incubated with the CDDP-treated DNA. The relative intensity of the bands can be used to quantify differences in these proteins between cell lines.

SITE oF ACTION oF *R. Brown and D. Bissett*
ETOPOSIDE.

plexes are lethal to proliferating cells and are believed to be responsible for the anti-tumour activity of these drugs. If topoisomerases are a target for anti-tumour drugs, obviously reduction in topoisomerase levels or in enzyme affinity for the drug may lead to resistance. Indeed cell lines selected for resistance to chemotherapeutic drugs have been shown to have reduced levels of topoisomerase activity.

How does DNA damage lead to cell death?

Unrepaired drug-induced DNA damage may result in:

(i)   inhibition of DNA synthesis;
(ii) alteration in gene expression;
(iii) induction of apoptosis;
(iv) loss of genetic information.

How an unrepaired DNA lesion eventually leads to cell death is an area of increasing interest. Generally it has been assumed that DNA lesions block DNA synthesis and prevent cellular replication. Even if the lesion can be by-passed, persistence of the lesion in one cell and possible loss of genetic information in both daughter cells may eventually be lethal. DNA lesions may also have pronounced effects on gene expression, affecting RNA elongation or even the binding of polymerases and transcription factors. Using the gel retardation assay systems previously described, it has been possible to show that drug DNA lesions interfere with binding of transcription factors [22]. Differential tolerance of DNA damage could account for differences in sensitivity to a chemotherapeutic drug, either between normal and neoplastic cells, or between sensitive and resistant tumours. Many features of cell death induced by anti-cancer drugs, such as chromatin condensation and activation of DNA endonucleases, are similar to apoptosis, often called programmed cell death [23]. Protein synthesis is required for apoptosis, suggesting that DNA damage induced by an anti-cancer drug may initiate a signal transduction pathway which leads to cell death.

Blocking DNA repair may sensitise cells to cytotoxics

If increased DNA repair plays a role in drug resistance of tumours, inhibitors of DNA repair may have potential to improve responses to chemotherapy [24]. Methylxanthines such as caffeine have been shown to inhibit DNA excision repair but also exert other effects which enhance the cytotoxicity of DNA damaging agents: cells are prevented from arresting in the G_2 phase of the cell cycle, possibly reducing the time available for DNA repair, and post-replication DNA repair is inhibited. These agents can enhance the cytotoxicity of DNA-damaging agents in vitro, but at present there is nothing to suggest that they will improve the therapeutic ratio of drugs such as CDDP. Nonetheless combinations such as pentoxifylline and CDDP have already entered phase I clinical trials.

Excision repair can also be inhibited by blocking the repair polymerisation step. Two agents that may inhibit DNA repair polymerases are hydroxyurea and cytosine arabinoside (ara-c). Hydroxyurea inhibits ribonucleotide reductase, leading to depletion of deoxyribonucleotide triphosphates, the precursors required for DNA synthesis. Ara-C causes chain termination when incorporated into DNA. The combination of hydroxyurea and ara-C has been shown to enhance the cytotoxicity of CDDP in a human colon cell line. Aphidocolin, an inhibitor of DNA polymerase-α and -δ, has been shown to inhibit repair of CDDP–DNA adducts and to potentiate the toxicity of CDDP in a human ovarian cell line.

Post-translational modification by ADP ribosylation is a ubiquitous mechanism to regulate enzyme activity. It has been shown that ADP ribosylation participates in excision repair. DNA ligase activity, involved in the final step in excision repair, is in part dependent upon post-translational modification by ADP ribosylation, and many other proteins involved in excision repair may also undergo ADP ribosylation. Inhibition of ADP-ribosylation, with 3-aminobenzamidine or nicotinamide, can sensitise tumours in mice to CDDP, although no effect is observed in cultured ovarian carcinoma cells.

Recent evidence has suggested a role for the p53 gene in response to DNA damage [26]. Genetic alteration of p53 is one of the most frequently observed changes in human neoplasia and the p53 gene has been implicated both as a proto-oncogene and as a tumour suppressor gene. Levels of p53 have been shown to be increased in cells after treatment with a variety of DNA damaging agents, including chemotherapeutic agents. In the case of ionising radiation, p53 levels alter in temporal association with a G_1 cell cycle arrest in cells with wild-type p53; cells with mutant p53 do not arrest at G_1 [27]. Hence, it has been suggested that wild-type p53 has a role in the inhibition of DNA synthesis following DNA damage, and alteration of p53 may cause a failure of this inhibition, an increased sensitivity to DNA damage, and an increased mutation rate in the genome. Modulation of the DNA damage-p53 pathway could potentially increase the effectiveness of both radiotherapy and chemotherapy.

While there remains uncertainty as to the role of DNA repair in determining the response of tumours to chemotherapy, it seems likely that an increased understanding of the biochemistry of DNA repair and its regulation will facilitate the design of novel and effective chemotherapeutic strategies.

References

1   Powis G. Signalling targets for anticancer drug development. Trends Pharmacol Sci 1991, **12**, 188–194.
2   Driscoll JS. The preclinical new drug research program of the National Cancer Institute. J Natl Cancer Inst 1984, **68**, 63–76.
3   Chabner BA, Collins JM. Cancer Chemotherapy - Principles and Practice. Lippincott, Philadelphia, 1990.
4   Epstein RJ. Drug-induced DNA damage and tumour chemosensitivity. J Clin Oncol 1990, **8**, 2062–2984.
5   Bocian E, Laverick M, Nias AHW. The mode of action of *cis*-dichloro-bis(isopropyl-amine)*trans*-dihydroxyplatinum IV (CHIP) studied by the analysis of chromosome aberration production. Br J Cancer 1983, **47**, 503–510.
6   Kohn KW. Principles and practice of DNA filter elution. Pharmacol Therapeut 1991, **49**, 55–77.
7   Anand R, Southern EM. Pulsed field gel electrophoresis. In: Gel Electrophoresis of Nucleic Acids – a Practical Approach. IRL Press, Oxford, 1990, pp 101–123.
8   Fichtinger-Schepman AMJ, van der Veer JL, den Hartog JHJ et al. Adducts of the anti-tumour drug *cis*-diamminedichloroplatinum(II) with DNA: formation, identification, and quantitation. Biochemistry 1985, **24**, 707–713.
9   Terheggen PMAB, Floot BGJ, Scherer E et al. Immunocytochemical detection of interaction products of *cis*-diamminedichloroplatinum(II) and *cis*-diammine (1,1-cyclo-butanedicarboxylato)platinum(II) with DNA in rodent tissue sections. Cancer Res 1987, **47**, 6719–6725.
10 Swift M, Retnauer PJ, Morrell D, Chase CL. Breast and other cancers in families with ataxia telangiectasia. N Engl J Med 1987, **316**, 1289–1294.
11 Goth R, Rajewsky MF. Persistence of O^6-ethylguanine in rat brain DNA: correlation with nervous system-specific carcinogenesis by ethylnitrosurea. Proc Natl Acad Sci USA 1974, **71**, 639–643.
12 Bohr VA, Phillips DH, Hanawalt PC. Heterogeneous DNA damage and repair in the mammalian genome. Cancer Res 1987, **47**, 6426–6436.
13 Masuda H, Ozols RF, Lai G et al. Increased DNA repair as a mechanism of acquired resistance to *cis*-diamminedichloroplatinum(II) in human ovarian cancer cell lines. Cancer Res 1988, **48**, 5713–5716.
14 Chao CC-K, Lee Y, Cheng P, Lin-Chao S. Enhanced host cell reactivation of damaged plasmid DNA in HeLa cells resistant to *cis*-diamminedichloroplatinum. Cancer Res 1991, **51**, 601–605.
15 Wood RD, Coverley D. DNA excision repair in mammalian cell extracts. Bioessays 1991, **13**, 447–453.
16 Shivji MKK, Kenny MK, Wood R. Proliferating cell nuclear antigen (PCNA) is required for DNA excision repair. Cell 1992, **69**, 1–20.
17 Chao CC-K, Huang S-L, Huang H, Lin-Chao S. Cross-resistance to uv radiation of a cisplatin-resistant human cell line: overexpression of cellular factors that recognise UV-modified DNA. Mol Cell Biol 1991, **11**, 2075–2080.
18 Donahue BA, Augot M, Bellon SF et al. Characterisation of a DNA damage-recognition protein from mammalian cells that binds specifically to intrastrand d(GpG) and d(ApG) DNA adducts of the anticancer drug cisplatin. Biochemistry 1990, **29**, 5872–2880.

19 Lindahl T, Sedgewick B. Regulation and expression of the adaptive response to alkylating agents. Annu Rev Biochem 1988, **57**, 133–157.
20 Linn S. How many pols does it take to replicate nuclear DNA? Cell 1991, **66**, 185–187.
21 Liu LF. DNA topoisomerase poisons as antitumour drugs. Annu Rev Biochem 1989, **58**, 331–375.
22 Broggin M, Ponti M, Ottolenghi S et al. Distamycin inhibits the binding of OTF-1 and NFE-1 transfactors to their conserved DNA elements. Nucleic Acids Res 1989, **17**, 1051–1059.
23 Eastman A. Activation of programmed cell death by anticancer agents: cisplatin as a model system. Cancer Cells 1990, **2**, 275–280.
24 Stewart DJ, Evans WK. Non-chemotherapeutic agents that potentiate chemotherapy efficacy. Cancer Treat Rev 1989, **16**, 1–40.
25 Lambert B, Jones BK, Roques BP et al. The noncovalent complex between DNA and the bifunctional intercalator ditercalinium is a substrate for the UvrABC endonuclease of *E. coli*. Proc Natl Acad Sci USA 1989, **86**, 6557–6559.
26 Lane D. p53, Guardian of the genome. Nature 1992, **358**, 15–16.
27 Kastan MB, Onyekwere O, Sidransky D et al. Participation of p53 protein in the cellular response to DNA damage. Cancer Res 1991, **51**, 6304–6311.

Molecular Biology for Oncologists
Edited by John Yarnold, Michael Stratton, Trevor McMillan
© *1993, Elsevier Science Publishers B.V. All rights reserved*

CHAPTER 20

The molecular biology of drug resistance

Daniel Hochhauser

Imperial Cancer Research Fund, University of Oxford, Institute of Molecular Medicine,
John Radcliffe Hospital, Oxford, OX3 9DU, UK

The failure to cure cancer with current chemotherapeutic regimes is largely due to drug resistance

The mechanisms of resistance to chemotherapeutic drugs may be considered to operate from the point of administration (e.g. failure in absorption), to distribution and metabolism. Factors affecting these processes include considerations such as poor absorption and decreased tumour blood supply. However, this review concentrates on some recent advances in understanding resistance at the molecular level.

Most studies in the laboratory on resistance mechanisms have used cell lines made resistant by gradual increases in drug concentration. The resultant resistant cell lines may have several hundred-fold resistance to drug and the relevance of such changes to the clinical situation is unclear. Furthermore, the term drug resistance is misleading in implying an abnormal tumour phenotype insofar as most normal cells are also relatively drug resistant.

Cells exposed to a single drug may develop resistance to unrelated compounds

It has been known for some time that exposure of tumour cells to one drug may induce cross-resistance to a variety of unrelated drugs to which the cell has not been exposed previously. This phenomenon is known as multidrug resistance

(MDR). The drugs involved include adriamycin, vinca alkaloids, mitomycin C but not platinum, bleomycin or alkylating agents. There has been extensive work on the development of resistance to platinum particularly with regard to enhanced DNA repair which is reviewed in Chapter 19 and is not considered further here.

P-glycoprotein is the main mechanism of multidrug resistance

The main mechanism of MDR is through the expression of P-glycoprotein (PGP), a 170 kDa molecule with six transmembrane domains and an ATP binding cassette (reviewed in [1]). PGP is related to a family of energy-dependent transporters, the ATP binding cassette superfamily, in both prokaryotic and eukaryotic species. Other members of this superfamily include a pigment transporter in *Drosophila*, a protein mediating chloroquine resistance in *Plasmodium falciparum* and, more recently, the product of the cystic fibrosis gene (CFTR) and two genes associated with the presentation of class I antigens by transport of peptides into the endoplasmic reticulum.

P-glycoprotein functions as a channel and a pump

The ubiquitously found ATP binding cassette proteins have been shown to be transporters but PGP has been demonstrated to have the properties of both a transporter and a channel. It appears that PGP is an active transporter which pumps hydrophobic drugs out of cells and hence reduces cytotoxicity. Recently, PGP has been shown to be associated with a volume regulated chloride channel activity. These two properties of a chloride channel and of a drug pump are separable, indicating that they represent two distinct and independent functions of the protein [2].

The MDR gene can be cloned in several ways

Cytogenetic analysis of drug resistant cells has revealed homogeneously staining regions and double minute chromosomes which are features indicating gene amplification. By use of a novel technique known as "in gel renaturation", these amplified regions were isolated and cloned. The cloned sequences were then used as radiolabelled probes on cDNA libraries made from resistant cell lines and the isolated clones were then sequenced. The MDR gene was also cloned by subtractive hybridisation of cDNA libraries from sensitive and drug resistant lines and by using specific monoclonal antibodies made from drug resistant cells to screen expression cDNA libraries [3].

PGP is the product of the MDR gene. In man, the MDR1 gene is located on chromosome 7q21-31 and consists of an open reading frame of 1280 amino acids. There are two human and three rodent MDR genes.

The MDR gene has been implicated in multidrug resistance by transfection experiments

Transfection experiments in which a cDNA encoding the human MDR1 gene was introduced into sensitive cell lines have demonstrated that the MDR1 gene confers MDR [4]. The other MDR genes (two altogether in humans) presumably evolved as gene duplications and their role in vivo is unclear although the MDR3 gene has been found to be overexpressed in the absence of MDR1 in B-cell prolymphocytic leukemia. In mouse, two MDR forms have been shown to be induced in drug resistant cells although only MDR1 has been shown to be active in transfection experiments.

Interestingly, mutations within the MDR1 gene may alter patterns of cross resistance [5]. In vinblastine resistant KB cells, a glycine to valine mutation at codon 185 resulted in a changed preferential resistance pattern from vinblastine to colchicine. This might indicate the key role of amino acid 185 in PGP drug interaction; this amino acid is located within the first hydrophobic region on the cytoplasmic side of the membrane and may be part of a drug binding site.

P-glycoprotein may be increased in several different ways

Studies on cell lines have revealed that up-regulation of PGP may be achieved in several different ways. Control of expression of the human MDR gene product occurs at the level of gene copy number, transcription, translation and post-translationally. As previously mentioned, there is frequent amplification of the genomic region containing the MDR1 gene in drug resistant cells leading to increased expression of the protein. Amplification through autonomously replicating extrachromosomal sequences has also been seen. In other cell lines, there is a stabilisation of the MDR mRNA which effectively increases cellular PGP, while in other cell lines post-translational changes such as phosphorylation lead to stabilisation of the protein itself or altered activity.

P-glycoprotein is widely distributed in tissues

The distribution of MDR in normal and tumour cells has been clarified using several monoclonal antibodies [6]; the most widely used is the C219 antibody which binds to an epitope near the ATP binding site and can be used for both immunoprecipitations and Western blotting. Unfortunately, C219 also stains the MDR3 product. Tissue distribution has also been studied using Northern blots of isolated mRNA, RNA slot blots, in situ mRNA hybridisation, ribonuclease protection and semiquantitative measurement of mRNA levels using the polymerase chain reaction (PCR).

The highest levels of PGP have been found in kidney (lumina of the proximal tubules), adrenal cortex (zona fasciculata and reticulata), stomach, duodenum, colon and placenta. High levels of expression are also found in endothelial cells of capillary blood vessels at blood-brain barrier sites and in the testes [7]. This has significance insofar as these constitute sanctuary sites in which relapse occurs in conditions such as acute lymphoblastic leukemia where there is presumably a failure of drug penetration.

Interestingly, it has recently been found that MDR1 and the cystic fibrosis gene product have complementary patterns of expression. A switch of expression was found in several tissues; for example in the intestine expression switches from CFTR to MDR1 as the cells migrate across the crypt-villus boundary [8].

P-glycoprotein is regulated as a stress protein

The renal adenocarcinoma cell line HDB46 shows an eightfold increase in MDR1 RNA levels in response to heat shock, ethanol, sodium arsenite and cadmium consistent with the role of MDR1 as a stress-inducible gene responsive to environmental insults [9]. The induction of MDR1 mRNA occurs concomitantly with development of a two- to sixfold increase in drug resistance to vinblastine following heat shock. This provides a link between the expression of PGP in response to cellular environmental shocks and the concomitant development of drug resistance.

MDR expression is often increased in tumours

Studies of the significance of MDR in human tumours have recently become available. An extensive study reported on levels of MDR1 mRNA in over 400 human cancers [10]. It was found that there is an inverse correlation of levels of MDR expression with chemosensitivity of the tumour, although numerous exceptions were found. Thus, for example breast and ovarian cancers had lower levels of MDR expression than colon and renal cell carcinomas. However, sarcomas and non-small cell lung cancers had low levels of MDR and a low frequency of expression despite showing a resistant phenotype.

Recent reviews confirm the general linkage between low levels of MDR and chemosensitivity with some studies showing an increase in MDR as the tumour becomes resistant [11].

Some resistant cancers show no pattern of increased MDR expression

Renal adenocarcinomas arise from the proximal kidney tubules and are generally refractory to chemotherapy. It was found that the levels of MDR in tumours and surrounding tissues is generally elevated. There is no correlation between the level of expression of MDR in tumour and surrounding tissue [12].

It is too early to allow the full significance of P-glycoprotein expression to be assessed

The majority of studies thus far have included too few patients to allow the clinical significance of MDR expression to be clarified. Prospective studies will help to clarify whether identifying subgroups of tumour patients expressing MDR helps to predict response to chemotherapy and might therefore be appropriate for the use of MDR modulators. Clearly, the expression of MDR will be insufficient to explain the development of steroid and antimetabolite resistance in many neoplasms.

Many agents can reverse multidrug resistance

A variety of pharmacological agents reverse MDR and this has been extensively reviewed [13]. These range from the calcium blockers, particularly verapamil and nifedipine, to tamoxifen, phenothiazines and cyclosporin, among others. The concentrations of some drugs required to reverse MDR in vitro would produce toxicity in vivo and this has been a major limitation to the effective clinical use of, for example, verapamil. Some of these agents probably act by binding to PGP and thereby cause intracellular accumulation of drug. Others, including phenothiazines, may act by binding calmodulin or do not affect efflux but cause redistribution within the cell.

The most extensive studies on MDR reversal occurred with verapamil. There are reports that verapamil may improve the rate of remission in multiple myeloma and non-Hodgkin's lymphoma [14].

MDR expression does not account for all cases of multidrug resistance

The phenomenon of multidrug resistance mediated through PGP has not, however, accounted for many other instances of cross-resistance found in many situations including both cell lines and clinical material. The phenotype in these cases differs in that although cross-reactivity occurs among different drug types, there is no resistance to vincristine.

Atypical multidrug resistance is associated with drugs interacting with topoisomerase II, a vital cellular enzyme

The phenomenon of this atypical multidrug resistance (at-MDR) is associated with drugs which interact with topoisomerase II [15]. Topoisomerase II is a 170 kDa protein which is required by the cell to separate intertwined DNA and cause re-laxation of supercoiling. This involves the cutting of both DNA strands, passage of the strands through the break and subsequent resealing of the break.

Topoisomerase II is required for several essential cellular processes including DNA replication, transcription, chromosomal segregation and possibly recombi-nation [16]. Recently, it has become clear that there are two forms of topoisom-erase II (termed α and β) in vertebrates which are located on different chromo-somes. They are regulated differently and may have differing functions [17]. The relative contribution of each of these remains unclear.

Many significant chemotherapeutic agents are topoisomerase II poisons

Several drugs interact with topoisomerase II by stabilising the cleavable complex in which the drug binds to the enzyme–DNA complex and alters the reaction equilibrium to prevent resealing of the nicked DNA. However, the exact mode of drug action is unclear and may differ between different agents. Such drugs include the epipodophyllotoxins etoposide (VP16) and teniposide (VM26), adriamycin and intercalating agents such as mAMSA. Neither the alkylating agents (e.g. cyclo-phosphamide) nor platinum are involved in the at-MDR phenotype.

Few studies of topoisomerase II expression have been carried out on tumour tissue

The majority of studies on the relationship of topoisomerase II content and drug sensitivity have been performed on cell lines rather than with clinical material. It is frequently found using cell lines that on removal of drug, there is rapid loss of resistance and reversion to parental line sensitivity. This does not occur clinically, where the development of resistance to topoisomerase II inhibitors persists long after the effective elimination of drug. Also, it should be noted that the degree of resistance found in many of these lines, often above 200-fold, and even exceeding 700-fold in one study, far exceeds that found in clinical specimens and it may be that cell lines with more moderate degrees of resistance more accurately resemble the clinical situation.

Cytotoxicity of topoisomerase poisons is related to the amount of enzyme present

Generally, it may be stated that the more topoisomerase II there is within the cell, the more sensitive the cell will be to topoisomerase II poisons as a higher level of enzyme will result in more DNA lesions. This has been confirmed in the relatively few studies in which tumor cell lines have been analysed to correlate enzyme content and chemosensitivity. In a study comparing doxorubicin and etoposide sensitivity in small-cell lung cancer (SCLC) and non-small-cell lung cancer (NSCLC) lines, the former were significantly more sensitive to both agents [18]. As expected, the topoisomerase II activities in nuclear extracts from SCLC lines were two-fold higher than with NSCLC lines. Western blotting revealed higher topoisomerase II content in the SCLC lines. Such studies confirm the clinical picture where SCLC are chemosensitive at presentation as compared with NSCLC. The frequent relapses which occur in SCLC may be a fruitful area to investigate the clinically relevant onset of at-MDR.

Similarly, topoisomerase II content was found to correlate well with chemosensitivity in a study of lymphocytes from patients with chronic lymphocytic leukemia (CLL) [19]. Such cells are arrested in the G_0/G_1 cell cycle phase and topoisomerase II was undetectable immunologically. Detectable topoisomerase II was found in the lymphocytes from a variety of other lymphoproliferative diseases such as Burkitt's lymphoma and acute lymphoblastic leukemia. The link between topoisomerase II amount and drug sensitivity has also been investigated in bladder and testis tumour cell lines in which the in vitro drug sensitivities of these lines mirrored the original tumour sensitivity with respect to adriamycin and platinum [20].

Many alterations in topoisomerase II have been found in studies on cell lines

Cell lines resistant to topoisomerase II poisons may show several different changes. These include decreased mRNA levels, alteration in the topoisomerase II isoform expressed, point mutations in the gene and decreased enzyme phosphorylation. The common factor is a reduction in cellular topoisomerase II activity presumably leading to less DNA strand breaks and hence decreased cytotoxicity. Whether these in vitro changes correspond to the situation clinically is now under investigation.

Glutathione transferases may be involved in drug resistance

The glutathione transferases are enzymes synthesised in the liver and kidney primarily but also found in all organs. They are involved in detoxification principally through drug conjugation to glutathione [21]. Although changes in GST have been found in many instances in drug resistant cells and they are plausible candidates playing a role in metabolising drugs [22], there is no definitive evidence that changes in their amount directly result in resistance. Experiments transfecting the GST cDNAs into sensitive cells failed to produce a resistant phenotype in many cases and only a modest 1.5–2-fold increase in some studies.

Future trends

The expression of PGP in tumour cells has now been demonstrated to be a significant, if not the sole, factor in multidrug resistance. Recent advances in research include the development of a transgenic mouse in which the human MDR1 gene is expressed in bone marrow. This may be used as a model for testing agents which reverse MDR. Conceivably, the expression of MDR in bone marrow stem cells might allow the use of higher doses of chemotherapeutic agents clinically. The major advances clinically in circumventing MDR will come from closer understanding of the molecular structure of PGP which will allow the development of more effective inhibitors.

Currently, studies are also under way to investigate the possibility of up-regulating topoisomerase II activity and hence increasing cytotoxicity. Oestrogens and cytokines such as GM-CSF which increase the topoisomerase II levels might be useful in achieving this. Similarly, agents potentiating topoisomerase II activity by phosphorylation such as protein kinase agonists might increase cytotoxicity.

It has also been found that the relative expression of the two topoisomerase II isoforms differs among cell types and that the isoforms differ in their drug sensitivity profiles. Further studies may allow the synthesis of isoform-specific drugs and determine the feasibility of the use of topoisomerase II poisons on the basis of measuring topoisomerase II activity in tumours. Another route could be through the development of competitive topoisomerase inhibitors rather than the current stoichiometric agents, i.e. competition with topoisomerase II for binding to DNA would be more effective in low topoisomerase II expressers.

References

1   Higgins CF. ABC transporters: from microorganisms to man. Annu Rev Cell Biol 1992, **8**, 67–113.
2   Gill DR, Hyde SC, Higgins CF et al. Separation of drug transport and chloride channel functions of the human multidrug resistance P-glycoprotein. Cell 1992, **71**, 21–32.
3   Endicott JA, Ling V. The biochemistry of P-glycoprotein mediated multidrug resistance. Annu Rev Biochem 1989, **58**, 136–171.
4   Gros P, Ben-Nariah Y, Croop JM, Housman DE. Isolation and expression of a complementary DNA that confers multidrug resistance. Nature 1986, **323**, 728–723.
5   Safa AR, Stern RK, Choi K et al. Molecular basis of preferential resistance to colchicine in multidrug-resistant human cells conferred by GLY185 Æ VAL-185 substitution in P-glycoprotein. Proc Natl Acad Sci USA 1990, **87**, 7225–7229.
6   Georges E, Bradley G, Gariepy J, Ling V. Detection of P-glycoprotein isoforms by gene-specific monoclonal antibodies. Proc Natl Acad Sci. USA 1990, **87**, 152–156.
7   Cordon-Cardo C, O'Brien JP, Casals D et al. Multidrug resistance gene (P-glyco-protein) is expressed by endothelial cells at blood-brain barrier sites. Proc Natl Acad Sci USA 1989, **86**, 65–68.
8   Trezise AEO, Romano PR, Gill DR et al. The multidrug resistance and cystic fibrosis genes have complementary patterns of epithelial expression. EMBO J **11**, 4291–4303.
9   Chin KV, Tanaka S, Darlington G et al. Heat shock and arsenite increase expression of the multidrug resistance (MDR1) gene in human renal cell carcinoma cells. J Biol Chem 1990, **275**, 221–226.
10 Goldstein LJ, Galski H, Fojo A et al. Expression of a multidrug resistance gene in human cancers. J Natl Cancer Inst 1989, **8**, 116–124.
11 Gottesman MM, Goldstein LJ, Fojo A et al. Expression of the multidrug resistance gene in human cancer. In: Molecular and Cellular Biology of Multidrug Resistance in Tumor Cells, Roninson IB ed. Plenum, New York, 1991.
12 Fojo AJ, Ueda K, Slamon et al. Expression of a multidrug resistance gene in human tumours and tissues. Proc Natl Acad Sci USA 1987, **84**, 265–269.
13 Stewart DJ, Evans WK. Non-chemotherapeutic agents that potentiate chemotherapy efficiency. Cancer Treat Rev 1989, **16**, 1–40.
14 Dalton WS, Grogan JM, Meltzer PS et al. Drug resistance in multiple myeloma and non-Hodgkin's lymphoma: detection of P-glycoprotein and potential circumvention by addition of verapamil in chemotherapy. J Clin Oncol 1989, **7**, 415–424.
15 Liu LF. DNA topoisomerase poisons as antitumour drugs. Annu Rev Biochem 1989, **58**, 351–375.
16 Hsieh T. Mechanistic aspects of type II DNA Topoisomerases. In: DNA Topology and its Biological effects 1990, Cozzarelli N, Wang, JC eds. Cold Spring Harbour Laboratory Press.

17 Chung TDY, Drake FH, Tan KB et al. Characterisation and immunological identification of cDNA clones encoding 2 human DNA topoisomerase II isozymes. Proc Natl Acad Sci USA 1989, **86**, 9431–9435.
18 Kasahara K, Fujiwara Y, Sugimoto Y et al. Determinants of response to the DNA topoisomerase II inhibitors doxorubicin and etoposide in human lung cancer cell lines. J Natl Cancer Inst 1992, **84**, 113–118.
19 Potmesil M, Hsiang YH, Liu L.F et al. Resistance of human leukemic and normal lymphocytes to drug-induced DNA cleavage and low levels of DNA topoisomerase II. Cancer Res 1988, **48**, 3537–3543.
20 Fry AM, Chresta CM, Davies SM et al. Relationship between topoisomerase II level and chemosensitivity in human tumor cell lines. Cancer Res 1991, **51**, 6592–6595.
21 Pickett CB, Lu AYH. Glutathione S-transferases: gene structure, regulation and biological function. Annu Rev Biochem 1989, **58**, 58743–58764.
22 Morrow CS, Cowan KH. Glutathione S–transferases and drug resistance. Cancer Cells 1990, **2**, 15–22.

Molecular Biology for Oncologists
Edited by John Yarnold, Michael Stratton, Trevor McMillan
© 1993, Elsevier Science Publishers B.V. All rights reserved

CHAPTER 21

Antibody technology is being transformed

Robert E. Hawkins[a] and Stephen J. Russell[b]

[a]MRC Laboratory of Molecular Biology, Hills Road Cambridge, CB2 2QH, UK
[b]Cambridge Centre for Protein Engineering, Cambridge, UK

Targeted therapy with antibodies poses a number of problems

Antibodies are versatile naturally occurring proteins which are potentially very useful for targeted therapy of a variety of malignancies. They are widely used in oncology to provide diagnostic and prognostic information and some successes have already been achieved using antibodies for targeted cancer therapy. However, progress has been hampered by a number of difficulties.

The ideal targeted reagent would fulfil a number of criteria (Table 1). In practice, these ideals are not achievable. Most cancer associated antigens are not entirely specific to the tumour, although many still do allow good discrimination between normal and malignant tissues. Cancer cells are heterogeneous and frequently only a subset of the cells in a tumour deposit express any one antigen. It is therefore desirable to use either a cocktail of targeting reagents or killing mechanisms that have some bystander effects. For the reagent to diffuse throughout the tumour, it should be small (preferably less than 30 kDa), but with a circulating half-life long enough to allow tissue diffusion to occur. The result of all these factors is suboptimal targeting with some degree of toxicity to normal tissues. Prolonged or repeated therapy is therefore necessary for real therapeutic benefit, so it is desirable that an immune response to the targeting reagent should not occur.

Treatment of cancer with polyclonal antisera can be effective [1], but since the arrival of monoclonal antibodies (MAb) [2] there has been an explosion of activity attempting to use them for targeted therapy. There are now many rodent MAbs with useful specificities, but most are not ideal for therapy because they recruit

Table 1.   The ideal targeted reagent for therapy.

The target antigen should be entirely specific
The reagent should reach 100% of its target population
The reagent should be non-toxic until bound to its target
Killing of the target should be highly efficient

---

human effector functions poorly (rat IgG2b is an exception) and are treated as in-
vading foreign proteins by the human immune system. This limits the effectiveness
of therapy and can lead to potentially damaging side effects. There has thus been a
large research effort into ways of generating human monoclonal antibodies. Some
useful human MAbs have been made from immunised patients by B-cell
immortalisation using Epstein Barr virus [3], but it is not possible to immunise
humans against most cancer antigens. In vitro immunisation [4] facilitates the
production of human MAbs to such antigens, but so far this approach is not very
efficient. Humanisation of rodent MAbs by genetic engineering is one alternative
to the direct production of human monoclonals. More recently, methods have begun to emerge for generating and screening large libraries of human antibodies in
vitro. The generation of an antibody is only the start of the process of developing
the optimal therapeutic reagent. The new genetic technology also allows us to
modify affinity, to produce different antibody fragments and to link them to a variety of effector functions. Some of these methods are outlined below.

The enormous diversity of antibodies results from the joining of a limited number of germline genes

The basic unit of an antibody is a Y-shaped molecule (epitomised by IgG) with the
antigen binding sites being located at the tips of the outstretched arms (Figure 1).
The stem of the Y, the Fc region, provides the effector arm of the molecule. This
region binds and activates complement and is recognised by the Fc receptors on
phagocytic effector cells such as macrophages, neutrophils and on NK cells that
mediate antibody dependent cell mediated cytotoxicity (ADCC). The IgG molecule
(150 kDa) comprises two heavy chain–light chain heterodimers linked together by
a disulphide bridge. Each heavy chain (50 kDa) is composed of three constant domains and one variable (V) domain, whereas each light chain (25 kDa) has one
constant and one V domain.

   Detailed analysis of the antigen binding sites of many MAbs shows that they are
very diverse, whereas the rest of the molecule is relatively constant (Figure 1).

Even within the V domains, there are relatively constant framework regions and hypervariable loops of amino acids known as complementarity determining regions (CDRs). X-Ray crystallographic analysis has demonstrated that V-regions conform to a common pattern. The framework regions are folded into two β-sheets which are sandwiched together and the hypervariable loops (three for each V domain) link the tips of individual β-strands to shape the antigen binding surface. The heavy chains and light chains pair so that a total of six CDRs come together to form the antigen binding surface. These CDRs interact not only with the antigen but also with each other.

At the genetic level, the enormous diversity required to make antibodies to any foreign material is supplied by remarkably few gene segments. This is possible because a number of gene segments recombine to form each expressed V-gene (Figure 2). In the case of the heavy chain, there are around 75 V-segments, 30 D-segments and 6 J-segments to choose from. These gene segments can recombine at random and thus the number of V-genes that can be formed is much larger: $75 \times 30 \times 6 = 13\ 500$. In fact, the total number is larger still because there is additional (junctional) diversity introduced during the process of recombination. Variable numbers of extra nucleotides are added and/or taken away at the junctions of the gene segments (both V–D and D–J joining). Similar processes operate in the light chains, although only V and J genes exist. The final method for generating diversity is a consequence of the fact that any light chain can combine with any heavy chain to form an expressed antibody. This remarkable system of recombination has the potential to generate more than 10^{12} possible antibodies from approximately 200 germline genes. This huge primary antibody repertoire can be further diversified. During antigen-driven B-cell proliferation, random somatic mutation of

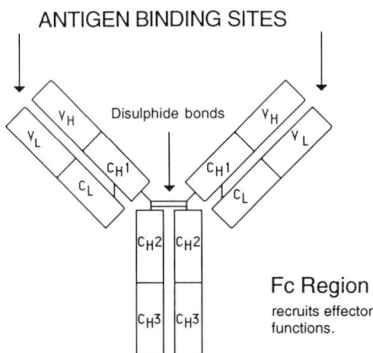

Figure 1. Antibody structure. The basic unit of an antibody is depicted. The antigen binding site is formed by six loops comprising the CDRs from the VH and VL domains. From Hawkins RE, Russell SJ. Br Med J 1992, **305**, 1270–1271. Reprinted with permission.

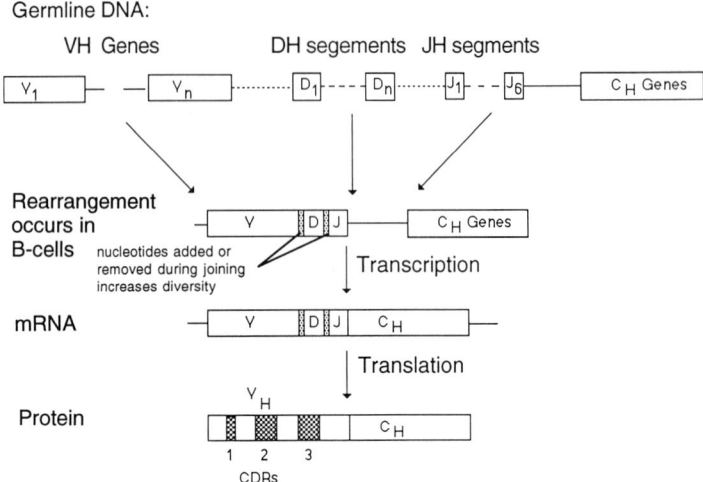

Germline DNA:

Figure 2. Genetic basis of antibody diversity. Recombination events in the production of immunoglobulin heavy chains. The germline genes randomly recombine during B-cell development to form the functional VH gene. The process of recombination and the joining errors give rise to the enormous antibody diversity. From Hawkins RE, Russell SJ. Br Med J 1992, **305**, 1270–1271. Reprinted with permission.

the rearranged V-genes occurs. This occasionally generates B-cells which secrete antibodies with improved affinity for the target antigen and by poorly understood mechanisms, such B-cells are selectively recruited by the immune system. This is the basis of improvements in antibody affinity (affinity maturation) which are associated with prolonged or repeated exposure to a single antigen [5].

Knowledge of the structural and genetic basis of antibody diversity outlined above forms the basis of many of the technological advances in making or improving antibodies which are discussed next.

Chimeric antibodies are composed of rodent variable regions and human constant regions

Transplanting the variable domains of a rodent MAb onto human constant regions creates a chimeric molecule [6] that is largely human but with the specificity and affinity of the parent antibody. This approach optimises effector functions [7] but does not overcome the problem of immunogenicity because one-third of the chimeric MAb is still of rodent origin. Recent trials of such antibodies have revealed that over 50% of humans mount an anti-mouse response after only one treatment [8].

Fully reshaped antibodies retain only the CDRs from the original rodent antibody

Rather than transferring the whole V domain, it has proved possible to transplant the hypervariable loops from a rodent MAb onto a human framework. This produces an antibody with the same specificity [9]. The process usually results in some loss of affinity, but mutations can be made in the framework regions to adjust the orientations of the transplanted hypervariable loops and restore full binding. Several MAbs have now been humanised by CDR-grafting [10] and one has already been used with clear therapeutic benefit [11]. Hopefully, CDR-grafted MAbs will be less immunogenic in humans than rodent or chimeric MAbs, although there is still some potential for immunological responses against the rodent-derived CDRs.

Expression of antibody fragments in bacteria is rapid and allows production in many forms

Chimeric and CDR-grafted antibodies are produced by mammalian cells transfected with the appropriate immunoglobulin genes. This is still the only way to make complete functional antibody molecules, although it is possible to make antibody fragments in bacteria. Bacterial production [12] is faster and cheaper than mammalian tissue culture systems. In *E. coli*, translation of the genes encoding the variable regions occurs in the cytoplasm. Specific signal sequences then direct the protein product to the periplasmic compartment where the environment is similar to that in the endoplasmic reticulum of a eukaryotic cell and the antibody fragment folds into a functional form. Figure 3 illustrates the antibody fragments produced in *E. coli*. Some of these can be expressed as fusion proteins linked to toxins (e.g. pseudomonas exotoxin) or enzymes to provide alternatives to the natural effector functions (see below).

Polymerase chain reaction allows the rapid cloning of immunoglobulin genes

One goal of antibody engineering is to generate antibodies entirely in vitro. Key technological advances are important in making this a reality. First, the polymerase chain reaction (PCR) allows amplification and cloning of large numbers of antibody V genes [13]. This method is very efficient and works from the RNA or DNA of B-cells at all stages of growth or differentiation. Heavy and light chain V-genes are cloned separately and information about the original pairing of the chains is lost. The V-genes are then randomly recombined to create a very large library of

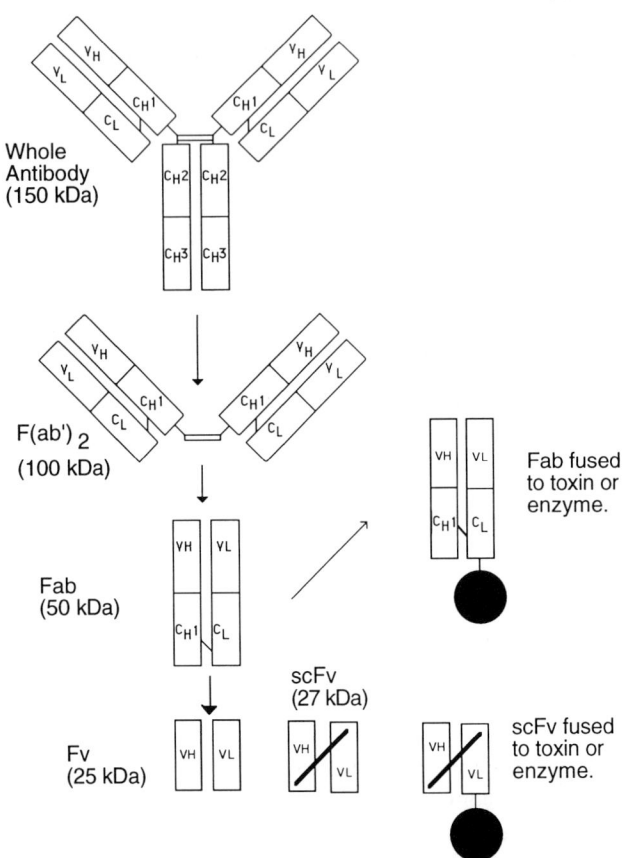

Figure 3. Antibody fragments. The available antibody fragments and their molecular weights are shown. The native antibody can be broken down into various fragments by proteolysis. More recently all these fragments have been expressed in bacteria both alone and as fusion proteins. From Hawkins RE, Russell SJ. Br Med J 1992, **305**, 1348–1350. Reprinted with permission.

paired heavy and light chains. In a mouse there are approximately 10^8 B-cells which collectively carry in the region of 10^5 different VL genes and 10^7 different VH genes. To cover every possible combination of heavy and light chains after PCR cloning and random recombination, 10^{12} clones would have to be produced and screened. Until recently, there were no methods that allowed screening of more than a million clones [14] and it was anticipated that much more powerful methods would be required to recover antibodies with optimal characteristics [15]. A technique that allows the screening of larger numbers is the display of antibody fragments on the surface of bacteriophage in fusion with one of the phage coat proteins [16].

Bacteriophage display of antibody fragments allows efficient selection for the required antibody

Filamentous bacteriophages (phage) are the bacterial equivalent of mammalian viruses and they infect bacteria but do not kill them. After infection with phage, the bacteria continue to grow and produce a large number of copies of the infecting phage. At one tip of the phage, there are a few (probably three) copies of the phage attachment protein, encoded by gene III. Antibody genes can be linked to the N-terminal codon of gene III and the modified phage genomes expressed in *E. coli*. The resulting phages display correctly folded functional antibody fragments fused to the N-terminus of the gene III protein (Figure 4a). Fortunately, the antibody fragments do not seriously compromise the infectivity of the so-called phage antibodies, which can be re-amplified in *E. coli*. Specific phage antibodies can be selected from a large diverse population by virtue of their ability to bind a particular target antigen (Figure 4b). This is similar to the way B-cells are selected when they encounter an antigen that binds their surface immunoglobulin. Phage antibodies can be expanded after one round of selection by re-infection of a bacterial culture and can then be subjected to further cycles of antigen selection and amplification. After several cycles, a relatively pure population of phage displaying antibodies of the desired specificity is obtained.

The phage system provides a rapid and effective way of producing large numbers of MAbs from immunised mice [17] or humans [18]. In the future, the development of mice transgenic for the whole human immunoglobulin locus may allow the development of human antibodies utilising immunisation. At present the most attractive way of developing human antibodies to "self" antigens is to develop an entirely in vitro system.

Using these technologies allows us to mimic the humoral immune system in vitro

Many MAbs with potential for human therapy are against self-antigens for which immunisation is not effective because controls inherent in the immune system prevent the production of high affinity anti-self antibodies. In vitro immunisation [4] is one way to avoid this, but a more attractive method is to mimic the entire natural process of antibody production in vitro. The B-cell repertoire in vivo is essentially a large random selection of paired heavy and light chains expressed on the surface of resting B-cells. Cells displaying specific antibody are selected when they encounter a foreign antigen that binds to their surface immunoglobulin. This is followed by differentiation into antibody-secreting plasma cells and memory B-cells

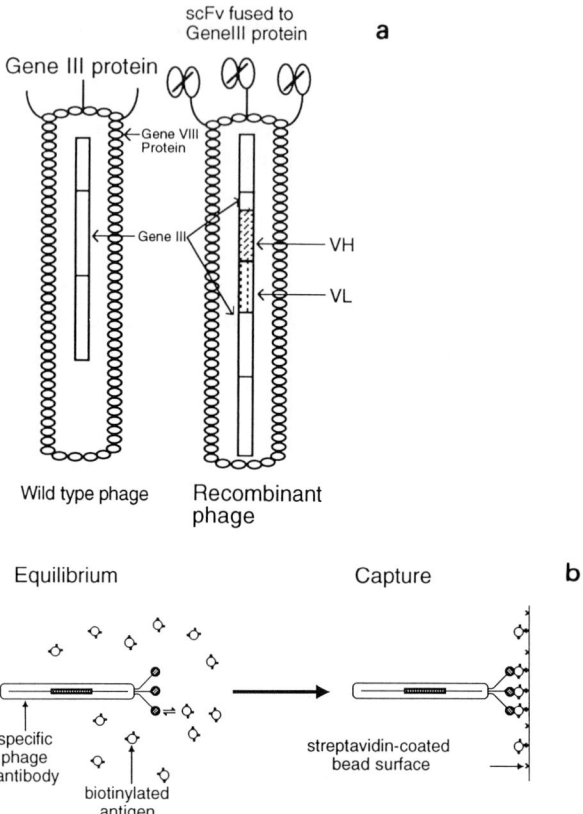

Figure 4. Phage system. (a) The wild type phage has multiple copies of the major coat protein (gene VIII protein) and usually has three copies of the gene III protein. The antibody genes (in this case an scFv) are inserted at the 5' end of gene III and encode a fusion protein with the antibody at the N-terminus. (b) The antibody is functional and can be selected according to its affinity for antigen. In this system [21], the biotinylated antigen and antibody react in solution and the phages that have bound antigen are subsequently captured using streptavidin conjugated magnetic beads. From [21]. Reprinted with permission.

which give rise to the secondary response if the same antigen is encountered again. To mimic this system in vitro, the V-genes from the peripheral blood B-cells of two healthy blood donors were cloned, randomly recombined and expressed as a phage antibody library containing 2×10^7 clones. MAbs specific for a variety of test antigens were selected from this human antibody library [19]. More importantly, the library yielded a number of human MAbs against human "self" antigens.

To date, phage antibody libraries containing up to 10^8 members have been generated but the theoretical maximum library size is much larger still. It should therefore be possible to develop phage libraries that will contain large numbers of easily accessible human antibodies specific for any antigen (Figure 5).

Bacteriophage display can be used to improve the affinity of existing antibodies

It has been shown that high affinity MAbs are more effective for cancer therapy [20]. Phage antibody libraries can be employed to mimic the natural process of affinity maturation; this can be applied to making antibodies entirely in vitro (Figure 5) or to improve an antibody made by other means. Selection of higher affinity variants from a library of randomly (PCR) mutated forms of an existing antibody [21] may thus have important therapeutic implications. MAb genes may be diversified in several other ways and used to create secondary phage antibody libraries containing higher affinity variants. For example, remixing heavy chains with new light chains to form a chain-shuffled library [18] can generate MAbs with increased affinity [22]. This process has some similarities to the repertoire shift that occurs naturally [5]. Thus, we can imitate the entire natural system of antibody production in vitro (Figure 5). Other methods of mutagenesis are available in vitro and thus there are alternative ways to try and improve the affinity or specificity of the initial antibody and make the optimal reagent for therapy. As a result of these advances, it may soon be a relatively simple matter to generate useful human antibodies against "self" antigens.

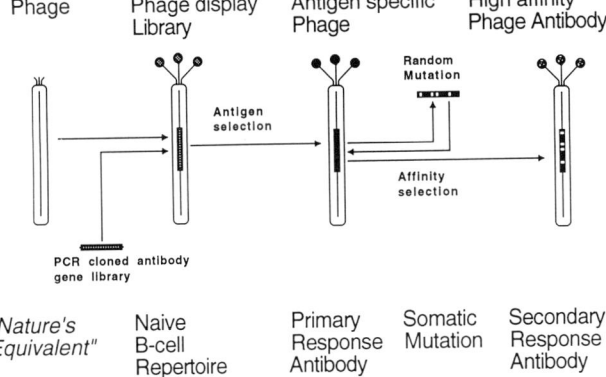

Figure 5. Antibodies in vitro. An overview comparing the phage system with that of the immune system. From Hawkins RE, Russell SJ. Br Med J 1992, **305**, 1348–1350. Reprinted with permission.

Genetic construction of antibody fusion proteins facilitates the production of targeted molecules with novel effector functions

Antibodies normally mediate their effects through the Fc portion, but fragments produced in *E. coli* cannot use this mechanism as the Fc portion expressed in bacteria is non-functional. Fab fragments expressed in bacteria can be conjugated chemically to purified Fc regions to recreate a fully functional antibody.

Bispecific antibodies, where one arm binds the target cell and the other binds the effector cells, have been used to target cytotoxic T-cells more effectively in the treatment of glioma [23]. Extensive testing of such constructs in vitro demonstrates their potential against other tumours but difficulties in large-scale production have limited their use. The recent development of simplified methods of production of bispecific antibodies [24] promise to change this. Alternatively, the introduction of (tumour-specific) antibody genes into T-cells for expression on the cell surface [25] may be an effective way to target T-cells. Furthermore, natural effector mechanisms can be boosted to improve the efficacy of an antitumour antibody. For example, antibody dependent cell mediated cytotoxicity (ADCC) can be enhanced by concomitant cytokine therapy or reinfusion of in vitro lymphokine activated killer (LAK) cells (see [26] for review).

For a variety of reasons, cancer cells can be resistant to natural antibody-mediated killing in vivo. Attempts have therefore been made to arm MAbs with alternative killing mechanisms. Radioimmunoconjugates have been used extensively for imaging purposes and in therapeutic trials. It is possible to radiolabel antibody fragments expressed in bacteria and these may have some advantages over the use of the whole molecule. For radioimmunodetection, the rapid clearance of smaller antibody fragments from the circulation may reduce background sufficiently to give a clear image of the tumour within hours rather than days. For therapy, the attraction of radioimmunoconjugates is their ability to kill tumour cells from a distance of several cell diameters. Their major limitations are that many cancers are relatively insensitive to radiation and free circulating antibody can cause significant bone marrow toxicity before binding to the tumour.

A variety of immunotoxins have been tested for anti-cancer activity. The most commonly used are plant or bacterial ribosomal toxins such as ricin or pseudomonas exotoxin. Alternatives include the use of human ribonuclease which digests RNA once internalised, and thus inhibits cell growth [27]. Chemical coupling of toxins to antibodies is straightforward but difficulties in large-scale production and poor in vivo stability limit their use. It is more convenient to link antibody fragments genetically to toxins and express them as functional fusion proteins in *E. coli* [27,28]. Preclinical tests of an anti-cancer scFv antibody fragment linked to pseudomonas exotoxin look promising [28].

Another approach is the use of antibody enzyme conjugates for prodrug activation (ADEPT) [29]. After the antibody-enzyme conjugate has reached the tumour, an inactive cytotoxic reagent is given which is converted locally (by the enzyme) to an active drug. Initial clinical testing of this concept demonstrated potential, but also showed the problems of immunogenicity of the enzyme (a bacterial enzyme was used) [30]. More recently antibody-(human) enzyme fusions have been produced and offer great promise if suitable prodrugs can be produced [31].

Combining antibodies in various ways should increase their effective target specificity

One problem with targeted therapy for cancer is that the targets are often not entirely specific to the tumour. The search continues for truly tumour specific antigens, but there are other solutions to this problem. The first is to use a cocktail of MAbs against several tumour antigens each of which is weakly expressed on a different host tissue. The overall effect will be to increase specificity but with low levels of toxicity in several tissues. The more toxic the effector mechanism, the more important it is that the targeting is precise. A second approach which could be used to improve specificity is to construct bispecific antibodies that recognise two distinct tumour antigens. Provided simultaneous expression of the two antigens is not observed in any single normal host tissue, the bispecific antibody should bind avidly to tumour cells alone, giving greatly enhanced specificity. The combination of increased specificity with very effective cytotoxic mechanisms may allow greatly improved tumour destruction with reduced toxicity to normal tissues.

The initial tissue distribution of macromolecules is critically affected by the vascular permeability

For therapy with macromolecules, it is frequently not appreciated that the amount that penetrates the tumour is dependent more on the size of the molecule, the blood flow to the tumour and the permeability of the blood vessels than on the binding characteristics of the antibody [32]. The time of exposure will also affect tissue penetration and the serum half-life is thus important; days for an IgG molecule and minutes for an Fv fragment. The optimal size is thus a trade-off between speed of tissue penetration and time of exposure. The speed at which a bound MAb dissociates from the surface of the tumour cell is an important determinant of its clinical potential and this is determined by a number of factors. Certainly its kinetic off-rate constant (K_{off}) is important. In addition, for cell surface antigens, avidity effects of bivalent or bispecific antibodies will selectively slow dissociation from cells expressing high levels of the target antigens. This results in increases in the tumour/normal tissue ratios with time after giving the antibody. Internalisation of some target antigen–antibody complexes occurs and thus effectively stops dissociation from the cell surface. This is not desirable for some effector functions (e.g. natural/ADEPT), but for immunotoxins it is a necessary prelude to cell killing.

The new techniques of antibody production solve some of the problems of antibody targeted therapy, but their clinical value has yet to be tested

It is clear from the preceding discussion that with the means to manipulate antibody genes, we are entering an era when large numbers of MAbs will be readily available in forms that until recently were hard or impossible to make. We should not forget that natural antibody responses are polyclonal and are supported by a variety of cellular functions (T-cells, phagocytic cells, etc.). This suggests that MAb cocktails will be of more value than single antibodies and that they may need to be used in conjunction with other available therapeutic modalities.

Table 2. Comparison of antibody effector mechanisms.

Feature	Effector mechanism				
	Natural antibody	Radiation	Toxin conjugate	Drug conjugate	ADEPT
Minimum size (kDa)[a]	150	25	65	25	<1
Bystander effects	Yes	Yes	No	No	Yes
Immunogenic	No	No[b]	Yes[c]	No	Yes[c]
Toxicity of unbound form	Low	High	Low[d]	Low[d]	Low[e]

All antibody components are human or humanised.
[a]Penetration of large tumours is largely a function of the size of the molecule. It is certainly poor for any molecule of 30 kDa or more [31].
[b]Antibody responses can occur to the chelating agent when this method of coupling radiation is used.
[c]Current constructs involve toxins/enzymes that are non-human and thus immunogenic, but newer forms may be made from human proteins and thus be non-antigenic.
[d]Toxicity partly depends on the method of linkage used. In vivo instability has been a problem with some methods. For toxins this can be overcome by using fusion proteins [27].
[e]The toxicity depends on the prodrug. There may be some activation of drugs in normal tissues and thus some toxicity.

If we look back at the original ideals for targeted therapy, the advantages and disadvantages of the various forms of antibodies can be examined. Many methods have not been clinically evaluated with the newly available methods of production, but a summary of the (largely) theoretical advantages and disadvantages of each is given in Table 2.

Targeted therapy for the treatment of malignant disease has been a goal for many years and MAbs currently provide the most convenient format for testing the possibility. The ultimate value of MAbs for therapy and the best format for their use will become clearer with further clinical testing, but the outlook remains optimistic.

References

1   Vreisendorp HM, Herpst JM, Germack MA et al. Phase I–II studies of yttrium-labelled antiferritin treatment for end-stage Hodgkin's disease, including radiation therapy oncology group 87-01. J Clin Oncol 1991, **9**, 918–928.
2   Köhler G, Milstein C. Continuous cultures of fused cells secreting antibody of predefined specificity. Nature (London) 1975, **256**, 495–497.
3   Crawford DH, Barlow MJ, Harrison JF et al. Production of human monoclonal antibodies to Rhesus D antigen. Lancet 1983, **i**, 386–388.
4   Borreback CAK, Danielsson L, Möller SA. Human monoclonal antibodies produced by primary in vitro immunization of peripheral blood lymphocytes. Proc Natl Acad Sci USA 1988, **85**, 3995–3999.
5   Berek C, Milstein, C. Mutation drift and repertoire shift in the maturation of the immune response. Immunol Rev 1987, **96**, 23–41.
6   Morrison SL, Johnson MJ, Herzenberg LA, Oi VT. Chimeric human antibody molecules: mouse antigen-binding domains with human constant region domains. Proc Natl Acad Sci USA 1984, **81**, 6851–6855.
7   Brüggemann M, Williams GT, Bindon CI et al. Comparison of the effector functions of human immunoglobulins using a matched set of chimeric antibodies. J Exp Med 1987, **166**: 1351–1361.
8   Meredith RF, Khazaeli MB, Plot WE et al. Phase I trial of iodine-131-chimeric B72.3 (human IgG4) in metastatic colorectal cancer. J Nucl Med 1992, **33**, 23–29.
9   Jones PT, Dear PH, Foote J et al. Replacing the complementarity-determining regions of a human antibody with those from a mouse. Nature (London) 1986, **321**, 522–525.
10 Russell SJ, Llewelyn MB, Hawkins RE. The human antibody library: Entering the next phage. Br Med J 1992, **304**, 585–586.
11 Hale G, Clark MR, Marcus R et al. Remission induction in non-Hodgkin lymphoma with reshaped monoclonal antibody CAMPATH-1H. Lancet 1988, **ii**, 1394–1399.
12 Skerra A, Pluckthün A. Assembly of a functional immunoglobulin Fv fragment in *Escherichia coli*. Science 1988, **240**, 1038–1041.
13 Ward ES, Güssow D, Griffiths AD et al. Binding activities of single immunoglobulin variable domains secreted from *Escherichia coli*. Nature (London) 1989, **341**, 544–546.
14 Huse WD, Sastry L, Iverson S et al. Generation of a large combinatorial library of the immunoglobulin library in phage lambda. Science 1989, **246**, 1275–1281.

15   Winter G, Milstein C. Man-made antibodies. Nature (London) 1991, **349**, 293–299.
16   MacCafferty J, Griffiths AD, Winter G, Chiswell DJ. Phage antibodies: filamentous phage displaying antibody variable domains. Nature (London) 1990, **348**, 552–554.
17   Clackson T, Hoogenboom HR, Griffiths AD, Winter G. Making antibody fragments using phage display libraries. Nature (London) 1991, **352**, 624–628.
18   Burton DR, Barbas CF, Persson MAA et al. A large array of human monoclonal antibodies to type 1 human immunodeficiency virus from combinatorial libraries of asymptomatic seropositive individuals. Proc Natl Acad Sci USA 1991, **88**, 10134–10137.
19   Marks JD, Hoogenboom HR, Bonnert TP et al. By-passing immunization: human antibodies from V-gene libraries displayed on bacteriophage. J Mol Biol 1991, **222**, 581–597.
20   Schlom J, Eggensperger D, Colcher D et al. Therapeutic advantage of high-affinity anticarcinoma radioimmunoconjugates. Cancer Res 1992, **52**, 1067–1072.
21   Hawkins RE, Russell SJ, Winter GJ. Selection of phage antibodies by binding affinity: mimicking affinity maturation. J Mol Biol 1992, **226**, 889–896.
22   Marks JD, Griffiths AD, Malmqvist M et al. By-passing immunisation: improving the affinity of a human antibody by chain shuffling. Bio/Technology 1992, **10**, 779–783.
23   Nitta T, Sato K, Yagita H et al. Preliminary trial of specific targeting therapy against malignant glioma. Lancet 1990, **335**, 368–371.
24   Kostelny SA, Cole MS, Tso JY. Formation of a bispecific antibody by use of leucine zippers. J Immunol 1992, **148**, 1547–1553.
25   Gross G, Waks T, Eshhar Z. Expression of immunoglobulin-T-cell receptor chimeric molecules as functional receptors with antibody-type specificity. Proc Natl Acad Sci USA 1989, **86**, 10024–10028.
26   Rosenburg SA. The immunotherapy and gene therapy of cancer. J Clin Oncol 1992, **10**, 180–199.
27   Rybak SM, Hoogenboom HR, Meade HM et al. Humanization of immunotoxins. Proc Natl Acad Sci USA 1992, **89**, 3165–3169.
28   Pastan I, Fitzgerald D. Recombinant toxins for cancer treatment. Science 1991, **254**, 1173–1177.
29   Bagshaw KD. Towards generating cytotoxic agents at cancer sites. Br J Cancer 1989, **60**, 275–281.
30   Bagshaw KD, Sharma SK, Antoniw P et al. Antibody directed enzyme prodrug therapy (ADEPT). First clinical report. In: Antibody Immunoconjugates and Radiopharmaceuticals (1991) Vol 4, 2, Order SE ed. M.A. Leibert, New York, p 204, abstract 0-11.
31   Bosslet K, Czech J, Lorenz P et al. Molecular and functional characterisation of a fusion protein suited for tumour specific prodrug activation. Br J Cancer 1992, **65**, 234–238.
32   Dvorak HF, Nagy JA, Dvorak AM. Structure of solid tumors and their vasculature: implications for therapy with monoclonal antibodies. Cancer Cells 1991, **3**, 77–85.

Molecular Biology for Oncologists
Edited by John Yarnold, Michael Stratton, Trevor McMillan
© *1993, Elsevier Science Publishers B.V. All rights reserved*

CHAPTER 22

Prospects for gene therapy of cancer

Jonathan D. Harris[a] and Karol Sikora[b]

[a]Molecular Pathology Laboratory, ICRF Oncology Group,
Royal Postgraduate Medical School, London, UK
[b] Department of Clinical Oncology, Hammersmith Hospital, London, UK

Cancer: a target for novel therapies

It is known that cancer is a somatic cell disorder where the genetic make-up of these cells has been altered resulting in clones with abnormal patterns of growth control. Oncogenes and tumour suppressor genes are part of the normal genome and these are vital in the transfer and processing (transduction) of physiological signals from outside the cell to the nucleus [1]. Consequently, any alteration to these sensitive mechanisms may result in the conversion of a cell to a cancerous phenotype. However, this is an oversimplification of the cancer process, since for no type of cancer is the molecular pathogenesis precisely understood. It is likely that a series of cooperative events are required for tumour formation. During the coming decade and the next, it is hoped that much of the human genome will have been sequenced, allowing comparisons to be made regarding the genetic make up of normal cells and those cells with abnormal growth control. Furthermore, the mechanisms of the expression of certain genes (transcriptional control) responsible for differentiation and proliferation will hopefully be elucidated. A number of different targets for gene therapy are possible, both in normal and malignant cells, as shown in Table 1 [2].

Table 1. Potential targets for gene therapy.

Normal cells
    Growth factors
    Drug resistance genes (afford protection against cytotoxic drugs)
    Cytokines
    Cell surface antigens (evoke an immune response)

Malignant cells
    Anti-sense molecules (block transcription or translation of "malignancy" signals)
    Prodrug activation (cytotoxic drug is only produced in tumour cells)
    Transcription factors (reduced tumour invasiveness)

Novel genetic based therapies possess an inherent advantage over current therapies

Cancer therapy aims to destroy as much of the tumour as possible without damaging normal host tissue to any significant extent. However, conventional therapies present a variety of problems such as drug resistance in chemotherapy, inaccessible areas for tumour surgery, etc. Since malignancy is primarily a genetic process, novel genetically based therapies may be developed. These would have an advantage over current therapeutic regimens, namely that of high tumour selectivity. Prognostic applications of the genetic analyses of oncogene alterations have already begun, such as in breast and ovarian cancer (*erb*B2 amplification) [3], neuroblastoma (N-*myc* amplification) [4] and lung adenocarcinoma (*ras* mutations) [5]. Further analysis of these and other oncogenes, as well as tumour suppressor gene alterations, will almost certainly reveal alternative targets for different forms of therapy at the molecular level.

Gene therapy and gene transfer

Principles and techniques of gene therapy

This chapter is concerned with somatic gene therapy, that is the insertion of exogenous genes into somatic cells to correct abnormalities or deficiencies of specific proteins. This can be achieved by the process of transfection whereby new genetic material is acquired by the host cell through physical or virus-dependent methods. Thus, two possibilities may occur. The new gene may be present in addition to the defective gene (addition therapy), or the new gene may replace the defective one at the same site (homologous recombination). Since the gene transfer process has oc-curred in somatic cells, the genetic information will not be passed to future gen-erations. This is different to germline gene therapy whereby genes would be intro-duced into the germline and passed on to successive generations. This specific type of therapy probably has a smaller part to play in cancer therapy and also raises serious ethical questions which are addressed later in this chapter. However, at a future date, germline gene therapy might be used against cancers that involve defective tumour suppressor genes, for example the Rb gene in retinoblastoma and the p53 gene in Li–Fraumeni syndrome [2].

Retroviruses are a most efficient vehicle for gene transfer

A number of methods are available for in vivo gene transfer, although they fall into two main categories, physical and viral. The physical category includes liposomal transfer and tissue injection [6], whilst viral-mediated gene transfer uses various vectors to carry foreign genetic information into cells. Because the vector lacks the proteins required for the correct production of an intact viral particle (i.e. gag, pol and env proteins), these must be supplied by an alternative route. This is achieved by using special packaging cells which can make all the viral proteins necessary for replication. These cells are unable to make infectious viral particles themselves and hence rely on the vector for infectivity. Once the vector is introduced into the packaging cell, that cell becomes a producer cell, making a vector particle [7,8]. Since this vector particle has had the viral "replication" genes removed, it can infect a cell once only, and hence the risk of disseminated infection is minimised (Figure 1).

Viral-mediated gene transfer relies upon using vectors that show a tropism to-wards a specific tissue or cell. The retrovirus can efficiently integrate itself into the host cell and thus transfer its genetic information into the host genome. Since much information is available regarding retroviral infection and the integration process, these vectors can be modified in such a way to carry exogenous genes into cells. Most of the gene transfer techniques currently available show limited value in somatic gene therapy since only a low number of altered target cells are produced. Consequently, most of the current gene therapy trials and protocols involve viral transfer with retroviruses. These have proved useful as their infection can be highly efficient and the dose administered may be controlled more easily than by physical methods. Apart from retroviruses, modified adenovirus and herpes simplex virus have recently been studied since they can infect specific tissues such as epithelial and neuronal cells and may be useful in diseases that affect such tissues [9]. For ex-ample, adenovirus shows high tissue tropism for respiratory epithelium and hence may be ideal for cystic fibrosis gene therapy.

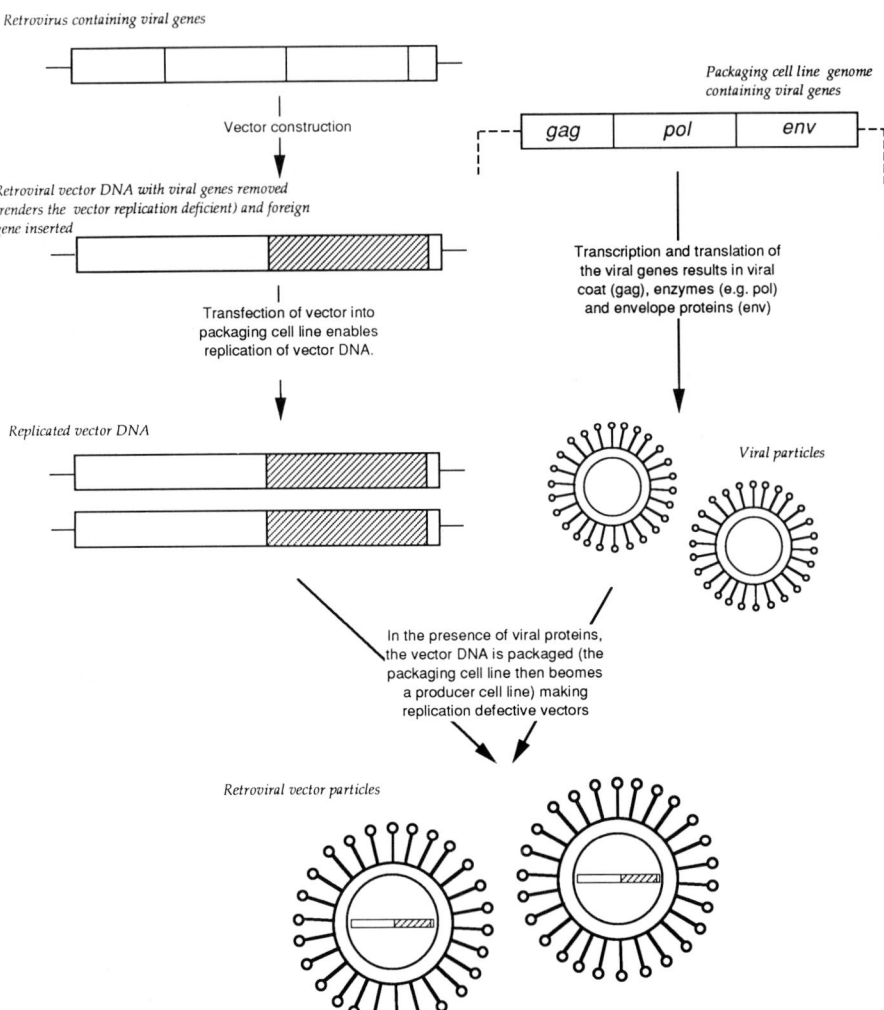

Retrovirus containing viral genes

Vector construction

Packaging cell line genome containing viral genes

| gag | pol | env |

Retroviral vector DNA with viral genes removed (renders the vector replication deficient) and foreign gene inserted

Transfection of vector into packaging cell line enables replication of vector DNA.

Transcription and translation of the viral genes results in viral coat (gag), enzymes (e.g. pol) and envelope proteins (env)

Replicated vector DNA

Viral particles

In the presence of viral proteins, the vector DNA is packaged (the packaging cell line then beomes a producer cell line) making replication defective vectors

Retroviral vector particles

Figure 1. Development and transfection of vector DNA constructs for retroviral mediated gene transfer. The vector is constructed so that it not only contains the foreign gene of interest, but also lacks viral proteins necessary for the production of intact viral particles (these are provided by the packaging cell line). This cell line lacks the "packaging signal" required to make infectious viral particles and so relies on the retroviral vector for this information. When the vector is transfected into the packaging cell line, replication of the vector DNA is performed by the cell line's replication machinery and viral particle proteins are transcribed and produced by the cell line itself. The vector DNA is then preferentially packed to form vector particles.

Expression of a foreign gene may be selectively increased

A number of techniques have been developed to improve the expression of genes once they have been inserted into their target cells. For example, the use of selective promoters can force the efficient transcription of transfected genes. The efficiency of the insertion process can usually be detected with a reporter gene, such as that which codes for neomycin phosphotransferase (confers resistance to the aminoglycoside antibiotic G418). The presence of a reporter gene in the vector allows for the selection of transduced cells and thus enhances the success of transfer [10]. Early trials using viral mediated gene transfer used such marker genes to track immune cells to assess the feasibility of gene transfer for malignant disease.

Gene therapy – past and present

The first experiment involving the manipulation of human genes began in 1989. This was not a gene therapy experiment as such, but involved the "marking" of immune cells to assess the possibility of gene transfer into human cells. Tumour infiltrating lymphocytes (TILs) are special immune system T-cells that migrate specifically to tumours. They can be isolated and grown in large numbers using interleukin-2 (IL-2), the growth factor for T-cells. In studies prior to 1989, approximately 40% of patients infused with these "anti-tumour" lymphocytes responded to such a treatment [11]. It was decided that it would be useful to follow the progress of the TILs once in the body and to calculate how many of these infused cells penetrated the tumour. The gene marking experiment involved using a retrovirus to transfer the neomycin resistance gene (*neo*-R) into TILs, previously isolated from patients with malignant melanoma [12]. These cells were re-infused into the same patient and blood samples taken at intervals after the re-infusion were analysed for TILs that showed resistance to neomycin. Although no clinical benefit was gained by the patients involved, significant information was produced regarding the gene transfer process. Since 1989, a number of other gene marking experiments have been approved or initiated to investigate gene transfer in TILs against melanoma (see below) and other cancers, for example neuroblastoma, chronic myeloid leukaemia and acute lymphocytic leukaemia [13,14].

The first human gene therapy trial was initiated on the 14th September 1990. Patients with adenosine deaminase (ADA) deficiency were chosen, since the disease results from a single gene defect. The disease primarily affects the T-cells of the immune system, resulting in a severe combined immunodeficiency (SCID) and is lethal unless corrected with a bone marrow transplantation (since the ADA enzyme is not present in patients with ADA deficiency, high levels of deoxyadenosine, toxic to both T- and B-cells, are produced) [15]. In the trial, T-cells were isolated from patients with ADA deficiency and were transduced with a retrovirus containing the ADA gene (and *neo*-R as a selectable marker gene). The antibiotic-resistant T-cells, containing the ADA gene, were isolated and grown in vitro (a process known as *clonal expansion*) and then were infused back into the patients. The results from the trial have proved to be encouraging. For example, a number of ADA+ T-cells (T-cells with the ADA gene) have remained in one of the patients for over 6 months since the infusions were stopped.

Variations on the theme

Cytokine therapy

There are a number of other techniques currently under development that may be considered as variations of classical gene therapy (i.e. gene replacement). It is known that cytokines can have a significant effect on tumour cell growth clinically, although they can cause unpleasant systemic side effects and have very short half-lives. Many current programmes involving gene therapy are using the vectoring of cytokine genes with cells that can home in specifically on tumours. A number of possible modifications are possible, including the transfection of cytokine genes into immune cells of the host normal or tumour cells (e.g. interleukin-2, in-terleukin-4, interferon and tumour necrosis factor (TNF)). The expression of cytokine genes in transfected cells may reduce their tumorigenicity and/or metastatic potential. Anti-tumour responses can be enhanced using cytotoxic lymphocytes (such as TILs), macrophages or antibodies. These could be induced by the expression of specific cytokines of tumour cells. Hence, it may be possible to convert a weakly immunogenic tumour, one that would only elicit a minor response, to a strongly immunogenic tumour that would result in a powerful destructive response [2].

The use of gene marking studies on TILs in melanoma patients has already been discussed. In addition, TNF genes have been transfected into TILs by Rosenberg and colleagues at the National Institutes of Health, USA. The rationale behind the experiments was that the secretion of TNF by the lymphocytes near the tumour site would enhance tumour destruction [16]. TNF has produced encouraging anti-tumour results in mice, although clinically the therapy has proved to be less successful. This is due to the large differential toxicity of the treatment between mice and humans. If it were possible to achieve a higher local concentration, the response might prove to be more promising. Clinical trials on 50 patients with melanoma using TNF-transfected TILs are currently in progress with the results expected by the middle of 1993.

Drug targeting and enzyme prodrug therapies

Perhaps the most promising of the gene therapies discussed in this chapter is VDEPT (virally directed enzyme prodrug therapy). This relies on a cellular gene being specifically expressed in cells of a particular tissue or tumour type, but not in normal cells. The vector carries a foreign gene coding for an enzyme that converts a harmless compound to a toxic drug. The transcription of this foreign gene is driven by the promoter of the cellular gene. Since the foreign gene is only expressed in tumour cells, the cytotoxic effects of the therapy on normal cells is minimised. Many tumours show transcription of certain host genes (markers) that are not required for survival, as outlined in Table 2. As more of these markers are discovered by molecular biologists, better VDEPT strategies may be developed. The best example of a current VDEPT system is that being developed by Brian Huber and colleagues in the United States. The model they chose was hepatocellular carcinoma since, in these tumours, the gene coding for α-fetoprotein is often expressed (normal liver cells, and most other cells of the body, do not express this gene). In the experiments, an artificial vector was produced that contained the α-fetoprotein promoter and the gene coding for thymidine kinase [17]. This enzyme was used since it can convert the relatively harmless compound 6-methoxypurine arabino-nucleoside, ara-M, to the phosphorylated cytotoxic agents, araAMP, ADP and ATP. When the liver cells were challenged with ara-M, only the hepatocellular carcinoma cells that expressed α-fetoprotein were destroyed.

Other promising investigations are currently under development, including utilising the conversion of the anti-fungal drug 5-fluorocytosine (5-FC) to the cytotoxic compound 5-fluorouracil (5-FU) by the enzyme cytosine deaminase (Figure 2A). This has the significant advantage over the previously described VDEPT stra-

Table 2. Examples of non-essential genes that are often overexpressed in tissues or tumours: selective targets for VDEPT (modified from [2]).

Selective gene marker	Tissue/tumour[a]
CEA (carcinoembryonic antigen)	GI, lung
α-Fetoprotein	Hepatocellular carcinoma
Villin	GI, pancreas
erb-B2	Breast, GI
erb-B3	Breast, GI
Dopa decarboxylase	SCLC, neuroectoderm
Calcitonin	Thyroid (medullary)
Thyroglobulin	Thyroid
Polymorphic epithelial mucin	Breast, pancreas

[a]GI, gastrointestinal; SCLC, small cell lung cancer.

tegy in that both 5-FC and 5-FU are in routine clinical use and hence much of their pharmacokinetics and other pharmacological data is known. A possible strategy using this system is shown in Figure 2B.

Anti-sense agents – the information blockers

Anti-sense oligonucleotides are small synthetic nucleotide molecules constructed so that they are complementary to specific DNA or RNA sequences. They are small, typically ranging in size between 15 and 20 nucleotides long, although some

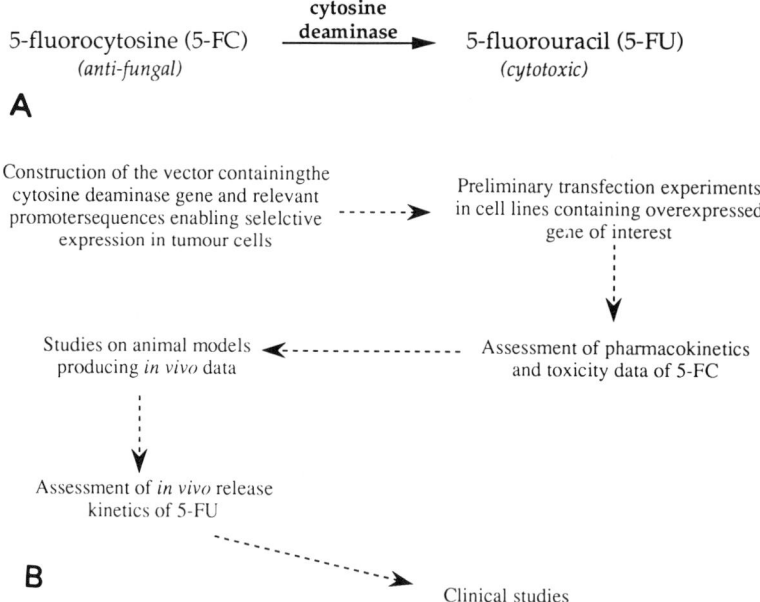

A

B

Figure 2. (A) Prodrug activation by cytosine deaminase. The non-mammalian enzyme cytosine deaminase normally converts cytosine to uracil via a deamination reaction. As a consequence, this enzyme will also deaminate 5-fluorocytosine (5-FC) to 5-fluorouracil (5-FU). Since 5-FC is relatively harmless to mammalian cells (it is clinically used as an anti-fungal), the conversion to the cytotoxic 5-FU represents a prodrug activation system that could be used in a VDEPT strategy. (B) A possible VDEPT strategy using the cytosine deaminase (CD) gene to convert 5-FC into the cytotoxic 5-FU might typically follow the above developmental programme. A retroviral vector would be constructed to deliver the CD gene and any relevant promoter sequences into the target cells (tumour). Various in vitro experiments concerning the efficacy of the vector and the conversion of 5-FC to 5-FU would be performed and subsequent in vivo models would provide valuable pharmaco-kinetic and toxicological information. If these experiments were successful, clinical studies would be initiated to assess the possible use of this VDEPT strategy in humans.

workers have reported successes with both smaller and larger oligomers. When these anti-sense molecules bind to their target sequences, they selectively inhibit the transcription or translation of those genes. If these inhibited genes are associated with disease states, then the subsequent down-regulation may prevent that disease and might result in a reversal of clinical symptoms. The location of mRNA molecules in the cytoplasm of cells provides an easier target compared to DNA molecules which would involve movement through the nuclear matrix. A number of anti-sense oligomers have been shown to have an anti-proliferative activity:

- c-*myc* in lymphoma cell lines [18]
- N-*myc* in neuroectodermal cell lines [19]
- c-*myb* in colon adenocarcinoma [20]
- *bcr-abl* in chronic myeloid leukaemia [21]
- c-*raf*1 in ras and raf-transformed NIH3T3 cells [22]

It has also been shown that these anti-sense agents can decrease the tumorigenicity and metastatic potential of cell lines. For example, Ki-*ras* proto-oncogene was transduced, in anti-sense orientation, into a small-cell lung cancer cell line expressing mutant Ki-*ras* (point mutation). This resulted in an inhibition of the mutant Ki-*ras* expression and the prevention of tumour growth in nude mice [23]. It has also been shown that the down-regulation of putative metastasis suppressor genes such as E-cadherin, by these anti-sense oligomers, results in the cells becoming invasive [24]. These agents thus represent a powerful method with which to analyse tumour invasion. Results have been published using a variety of cell lines and animal models and they indicate the potentially wide applications for the manipulation of abnormal growth and behaviour patterns of tumour cells.

Safety aspects and the ethics of gene therapy

In its simplest form, gene therapy is much like any other kind of therapy, and has its inherent risks. Any risk to the patient must be balanced by corresponding benefit. If no alternative method of therapy is available and the likelihood of gain is small, then it will still be possible to carry out the therapy on that patient (since some benefit is better than none). Therefore gene therapy must be assessed by the same criteria used for other clinical protocols. In addition, since this type of therapy is new, only clinical trial data can provide information regarding the efficacy of the treatment and may reveal any unexpected complications.

There is no doubt that gene therapy will remain a controversial therapy for some time. There are the patients for whom benefit may be gained that might not be available using conventional treatments. On the other hand, there is the sinister possibility of some workers abusing gene therapy to genetically engineer humans of a desired form, the "Brave New World" scenario. This is compounded with the controversy surrounding the potential for germline therapy, although no protocols have been proposed yet. It is possible that in the future one may alter the characteristics of the unborn child and this may lead to abuse of the method, for example changing desirable traits, such as hair colour, skin colour and possibly the personality [25]. There is much current debate surrounding this particular type of gene therapy and it is likely that this will continue for some time.

Future developments and prospects for gene therapy

In only a short time, the dream of gene transfer into humans has become a reality. Early clinical methods have depended on removing the cells from the body and performing in vitro transfection. Techniques are currently being developed that will carry out this process in vivo. A number of potential gene transfer and other gene therapy protocols will also be developed over the next few years, for example using alternative viral vectors in gene transfer or producing effective anti-sense therapies. As recombinant DNA technology advances and more is learnt of gene expression in human cells, additional targets for gene therapy may also be found.

Taking all the information described in this chapter into account, it is clear that although many different types of gene therapy are under way, this novel treatment is still in its infancy and has a long way to progress before fully effective therapies are implemented. As this new type of treatment begins its slow transition from infancy to adulthood, lessons will be learned and success may be coupled with disappointment. From the early clinical studies, however, it is clear that gene therapy is here to stay and future developments in this field may provide a cure for some types of cancer.

References

1 Bishop JM. Molecular themes in oncogenesis. Cell 1991, **64**, 235–248.
2 Gutierrez AA, Lemoine NR, Sikora K. Gene therapy for cancer. Lancet 1992, **339**, 715–721.
3 Borg A, Baldetrop B, Ferno M et al. *Erb-B*2 amplification in breast cancer with a high rate of proliferation. Oncogene 1991, **6**, 137–143.
4 Brodeur G, Seeger RC, Schwab M et al. Amplification of N-*myc* in untreated neuroblastoma correlates with advanced disease stage. Science 1984, **24**, 1121–1124.
5 Slebos RT, Kiberlaan RE, Paledio O et al. K-ras oncogene activation as a prognostic marker in adenocarcinoma of the lung. N Engl J Med 1990, **323**, 561–565.
6 Bevinisty N, Reshef L. Direct introduction of genes into rats and expression of the genes. Proc Natl Acad Sci USA 1986, **83**, 9551–9555.

7 Miller AD. Retrovirus packaging cells. Hum Gene Ther 1990, **1**, 5–14.
8 Cornetta K. Safety aspects of gene therapy. Br J Haematol 1992, **80**, 421–426.
9 Geller AI, Keyomarsi K, Bryan J, Pardee AB. An efficient deletion mutant packaging system for defective herpes simplex virus vectors: potential applications to human gene therapy and neuronal physiology. Proc Natl Acad Sci USA 1990, **87**, 8950–8954.
10 Kohn DB, Kantoff PW, Riglitis MA et al. Retroviral mediated gene transfer into mammalian cells. Blood 1987, **13**, 285–298.
11 Rosenberg SA, Packard BS, Aebersold PM et al. Use of tumor infiltrating lymphocytes and interleukin-2 in the immunotherapy of patients with metastatic melanoma: a preliminary report. N Engl J Med 1988, **319**, 1676–1680.
12 Rosenberg SA, Aebersold P, Cornetta K et al. Gene transfer into humans: immunotherapy of patients with advanced melanoma using tumour infiltrating lymphocytes modified by retroviral gene transduction. N Engl J Med 1990, **323**, 570–578.
13 Anderson WF. Human gene therapy. Science 1992, **256**, 808–813.
14 Autologous bone marrow transplant for children with AML in first complete remission: use of marker genes to investigate the biology of marrow reconstitution and the mechanism of relapse. Hum Gene Ther 1991, **2**, 137–159.
15 Miller AD. Human gene therapy comes of age. Nature 1992, **357**, 455–460.
16 TNF/TIL human gene therapy protocol. Hum Gene Ther 1990, **1**, 441–480.
17 Huber BE, Richards CA, Krenitsky TA. Retroviral-mediated gene therapy for the treatment of hepatocellular carcinoma: an innovative approach for cancer therapy. Proc Natl Acad Sci USA 1991, **88**, 8039–8043.
18 McManaway ME, Neckers LM, Loke SL et al. Tumour specific inhibition of lymphoma growth by an antisense oligodeoxy-nucleotide. Lancet 1990, **335**, 808–811.
19 Whitesell L, Rosolen A, Neckers LM et al. Episome-generated N-*myc* antisense RNA restricts the differentiation potential of primitive neuroectodermal cell lines. Mol Cell Biol 1991, **11**, 1360–1370.
20 Melani C, Rivoltini L, Parmiani G et al. Inhibition of proliferation by c-myb antisense oligodeoxynucleotides in colon adenocarcinoma cell lines that express c-myb. Cancer Res 1991, **51**, 2897–2901.
21 Szcylik C, Skorski T, Nicolaides NC et al. Selective inhibition of leukaemia cell proliferation by bcr-abl antisense oligodeoxynucleotides. Science 1991, **253**: 562–568.
22 Kolch W, Heidecker G, Lloyd P, Rapp UR. Raf-1 protein kinase is required for growth of induced NIH/3T3 cells. Nature 1991, **349**, 426–428.
23 Mukhopadhyay T, Tainsky M, Cavender AC, Roth JA. Specific inhibition of K-ras expression and tumorigenicity of lung cancer cells by antisense RNA. Cancer Res 1991, **51**, 1744–1748.
24 Vleminck K, Vakaet L, Mareel M et al. Genetic manipulation of E-cadherin expression by epithelial tumour cells reveals an invasion suppressor role. Cell 1991, **66**, 107–119.
25 Lebo RV, Golbus MS. Scientific and ethical considerations in human gene therapy. Baillieres Clin Obstet Gynaecol 1991, **5**, 697–713.

Molecular Biology for Oncologists
Edited by John Yarnold, Michael Stratton, Trevor McMillan
© *1993, Elsevier Science Publishers B.V. All rights reserved*

CHAPTER 23

Introduction to gene structure and expression

Richard Wooster

Section of Chemical Carcinogenesis, Institute of Cancer Research,
15 Cotswold Road, Sutton, Surrey, SM2 5NG, UK

Introduction

The aim of this chapter is to introduce aspects of gene structure and regulation. While it is not possible to give an exhaustive account, a number of topics are discussed that are pertinent to the current field of molecular oncology. The first of these is the structure of the DNA molecule and how it is organised into genes. The second concerns transcription and the factors responsible for its regulation. The third topic concerns the processing of transcribed RNA, and the final section discusses the topic of DNA polymorphism.

The structure of DNA

Nucleic acids consist of a sugar, a phosphate and a nitrogenous base

Deoxyribonucleic acid (DNA) is a chain of chemically linked nucleotides. The nucleotides consist of a sugar, a phosphate group and a nitrogenous base. The sugar and phosphate groups form the backbone of the DNA molecule. In DNA, the sugar is deoxyribose, while in RNA the sugar is ribose. In DNA, the nitrogenous bases are adenine (A), guanine (G), cytosine (C) and thymine (T). In RNA, uracil (U) replaces thymine (T). From their chemical structure, adenine and guanine are defined as purines, while cytosine, thymine and uracil are pyrimidines.

The DNA molecule is a double stranded helix

The DNA in a cell is double-stranded, forming a double helix where one nitroge-nous base forms weak bonds with another on the opposite strand to form a base pair (bp). In general, adenine pairs with thymine and cytosine pairs with guanine, and therefore, one strand is complementary to the other. By convention, one end of the molecule is defined as 5' while the other end is 3', and the sequence is written in the 5' to 3' direction (5' and 3' derive from the numbering of the carbon atoms in the sugar molecule). When a gene is present within the DNA sequence, the se-quence is written such that the DNA coding for the amino terminus of the protein is to the left of the page (the 5' end of the gene), while the carboxy terminus is to the right of the page (the 3' end of the gene).

DNA codes for proteins

The four bases that store the genetic information in DNA must code for the 20 amino acids that are used in proteins. A sequence of three bases, called a codon, is specific for a particular amino acid. There are a total of 64 different codons. Of these, three (TGA, TAG and TAA) encode signals which stop translation of mRNA into protein (stop codons). The remaining triplet codons encode the various amino acids. Some amino acids are coded by only one codon while others have up to six different codons. The codon ATG is unique in that it is present at the start of every coding region. This codon also codes for the amino acid methionine and is therefore also present within the coding regions.

The human genome contains 6×10^9 base pairs

The diploid human genome contains approximately 6 000 000 000 bp which are divided into 46 chromosomes. Each genome contains two copies of each autosome (numbered from 1 to 22) and two sex chromosomes (called X and Y). The haploid human genome of 23 chromosomes contains about 100 000 genes. The average size of a gene at the DNA level is about 10 000 bp, i.e. 10 kb (see Table 1). Therefore, only 30% of the genome is covered by genes. Furthermore, only 2–3 kb of the average gene codes for protein. Thus, only 5–10% of the human genome codes for proteins. The known roles of the remaining 90–95% of the genome in-clude the maintenance of the structure of the genome itself and the regulation of gene transcription (see below).

Genes contain exons and introns

Thousands of genes have been cloned and sequenced and it is possible to describe an idealised gene (Figure 1). The DNA 5' to the coding region usually contains se-quences that control the transcription of the gene (see below). The gene itself is

Table 1. A comparison of the size of genes, the number of exons in the gene and the size of the processed messenger RNA.

Gene	Number of exons	Gene size (kb)	mRNA size (kb)
Dihydrofolate reductase	6	31	2
DMD	>60	2000	14
RB1	27	250	4.6
WT1	10	50	3
β-Globin (mouse)	3	0.85	0.59

DMD, Duchesnes muscular dystrophy; RB1, retinoblastoma susceptibility gene; WT1, Wilm's tumour gene.

divided into two distinct elements, introns and exons. The exons contain the coding region of the gene. The introns are removed from the primary transcript (see RNA splicing) to leave the exons. The coding region has a start codon towards the 5' end of the gene with one stop codon at the 3' end of the coding region. The sequence between the start and stop codon constitutes an open reading frame (ORF), and this is flanked by 5' and 3' non-coding sequence. The number of exons and introns within a gene varies between 1 and over 60 (see Table 1).

Transcription

The transcription of DNA produces RNA

DNA is transcribed into RNA by DNA-dependent RNA polymerases (RNA pol). These polymerases are part of a larger aggregate of proteins that moves in a 5' to 3'

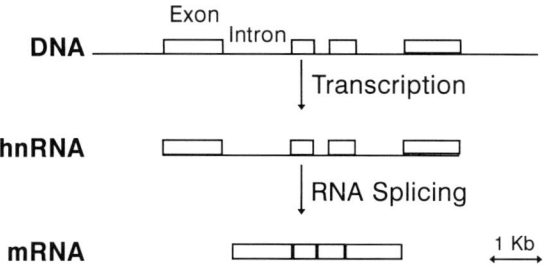

Figure 1. The intron/exon structure of a idealised gene. The alternating exon/intron structure of an idealised gene is shown at the top of the diagram. This gene is transcribed to produce the hnRNA shown and the introns are removed by RNA splicing.

280   R. Wooster

Table 2. The numbers of specific motifs in known promoters.

	TATA	CAAT	GC
SV40 early promoter	0	0	6
Thymidine kinase promoter	1	1	2
Histone 2B promoter	1	2	0

direction along the DNA from a point of initiation to a point of termination. There are three RNA polymerases called RNA pol I, II and III. RNA polymerase II transcribes genes that code for proteins (pol I transcribes ribosomal genes and pol III transcribes transfer RNA genes). The activity of RNA polymerase II, and hence the rate of transcription and expression, is tightly controlled. The control involves two major components: the presence of specific DNA sequences upstream or downstream of the gene, and proteins that recognise and bind to these specific sequences. The DNA sequences are called promoter and enhancer sequences. The proteins that bind to them are called transcription factors.

Promoter and enhancer regions are segments of DNA that are involved in the regulation of transcription

A promoter consists of a number of short conserved nucleotide sequences (motifs) located 5' (upstream) of a gene which functions by binding transcription factors (see below). The same DNA motif is often found in many different promoter sequences. For example, one of the motifs contains the consensus sequence TATA, called the TATA box. This four-nucleotide sequence is usually located 25 bp upstream (5') of the first base to be transcribed (the transcription start site). Other common nucleotide motifs include GC and CAAT boxes. While some of these sequences are required for transcription initiation, they are not all essential and the number and types of motifs adjacent to individual genes are variable (see Table 2).

An enhancer is a conserved sequence that may be positioned from a few hundred bases to tens of kilobases from the start point of transcription of a gene. In contrast to promoters, enhancer regions can be positioned either upstream or downstream of a gene (i.e. 5' or 3') and can function with the nucleotide sequence in either orientation.

Transcription factors are proteins that bind to specific DNA sequences within promoters and enhancers

Many proteins are capable of binding to DNA but not all these are transcription factors. For example, histone proteins bind to DNA throughout the genome and help to organise the architecture and packaging of the chromatin (see Chapter 24). By contrast, transcription factors bind specifically to DNA sequences within promoters and enhancers. Some transcription factors bind to most promoter regions and are essential for transcription. Others only bind to promoters and enhancers associated with particular genes and ensure tissue-specific expression of that gene. Examples of transcription factors and the DNA sequence to which they bind are given in Table 3.

Transcription factors often share short amino acid sequences that are important for binding to DNA

There are a number of DNA binding motifs in the protein sequences of transcription factors. The Wilm's tumour suppressor gene product contains four repeating protein sequences that bind specifically to DNA. Each protein sequence incorporates a zinc ion that is hydrogen-bonded to two cysteine and two histidine residues. This sequence is referred to as the zinc finger. Other transcription factors containing zinc fingers include the steroid hormone receptors, e.g. the oestrogen receptor protein.

Another motif is the leucine zipper. This consensus contains a leucine amino acid every seven residues along the polypeptide chain of the transcription factor. It is thought that the leucines allow the transcription factors to form heterodimers and homodimers. Dimerisation is thought to occur before the factors can bind to the DNA. One example of the leucine zipper structure is the formation of Jun–Jun homodimers and Jun–Fos heterodimers. Jun and Fos proteins are oncogene products that are components of an important complex of transcription factors known as AP1. Finally, the helix-turn-helix motif of the homeobox transcription factors is another common and important protein sequence which interacts specifically with regulatory sequences in DNA.

Table 3. A comparison of the DNA binding motifs in transcription factors and the sequences to which they bind.

Transcription factor	DNA binding motif	Module name	DNA consensus sequence
TFIID	Zinc finger	TATA	TATAAAA
SP1	Zinc finger	GC	GGGCGG
AP1	Leucine zipper	AP1	
p53	Acidic domain	PBS	(A/T)GPyPyPy

The processing of RNA

RNA splicing leads to the removal of introns from the transcribed RNA

The initial transcript of a gene is called heterogeneous nuclear RNA (hnRNA) and is composed of both introns and exons. During the process known as RNA splicing, the hnRNA is cut and subsequently rejoined, without the introns, to create messenger RNA (mRNA). The junctions between introns and exons contain short consensus signal sequence (see Figure 2). These sequences act as a recognition signal for proteins that bind to these sites and perform the DNA cutting and splicing reactions. Naturally occurring or artificially introduced mutations within these conserved splice signal sequences reduce or remove the ability of the hnRNA to undergo splicing. Furthermore, the introduction of new splice signal sequences can lead to a change in the natural pattern of splicing.

Alternative splicing produces several mRNAs from a single precursor hnRNA

Most hnRNAs are processed to produce a single-sized mRNA. However, some hnRNAs are processed to produce multiple mRNAs, where the difference between them is the presence or absence of certain exons. For example, there is only one WT1 (Wilm's tumour) suppressor gene. However, two alternative splice sites lead to the possibility of four distinct WT1 mRNAs coding for proteins with distinct functions. At one of these splice sites, the splicing reaction can lead to a 17 amino acid insert towards the N-terminus of the protein. At the other site, the splicing reaction can lead to the insertion of three amino acids between two of the zinc fingers in the DNA binding region. All four variants are found in the tissues in which WT1 is expressed, where they are thought to contribute to the normal function of the WT1 gene product.

The degradation of mRNA is prevented by sequences at the ends of the RNA molecule

mRNA molecules terminate in a GT region which is followed by the nucleotide sequence AAUAAA at the 3' end of the gene. The AAUAAA sequence is called a

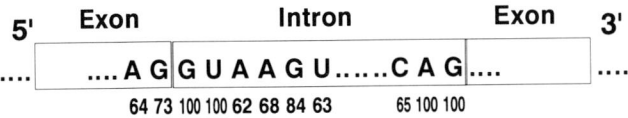

Figure 2. The consensus splice sequences. The intron/exon boundaries have been amplified to reveal the DNA sequence surrounding these junctions. The numbers below the DNA sequences shown the percentage of known sequences that have this consensus.

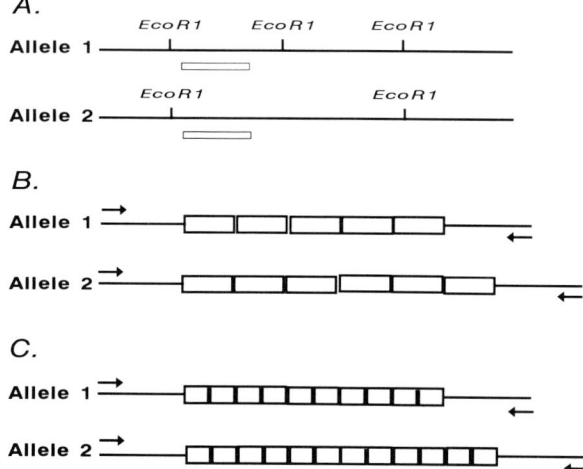

Figure 3. The structure of DNA polymorphisms. (A) RFLPs. The solid line represents DNA sequence. *Eco*R1 is the restriction endonuclease that is used in this example. The vertical lines indicate the positions at which this enzyme will cleave the DNA. The differential cleavage pattern can be visualised using the probe indicated by the open box. (B) Minisatellites. The solid lines represent "random" DNA sequences. The open box shows a repeat unit (6–50 bp) and this unit is present 5 times in allele 1 and 6 times in allele 2. These repeats can be amplified using the primers shown by the arrows and the PCR. (C) Microsatellites. As for (B) except that the repeat unit is smaller (2–5 bp) and the number of repeat units are 10 and 12 for alleles 1 and 2, respectively.

polyadenylation signal, and it stimulates a protein complex to add a chain of adenosines to the 3' end of the mRNA. This polyA tail is thought to prevent degradation of the molecule from the 3' terminus. The 5' terminus is protected by a "cap" that constitutes a 5'–5' bond structure (the usual DNA and RNA bonding structure is a 5'–3' bond). The centre of the RNA message is still susceptible to degradation and processing.

DNA polymorphisms

Although the human genome is almost identical from one person to another, there are small variations in DNA sequences between individuals. These are usually in non-coding DNA, either in the extensive regions of DNA between genes or within introns. Several types of sequence variation are recognised, including restriction site polymorphism, minisatellites and microsatellites (see below). They have an important role in genetic linkage analysis and in the study of DNA deletions in tumours (e.g. loss of heterozygosity)

Restriction endonucleases can be used to detect restriction fragment length polymorphisms

Over 100 enzymes called restriction endonucleases have been isolated from bacteria and are used in the laboratory to cleave DNA at defined nucleotide sequences. If a single base sequence variation changes a restriction enzyme cleavage site, the enzyme will not cut the DNA there (see Figure 3A). It is therefore possible to detect differences in homologous DNA sequences between two alleles if a single base change in one allele falls within a restriction site. This type of polymorphism is known as a restriction fragment length polymorphism (RFLP). A single base polymorphism occurs approximately every 200 base pairs, although not all fall within restriction enzyme recognition sites.

Minisatellites contain a variable number of tandem repeats

Minisatellites consist of short sequences of DNA repeated a number of times in tandem array (see Figure 3B). The basic sequence can be from 6 bp to over 50 bp in length. The number of times the basic sequence is repeated varies from allele to allele (from 10 to more than 60 times) and gives rise to a high level of polymorphism in the population. These repeats are called minisatellites, or variable number of tandem repeats (VNTRs). They may be analysed by Southern blotting and DNA hybridisation or by the polymerase chain reaction (see Figure 3B). These repeats occur throughout the genome, although they tend to cluster towards the telomeres of chromosomes. In general, VNTRs are more polymorphic than RFLPs because the latter generally have only two forms.

Microsatellites are minisatellites that contain small repeat units

Microsatellites are VNTRs in which the tandem repeat sequence is from 1 to 5 bp long. One of the most widely exploited is the dinucleotide GT repeat (Figure 3C). These are analysed using the polymerase chain reaction. It has been estimated that there are 50 000 to 100 000 microsatellites in the human genome, of which about 2000 have been characterised, i.e. located. As with VNTRs, microsatellites are generally more polymorphic than RFLPs.

References

General

Lewin. Genes IV. Oxford University Press, Oxford, 1990.
Watson JD, Hopkins NH, Roberts JW et al. Molecular Biology of the Gene. Benjamin/
    Cummings, Menlo Park, CA, 1987.

Specific

DNA structure: Watson JD, Crick FHC. Clinical implications of the structure of DNA. Nature 1953, **171**, 964–967.

Introns and exons: Breathnach R, Chambon P. Organization and expression of eucaryotic split genes coding for proteins. Annu Rev Biochem 1981, **50**, 349–383.

Enhancers and promoters: Muller MM, Gerster T, Schaffner W. Enhancer sequences and the regulation of gene transcription. Eur J Biochem 1988, **176**, 485–495.

Aoyama N, Nagase T, Sawazaki T et al. Overlap of the p53-responsive element and cAMP-responsive element in the enhancer of human T-cell leukemia virus type I. Proc Natl Acad Sci USA 1992, **89**, 5403–5407.

Transcription factors: Evans RM, Hollenberg SM. Zinc fingers: guilt by association. Cell 1988, **52**, 1–3.

Evans RM. The steroid and thyroid hormone receptor superfamily. Science 1988, **240**, 889–895.

Little MH, Prosser J, Condie A et al. Zinc finger point mutations within the WT1 gene in Wilms tumor patients. Proc Natl Acad Sci USA 1992, **89**, 4791–4795.

RNA splicing: Sharp PA. Splicing of messenger RNA precursors Science 1987, **235**, 766–771.

Niwa M, MacDonald CC, Berget SM. Are vertebrate exons scanned during splice-site selection? Nature 1992, **360**, 277–280.

Alternative splicing: Haber DA, Housman DE. Role of the WT1 gene in Wilms' tumour. Cancer Surv 1992, **12**, 105–117.

RFLPs: White R, Leppert M, Bishop DT et al. Construction of linkage maps with DNA markers for human chromosomes. Nature 1985, **313**, 101–105.

Minisatellites: Jeffreys AJ, Wilson V, Thein SL. Hypervariable 'minisatellite' regions in human DNA. Nature 1985, **314**, 67–73.

Microsatellites: Weber JL, May PE. Abundant class of human DNA polymorphisms which can be typed using the polymerase chain reaction. Am J Hum Genet 1989, **44**, 388–396.

Litt M, Luty JA. A hypervariable microsatellite revealed by in vitro amplification of a dinucleotide repeat within the cardiac muscle actin gene. Am J Hum Genet 1989, **44**, 397–401.

Molecular Biology for Clinical Oncologists
Edited by John Yarnold, Michael Stratton, Trevor McMillan
© 1993, Elsevier Science Publishers B.V. All rights reserved

CHAPTER 24

Chromatin structure and nuclear function

A. Bassim Hassan and Dean A. Jackson

Sir William Dunn School of Pathology, University of Oxford,
South Parks Road, Oxford, OX1 3RE, UK

Introduction

In diploid mammalian cells, 6×10^9 base pairs (bp) of DNA are folded to occupy a nucleus measuring roughly 10 µm across. To achieve this, DNA as a DNA/protein complex, referred to as chromatin, must fold into a series of higher order arrays. The degree of condensation varies during the cell cycle. During interphase, the chromatin is relatively decondensed, more decondensed chromatin correlating with gene expression. The only apparent features at this stage are variations in chromatin density and a small number of highly structured centres of ribosomal RNA synthesis and ribosome synthesis (nucleoli). During cell division, chromatin condenses into a compact form with a characteristic number of discrete units called chromosomes. The nucleus also contains the machinery required for transcription of genes and processing of the RNA products, the precise replication of the genome once per cell cycle and the processes of recombination and repair. In addition, complex interphase nuclear membrane components control the two-way communication of the nucleus with the cytoplasm and thus indirectly with other cells of a tissue, tumour or organism.

This review outlines advances in cell and molecular biology that have provided some insight into the immensely complex organizational features of eukaryotic nuclei. For this very reason, we attempt to introduce a conceptual framework but do so, we hope, without detracting from relevant experimental detail.

DNA wraps around histone core particles to form a 10 nm fibre of nucleosomes, the basic subunits of chromatin structure

The highly conserved fundamental subunit of chromatin structure, the nucleosome, consists of 146 bp of DNA wrapped, as 1.75 turns, around a protein core. Each core particle is an octamer of the four core histone proteins, H2A, H2B, H3 and H4 (Figure 1) [1]. Histones do not bind to specific DNA sequences; interactions between negatively charged phosphate groups on DNA and positive charges on the lysine and arginine-rich core histones are sufficient for stable nucleosome formation [2]. Nucleosomes are positioned at regular intervals along DNA and are separated by intervening linker DNA that varies from 0 to 60 bp in different cells. Transcriptionally active or competent (i.e. not active but capable of transcription given the appropriate signals) regions of the genome form euchromatin, typified by the open or extended nucleosome arrays visualized as beads on a string by electron microscopy.

Inactive genes, in the presence of histone H1, are further condensed into a 30 nm structure that resembles a solenoid

Active chromatin represents only 10%, or less, of the total present in most cells. Most chromatin is not expressed and is packaged into a relatively condensed, transcriptionally inert 30 nm chromatin fibre which is stabilized by a fifth histone, histone H1. This structure has been called the "solenoid" (Figure 2) [1,3].

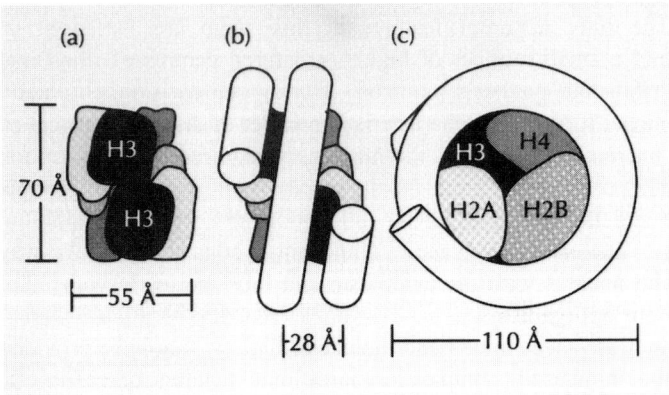

Figure 1. The nucleosome. A schematic illustration showing the locations of core histones in the histone octamer (a) and nucleosome core (b, end view; c, side view) with its 146 bp of DNA (1.75 turns). From [1]. Reproduced with permission.

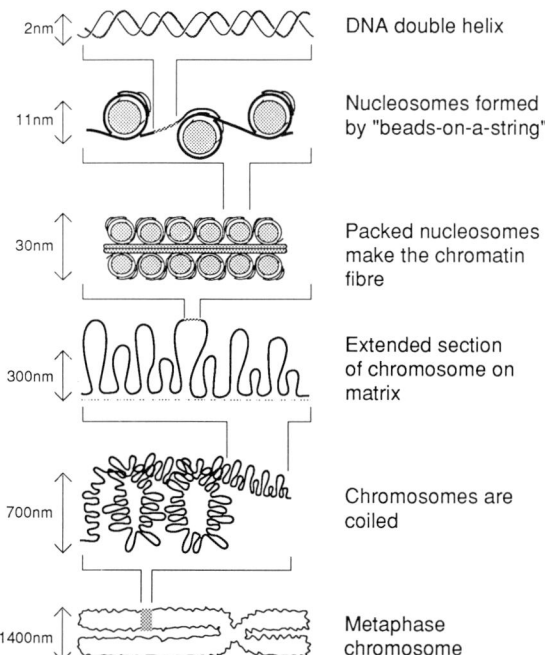

2nm — DNA double helix

11nm — Nucleosomes formed by "beads-on-a-string"

30nm — Packed nucleosomes make the chromatin fibre

300nm — Extended section of chromosome on matrix

700nm — Chromosomes are coiled

1400nm — Metaphase chromosome

Figure 2. Levels of chromatin condensation. A schematic illustration showing how the DNA duplex must fold in order to achieve the level of condensation seen in chromosomes. During interphase, chromatin loops form through their interaction with components of the nuclear matrix. Within these loops, chromatin might be folded into a condensed, 30 nm, fibre or the more open, 10 nm, structure characteristic of active genes. Dramatic architectural changes accompany chromosome formation, prior to mitosis. From Alberts et al. The Molecular Biology of the Cell, 2nd edition, Garland Publishing, 1989. Reproduced with permission.

These differences in chromatin structure are reflected by the sensitivity of different parts of the genome to digestion with nucleases. Pancreatic deoxyribonuclease (DNase I) is often used to demonstrate this. Transcriptionally active or competent regions, being relatively open, are preferentially "sensitive" to digestion with DNase I. In addition, specific regions acquire a more dramatic DNase I "hypersensitivity" that seems to correlate with the relative absence of nucleosomes [3,4]. These regions are commonly upstream of active genes and they may be indicative of different chromatin structures associated with the initiation of transcription.

The protein content of active chromatin itself differs from inactive chromatin in many ways [4–6]. For example, histone H1 is partially replaced by other proteins in active chromatin. The histones are also biochemically altered in active chromatin in ways that reduce nucleosome core stability.

A. Bassim Hassan and D.A. Jackson

Transcription and DNA replication requires the passage of RNA or DNA polymerases along the DNA in nucleosomes. It seems that the nucleosome core components can remain in place while this happens, but they may need to unfold to some degree [6].

The chromatin fibre interacts with a nuclear substructure to form chromatin loops

The nucleosomal fibre (10 or 30 nm depending on transcriptional status) is then thought to be arranged into chromatin loops containing, on average, 80–100 kb of DNA (Figure 2) [7,8]. This range of sizes has been defined using a variety of "nuclear derivatives" (i.e. nuclei isolated from cells in various ways), principally because there is no means of assessing loop size in an intact nucleus. For example, an analysis of nuclei extracted from cells in a buffer containing a high salt concentration (2 M NaCl) first suggested the presence of supercoiled DNA loops within interphase nuclei of eukaryotic cells. These DNA loops have been shown to be attached to structures called "nuclear matrices" and "nuclear scaffolds".

More recently, the use of encapsulated cells has allowed the analysis of chromatin domain structures under conditions approaching those found in the cell [7]. By encapsulating cells in agarose microbeads prior to permeabilization and analysis under physiological conditions, artefacts generated by the preparation are greatly reduced. Under these conditions, it has been demonstrated that the chromatin of HeLa cells (derived from a human cervical carcinoma) has an average loop size of 85 kb, with a broad size distribution covering 5–250 kb.

A higher level of chromatin condensation is required for chromosome formation prior to mitosis

If cells are to divide, chromatin duplicated during S-phase must separate so that exactly one chromosome set passes to each daughter. The condensation of chromatin to form chromosomes is achieved through additional levels of higher order packaging, with loops folding into 100 nm coils which are further coiled or stacked into compacted chromosome structures (Figure 2) [9]. This occurs in response to a complex series of events that are coupled to the cell cycle. Prior to mitosis, solubilization of the nuclear membrane and chromosome condensation occur. This occurs in response to the phosphorylation of proteins involved in chromatin structure (e.g. histone H1) that correlate with the appearance of specific kinase activities (e.g. cdc2, see Chapter 12).

Although much less complex in terms of their protein composition, chromosomes still maintain distinct structure; this is evident from the clearly recognizable banding patterns that can be seen after staining [10]. Defined "telomeres" (ends) and "centromeres" (sites of spindle attachment) each play important roles and have well characterized components [11,12].

The nuclear matrix is composed of proteins supported on a nucleoskeleton of core filaments

The nuclear matrix has been found to be composed of as many as 500 polypeptides. Some of the genes coding for these peptides have been cloned but their function is not yet clear [13]. Electron microscopy has allowed the visualization of some components of the nucleus. For example a 10 nm filamentous structure has been observed with a 23 nm repeat (Figure 3). This structure is characteristic of a group of fibres known as intermediate filaments. To date, however, the only members of intermediate filament family that are known to be present in eukaryotic nuclei are the lamins which form a filamentous network at the nuclear periphery.

As chromatin domains of mitotic cells or chromosome preparations are similar to those measured during interphase, it seems likely that the nucleoskeleton proteins will contribute to both chromatin organization during interphase and chromosome structure at mitosis.

Some chromatin loops form functional domains that represent independent units of genetic activity

At least a subset of the structural chromatin loops described above correlate with functional domains defined classically by their sensitivity to pancreatic deoxyribonuclease (DNase) (see Figure 4). This clearly reflects the chromatin structure of transcriptionally active genes but, perhaps surprisingly, is not confined to transcription units, with large regions 5' and 3' of a gene often displaying the same sensitivity [3,4]. The extent of DNase sensitivity can also correlate with levels of gene expression; decreasing DNase sensitivity across the entire β-globin domain during erythrocyte maturation correlates with a progressive condensation of chromatin and subsequent reduction in β-globin gene expression [4]. A clear example of correlations between chromatin structure and function is illustrated by studies on the chicken lysozyme gene [14]. Here, a region of DNase-I sensitivity has been well characterized. It extends over 21 kb, with the 4 kb transcription domain located centrally. Regions at both ends of this domain have been shown to attach to

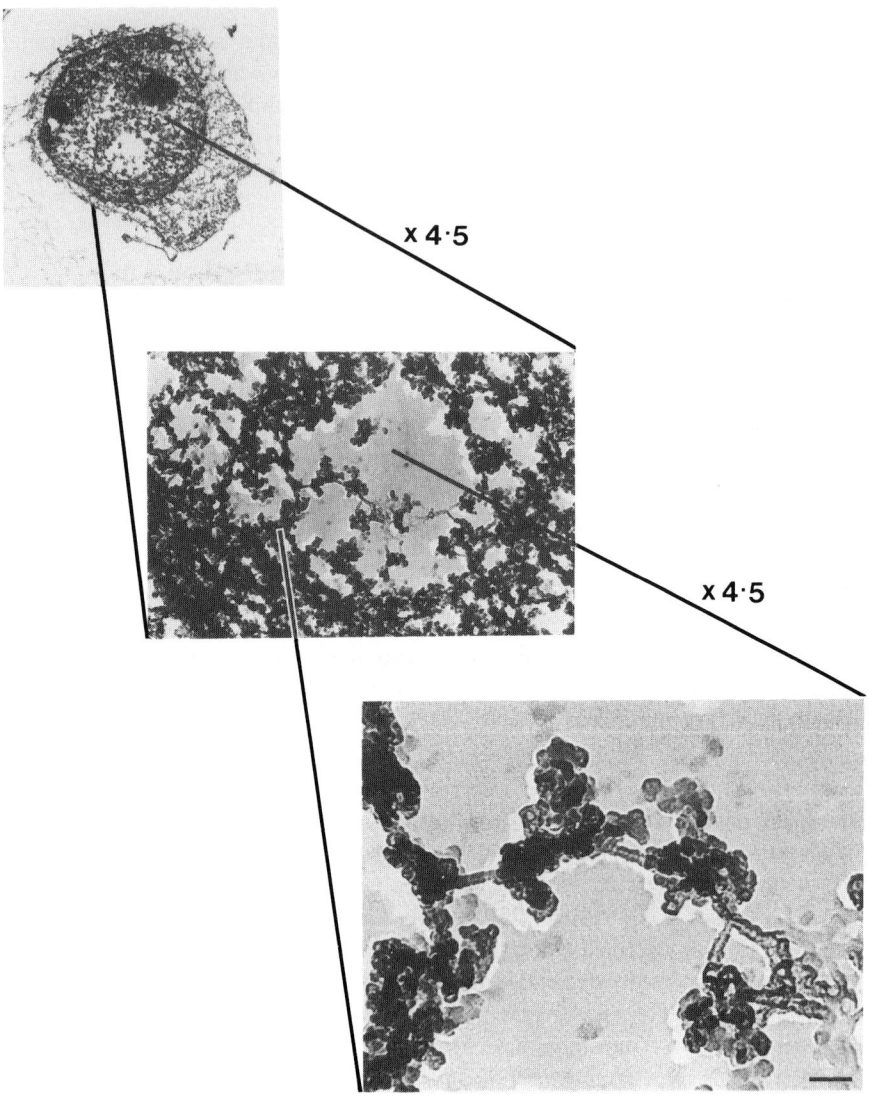

Figure 3. The nucleoskeleton. Electron micrographs of a thick resinless section of an extracted HeLa cell. Cells were encapsulated, lysed, chromatin cut with *Hae*III and 20% removed by electroelution (all procedures up to fixation took place in a physiological buffer). At low magnification, the nucleus, remnant cytoplasm (this collapses during EM processing) and surrounding agarose filaments are seen. At higher magnifications, residual chromatin clumps appear to be arranged along an underlying nucleoskeleton. Note that possible domain structures (individual chromatin clumps) and the skeleton are obscured in samples that retain all their chromatin. The bar is 100 nm (see [21] for further details.)

Inert   **Active**

Heterochromatin   **Euchromatin**

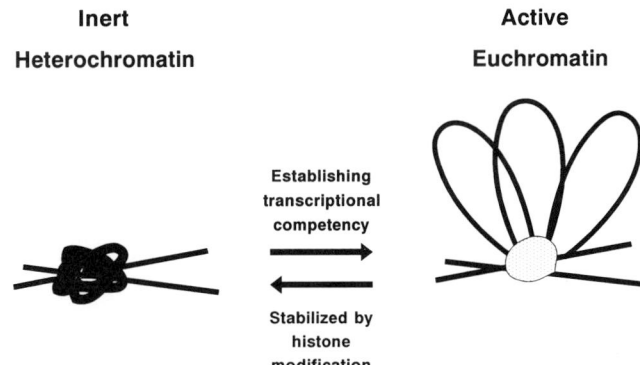

Establishing
transcriptional
competency

Stabilized by
histone
modification

Figure 4. Chromatin loops and gene expression. Structural chromatin loops, three are shown, form independent domains and so allow regions with different chromatin states to be established. In heterochromatin, domains fold into a 30 nm fibre (three are depicted on the left). As domains acquire a state of transcriptional competency they assume a more open, 10 nm structure. This decondensation allows access of activating factors that bind appropriate recognition motifs and locate RNA polymerases to the correct site on the promoter. In the model shown here, the polymerase is part of a much larger active centre (stippled) that forms part of the nuclear matrix - binding the gene as transcription proceeds.

the matrix and scaffold. These are called "A-elements". The significance of these elements is that they can separate areas of differing condensation [15]. Thus, active, decondensed areas are not inactivated by condensation induced by adjacent regions of heterochromatin. These "positional effects" need to be considered when genes are introduced into unnatural chromosomal sites.

Additional sequences within locus control regions may be required to establish transcriptional competency

Regions of DNase I hypersensitivity have been identified 20–50 kb upstream of some genes. These have the ability to affect expression of linked genes and so have been termed "locus control regions" (LCR) [16]. Unlike A-elements the locus control region appears to operate in isolation and so is unlikely to form domain borders. Subsequent investigations, describing numerous ubiquitous and tissue specific transcription factor binding sites, support the idea that the locus control region operates as a type of "super enhancer" and functions to maintain chromatin in a state of expressional competency across an entire gene domain.

Together these experiments support the idea that chromatin is arranged into structural loops which correlate in different ways with gene expression. Once formed, such domain structures can protect or isolate internal genes from the re-

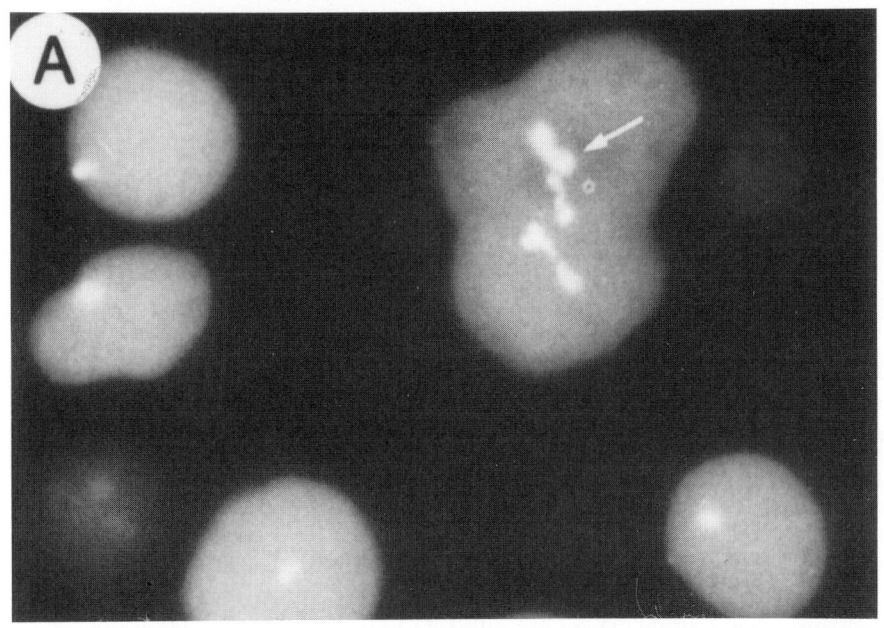

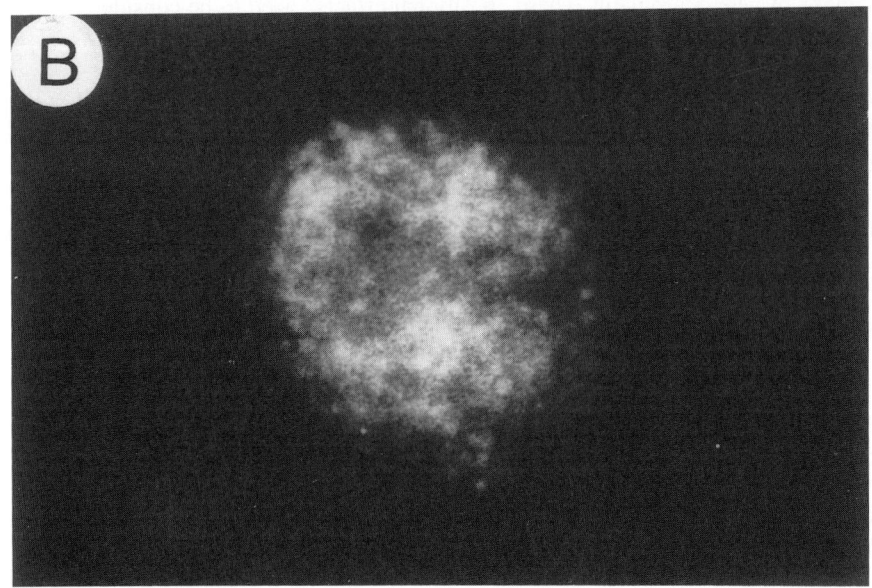

pressive influence of any local heterochromatin and so represent one important component on the pathway to establishing gene expression. Within domains, additional sequence elements (e.g. promoters and enhancers) form a hierarchy of interactions that eventually lead to transcription. Locus control regions appear to function early in the expression pathway, by providing signals that establish a transcriptionally competent chromatin structure.

Nuclear organization provides a means of developing specific nuclear subcompartments and nuclear context

It is obvious that a precisely structured chromatin domain can influence nuclear structure and function in many ways. For example, a nucleoskeleton-based domain organization might allow the genome to be arranged so that defined parts assume predetermined nuclear locations. Although this may not be absolutely precise, it now seems clear that chromosomes do occupy discrete territories at preferred sites. This could be important as it allows nuclei to be compartmentalized so that specific sites are rich in the component needed to perform different functions. Such an arrangement would undoubtedly reduce the demand for such components while providing one explanation for the complex spatial and temporal patterns of expression seen when genes are expressed from unnatural chromosomal sites in transgenic animals [17].

Many other lines of evidence support the view that host chromatin exerts a dominant position effect over other transcription controls. For example, in yeast (*Saccharomyces cerevisiae*) where origins of replication are known to operate at different times during the S-phase of the cell cycle, it can be demonstrated that origins situated close to teleomeric heterochromatin replicate later than they would at their natural site. This suggests that the suppressing effect of heterochromatin at the telomere can spread into adjoining regions and alter their function.

Figure 5. Nuclear structure influences RNA processing and DNA synthesis. (A) EBV transcripts emanating from two copies of the viral genome integrated at one site in Namalwa cells were visualized using biotinylated probes followed by fluorescein-avidin. RNA tracks remained when >85% of the DNA was removed. From [22]. Reproduced with permission. (B) Encapsulated HeLa cells were permeabilized, incubated in a replication mix supplemented with biotinyl-11-dUTP and sites of DNA synthesis visualized with FITC-streptavidin. Similar patterns were seen after removing at least 90% of the DNA.

Active RNA and DNA polymerase complexes are associated with the nucleoskeleton

Active genes show a strong preference for the nuclear matrix; they lie close to sites of attachment at the base of chromatin loops [18]. In many instances, protein factors necessary for gene activation have been shown to bind the nuclear matrix. Transcription also appears to correlate with active genes binding to the nuclear matrix [19].

The idea that sites of transcription can be determined by nuclear architecture is appealing because it begins to explain the observed variabilities when genes are expressed from unnatural chromosomal sites; the distribution of transcription complexes then becomes one important determinant of nuclear context. This structural theme continues as the nascent RNA passes from its site of synthesis towards the cytoplasm (Figure 5). Remarkably, in situ hybridization to nascent transcripts shows that RNA synthesized from a particular gene follows a preferred route to the cytoplasm [20], along "tracks" which persist even when most chromatin is removed.

Replication complexes are clustered into some 200 replication factories each of which contains 20–50 replication units

The most impressive evidence concerning the specificity of nuclear organization comes from the analysis of sites of DNA synthesis [7] (Figure 5). Using thymidine analogues, fluorescence microscopy has shown some 200 discrete sites of DNA replication activity in early S-phase nuclei of higher eukaryotic cells. Since at least 5000 replication complexes (replicons) must be active at this time, each replication factory must contain 20–50 active replication complexes, operating simultaneously. Furthermore, as labelled foci persist once replication is complete, their organization must reflect nuclear structure and not simply some property of the replication process. To support this view, replication has been shown to be intimately associated with the nuclear matrix and nucleoskeleton, with different patterns or shapes of foci characterizing different phases of the cell cycle.

Efficacy of expression from genes introduced into unnatural chromosomal sites: prospects for gene therapy

The ability to treat genetic defects using gene therapy now seems to be a serious goal. However, to achieve the desired result demands an intimate knowledge of all factors that might influence expression in situ. It may be assumed that protein factors present in particular cells will direct authentic expression. However, the position in the genome at which a gene is incorporated can have a dramatic effect on both the timing and extent of expression [7,17]. As the position effect factor is completely unpredictable, the ideal option would be "replacement" therapy, replacing the defective gene with its normal counterpart by homologous recombination. But this is not without its technical difficulties (which may prove insurmountable). Any approach that demands integration into the host genome has mutagenic potential and could well face ethical problems, so it would be valuable to develop a means of achieving stable expression for episomal (extrachromosomal) genes. Of course, this is not likely while we know so little of how gene expression is influenced by nuclear organization.

Conclusions

At two organizational extremes, chromatin structure is well established. At one extreme, the specificity of histone–DNA interaction in nucleosomes has been resolved to 7 Å using X-ray crystallography. At the other extreme, chromosomes form with a specificity allowing each to be recognized after staining from one cell generation to the next. During interphase, these chromosomal features are lost and any organizational specificity is difficult to assess. Even so, there is no shortage of evidence to indicate that nuclear architecture is far from ill-defined. Chromatin, for example, is packaged into arrays of higher-order loops that in some cases correspond to functionally discrete domains. Understanding the basis of genome organization in eukaryotic cells is of fundamental importance, because only then will transcriptional competency and the control of gene expression during differentiation, development and malignancy be fully understood.

References

1   Smith MM. Histone structure and function. Curr Opinion Cell Biol 1991, **3**, 429–437.
2   Travers AA. DNA bending and nucleosome positioning. Trends Biochem Sci 1987, **12**, 108–112.
3   Elgin SCR. Chromatin structure and gene activity. Curr Opinion Cell Biol 1990, **2**, 437–443.
4   Weisbrod S. Active chromatin. Nature 1982, **297**, 289–295.
5   Felsenfeld G. Chromatin as an essential part of the transcription mechanism. Nature 1992, **355**, 219–223.
6   Freeman LA, Garrard WT. DNA supercoiling in chromatin structure and gene expression. Crit Rev Eukaryot Gene Express 1992, **2**, 165–209.
7   Jackson DA. Structure-function relationships in eukaryotic nuclei. BioEssays 1991, **13**, 1–10.

8   Roberge M, Gasser SM. DNA loops: structural and functional properties of scaffold-attached regions. Molec Micro 1992, **6**, 419–423.

9   Manuelidis L. A view of interphase chromosomes. Science 1990, **250**, 1533–1540.

10 Bickmore WA, Sumner AT. Mammalian chromosome banding - an expression of genome organization. Trends Genet 1989, **5**, 144–148.

11 Blackburn EH. Telomeres. Trends Biochem Sci 1991, **16**, 378–381.

12 Earnshaw WC. When is a centromere not a kinetochore? J Cell Sci 1991, **99**, 1–4.

13 Berezney R. The nuclear matrix: a heuristic model for investigating genomic organization and function in the cell nucleus. J Cell Biochem 1991, **47**, 109–123.

14 Bonifer C, Hecht A, Saueressig H et al. Dynamic chromatin: the regulatory domain organization of eukaryotic gene loci. J Cell Biochem 1991, **47**, 99–108.

15 Kellum R, Schedl P. A position-effect assay for boundaries of higher order chromosomal domains. Cell 1991, **64**, 941–950.

16 Hanscombe O, Whyatt D, Fraser P et al. Importance of globin gene order for correct developmental expression. Genes Dev 1991, **5**, 1387–1394.

17 Palmiter RD, Brinster RL. Germ-line transformation of mice. Annu Rev Genet 1986, **20**, 975–985.

18 Getzenberg RH, Pienta KJ, Ward WS, Coffey DS. Nuclear structure and the three-dimensional organization of DNA. J Cell Biochem 1991, **47**, 289–299.

19 Cook PR. The nucleoskeleton and the topology of transcription. Eur J Biochem 1989, **185**, 487–501.

20 Carter KC, Lawrence JB. DNA and RNA within the nucleus: how much sequence-specific spatial organization. J Cell Biochem 1991, **47**, 124–129.

21 Jackson DA, Cook PR. Visualization of a filamentous nucleoskeleton with a 23 nm axial repeat. EMBO J 1988, **7**, 3667–3677.

22 Xing Y, Lawrence JB, Preservation of specific RNA distribution within the chromatin-depleted nuclear substructure demonstrated by in situ hybridization coupled with biochemical fractionation. J Cell Biol 1991, **112**, 1055–1063.

Molecular Biology for Oncologists
Edited by John Yarnold, Michael Stratton, Trevor McMillan
© *1993, Elsevier Science Publishers B.V. All rights reserved*

CHAPTER 25

The PCR revolution

Rosalind A. Eeles[a,b], William Warren[b] and Alasdair Stamps[c]

*[a]Academic Unit of Radiotherapy, Royal Marsden Hospital, and
Sections of [b]Molecular Carcinogenesis and [c]Pathology,
Institute of Cancer Research, 15 Cotswold Road, Sutton, Surrey, SM2 5PT, UK*

Introduction

The polymerase chain reaction (PCR) is an in vitro method which uses enzymatic synthesis to amplify, exponentially, specific DNA sequences (Figure 1). Devised by Mullis and refined by Saiki et al. in 1985 [1], it originally used a DNA polymerase that was not heat stable, so fresh enzyme had to be added before each cycle. The PCR revolution followed the development of computerised thermal cyclers which automatically heat and cool the samples, and the introduction of a thermostable Taq polymerase [2] isolated from algae (*Thermus aquaticus*) living in the hot springs of Yellowstone National Park.

The technique is powerful enough to amplify one copy of a specific DNA sequence millions of times. Prior to its introduction, amplification of a particular segment of DNA could only be achieved by labour intensive and time consuming cloning using bacteria. This could take several weeks, whereas PCR takes hours. The importance of the technique is reflected by the exponential increase in the number of publications relating to PCR, from 3 in 1986 to a staggering 1700 in 1990. These include several reviews [3–8].

Principles of the polymerase chain reaction

The polymerase chain reaction is a cyclical process of heating and cooling to denature, anneal and enzymatically amplify DNA

The standard reaction uses two oligonucleotide primers which are complementary

299

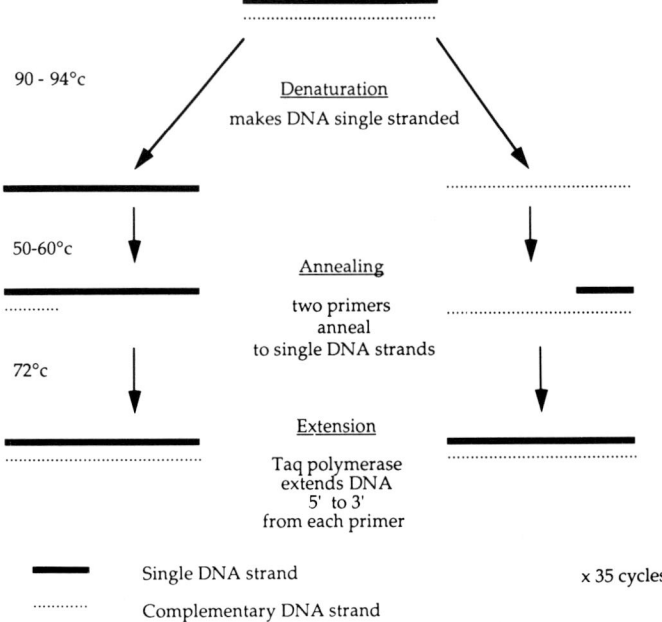

Figure 1. The polymerase chain reaction. Steps involved in one cycle. —, single DNA strand; ······, complementary DNA strand.

to and hybridise with opposite DNA strands flanking the region of interest in the target DNA. The primers are generally around 20 nucleotides in length, sufficiently long to be unique within the genome (Figure 2).

The reaction (shown diagramatically in Figure 1), consists of the following steps:

(i) *template denaturation*: at 90–94°C for 0.5–2 min; this separates the two DNA strands.

(ii) *primer annealing*: at 50–60°C for 0.5–1 min; the primers anneal to the template. (The annealing temperature depends on the nucleotide composition of the primers.)

(iii) *extension*: at 72°C; new DNA strands are synthesised from the primers, complementary to the single stranded template DNA to which the primer has hybridised.

These steps are then repeated. Thus, the newly synthesised DNA strands also become available as templates for a further round of DNA synthesis in the next cycle of the reaction. The DNA is therefore amplified exponentially, theoretically 2^n times where n is the number of cycles, although in practice the efficiency is not 100%. Typically 30–40 cycles are performed and although it is easier to amplify small fragments of a few hundred base pairs, up to 10 kb can be amplified. However, the larger the amplification product, the greater the number of shorter non-specific sequences that may also be amplified.

Details of the technical aspects can be found in several texts [7,8].

Problems: the major drawback is contamination

Because it is so sensitive, the greatest problem with PCR is contamination of new reactions by products from previous reactions. The potential scale of the problem is as follows: if the products of a 100 μl PCR were added to an Olympic-sized swimming pool, then a 0.1 ml aliquot from this pool would contain 40 amplifiable molecules [9]! Contamination can be minimized by aliquoting solutions and using separate pipettes and solutions for pre- and post-PCR experiments. Some experimenters use positive displacement pipettes or plugged pipette tips which are now available to eliminate contamination due to the production of aerosols. Many laboratories also have designated areas for handling PCR products. Negative controls should always be included in each PCR experiment to monitor potential PCR contamination, and the control reactions should be set up last (Figure 3).

```
ACACAACTGTGTTGACTAGCAACCCAG.......TGTGGGGCAAGGTGAACGTGGATGAAGTTG
                   ***********************CCACTTGCACCTACTTCAAC


The Primers are:

ACACAACTGTGTTGACTAGC --------->
5'end                3'end


                        <----------CCACTTGCACCTACTTCAAC
                            3'end              5'end

     When ordering the primers, write as:

     5'ACACAACTGTGTTGACTAGC

     5'CAACTTCATCCACGTTCACC
```

Figure 2. DNA sequence to be amplified is in bold type. ****, complementary DNA strand in the DNA to be amplified.

1 2 3 4 5 6 7 8 9 10

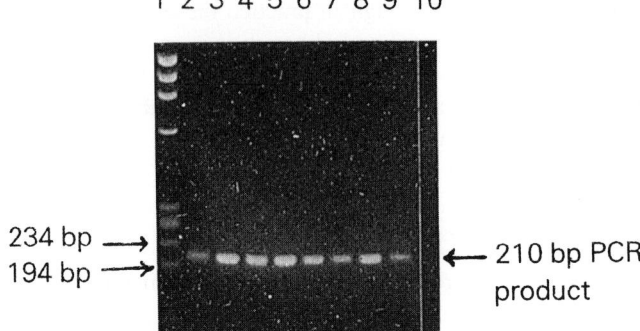

234 bp →
194 bp →   ← 210 bp PCR
                                          product

Figure 3. Ethidium bromide-stained 2% agarose gel. PCR products, each of 210 base pairs in lanes 2–9. Lane 10 is a negative control (no DNA to demonstrate that there is no PCR contamination). Lane 1: φ × 174 bacterial digested DNA, digested into pieces of known size to act as a size marker.

Applications of the polymerase chain reaction

Amplification of DNA segments for further analysis: a substitute for cloning

PCR provides a rapid method of acquiring large quantities of a DNA segment located between two known primer sequences. It can be applied to genomic DNA, cDNA or to various cloned sequences. Subsequently, the amplified segment can be subjected to a wide variety of procedures with many different end points [10–13]. In the context of cancer research, it is frequently sequenced or subjected to a number of procedures (e.g. single stranded conformational polymorphism analysis (SSCP) that allows detection of mutations that may be implicated in oncogenesis. Alternatively, the product can be blotted onto a filter and hybridised to probes. If the amplified segment contains a polymorphic region, the procedure can be adapted for linkage analysis of heritable predisposition to cancer or for studies of loss of heterozygosity.

Isolation of new genes using low stringency PCR for finding gene families

Related genes, both between species and within one species (e.g. the tyrosine kinase family of genes), have shared (conserved) sequences. To discover new, but related genes, degenerate primers can be designed which correspond to all possible nucleotide sequences that code for the protein in question [14], and further degeneracy can be introduced in order to encompass related protein sequences. Using a lower annealing temperature with conventional primers is another way of amplifying "related" sequences.

PCR can amplify specific DNA sequences from small amounts of material

Because of its huge amplification potential, the PCR may be used to isolate specific segments of DNA from tiny amounts of tissue. Indeed, at its most sensitive, sequences can be amplified from single cells, from single human hairs or blood spots [15,16]. More commonly in medical practice, it is applied to samples taken at amniocentesis or chorionic villous sampling [17–20] allowing antenatal diagnosis of a variety of diseases.

PCR can selectively detect DNA sequences not normally present in the tissue being tested

Viruses
Viral DNA can be detected by PCR for diagnosis, such as in cases of the acquired immunodeficiency syndrome [21] where the patient is still antibody negative, or in babies born to HIV positive mothers who will have acquired antibodies across the placenta, but may not necessarily be infected with the virus [22]. The presence of viral DNA in precancerous lesions has also been detected using PCR [23].

Translocations
PCR can be used for the diagnosis or detection of minimal residual disease. Certain chromosomal translocations commonly occur in follicular non-Hodgkin's lymphomas (t(14;18)) and chronic myeloid leukaemia (Philadelphia chromosome positive: t(9;22)). In follicular non-Hodgkin's lymphomas, the translocation involves the immunoglobulin heavy chain region on chromosome 14 and the BCL-2 region (a putative oncogene) on chromosome 18. The primers used hybridise to the regions flanking the translocation and will therefore only amplify the intervening DNA when the translocation is present. This can be used both for diagnosis and the detection of minimal residual disease, even in the presence of a histologically negative bone marrow [24]. Dilution experiments show that PCR assays can nonetheless detect a single lymphoma cell amongst one million normal cells. This is more than a 10^4-fold improvement in sensitivity over Southern blotting methods. In the case of chronic myeloid leukaemia, the breakpoint on chromosome 9 can occur over a large area, and the potentially large fragments cannot be amplified. This problem can be overcome by basing the amplification on mRNA, rather than genomic DNA. This is performed on cDNA synthesised from mRNA by reverse transcriptase [25].

PCR allows analysis of highly degraded DNA samples

Many DNA samples of forensic or anthropological interest are highly degraded, either because they are very old or have been maintained under suboptimal conditions. The PCR will, however, selectively amplify the minority of intact DNA fragments in such specimens that carry both primer sequences. Thus, one can sometimes salvage information from extremely degraded material that may be many thousands of years old [26]. In the context of forensic medicine, it has become possible to identify the individual from whom a degraded specimen originated by PCR across highly polymorphic regions known as microsatellites [27]. Using several of these segments one can build up a profile that is almost unique for any individual person. In the context of cancer research, DNA from histopathology archives has become available for analysis. The process of fixation in formalin and embedding in paraffin blocks for pathological examination degrades DNA considerably. However, the PCR often works effectively upon DNA extracted from this type of specimen and thus widens dramatically the range of cancers and other tissues available for study [28].

Gene expression

PCR can be used to investigate gene expression particularly when levels of mRNA are very low. Initially mRNA must be converted into cDNA by reverse transcription. Subsequently, the cDNA becomes the substrate for amplification. Careful use of the appropriate controls can make this procedure quantitative or semi-quantitative. One problem that arises is distinguishing PCR product amplified from the cDNA from product that may derive from residual traces of genomic DNA in the RNA preparation. This can be overcome by placing the two primer sequences in adjacent exons, which are separated by a large intron. In this situation, the PCR is unlikely to be effective on the genomic DNA because the segment to be amplified is too large. Because the intron is removed during the formation of mRNA, however, amplification from the cDNA which is synthesised directly from the mRNA will proceed unhindered.

Future developments

The PCR revolution has occurred because of the application of computer technology and a thermostable DNA polymerase to the cyclical amplification of DNA. Its power of amplification from minute samples has opened up enormous possibilities for rapid diagnosis, forensic science and research. Further refinements of existing diagnostic methods to incorporate the benefits of PCR will lead to even wider use of this technique in the next few years.

Acknowledgement

This work was supported by The Cancer Research Campaign and The Royal Marsden Hospital.

References

1 Saiki RK, Scharf S, Faloona F et al. Enzymatic amplification of beta-globin genomic sequences and restriction site analysis for diagnosis of sickle cell anaemia. Science 1985, **230**, 1350–1354.
2 Saiki RK, Gelfand DH, Stoffel S et al. Primer-directed enzymatic amplification of DNA with a thermostable DNA polymerase. Science 1988, **239**, 487–491.
3 Remick DG, Kunkel SL, Holbrook EA, Hanson CA. Theory and applications of the polymerase chain reaction. Am J Clin Pathol 1990, **93(4) suppl 1**, S49–S54.
4 Rodu B. The polymerase chain reaction: the revolution within. Am J Med Sci 1990, **299**, 210–216.
5 Wright PA, Wynford-Thomas D. The polymerase chain reaction: miracle or mirage? A critical review of its uses and limitations in diagnosis and research. J Pathol 1990, **162**, 99–117.
6 McCormick F. The polymerase chain reaction and cancer diagnosis. Cancer Cells 1989, **1**, 56–61.
7 Erlich HA ed. PCR Technology. Principles and Applications for DNA Amplification. Stockton Press, 1989.
8 Innis MA, Gelfand DH, Sninsky JJ, White TJ eds. PCR Protocols. A Guide to Methods and Applications. Academic Press, New York, 1990.
9 Kwok S, Higuchi R. Avoiding false positives with PCR. Nature 1989, **339**, 237; **339**, 490.
10 Verlaan-de Vries M, Bogaard H, Van den Elst JH et al. A dot-blot screening procedure for mutated ras oncogenes using synthetic oligodeoxynucleotides. Gene 1986, **50**, 313–320.
11 Orita M, Suzuki Y, Sekiya T, Hayashi K. Rapid and sensitive detection of point mutations and DNA polymorphisms using the polymerase chain reaction. Genomics 1989, **5**, 874–879.
12 Myers RM, Lumelsky N, Lerman LS, Maniatis T. Detection of single base substitutions in total genomic DNA. Nature 1985, **313**, 495–498.
13 Curiel DT, Buchhagen DL, Chiba I et al. A chemical mismatch cleavage method useful for the detection of point mutations in the p53 gene in lung cancer. Am J Respir Cell Mol Biol 1990, **3**, 405–411.
14 Wilks AF. Two putative protein-tyrosine kinases identified by application of the polymerase chain reaction. Proc Natl Acad Sci USA 1989, **86**, 1603–1607.
15 Higuchi R, von Beroldingen CH, Sensabaugh GF, Erlich HA. DNA typing from single hairs. Nature 1988, **332**, 543–546.
16 Witt M, Erickson RP. A rapid method for detection of Y-chromosomal DNA from dried blood specimens by the polymerase chain reaction. Hum Genet 1989, **82**, 271–274.

17 Gasparini P, Novelli G, Savoia A, Dallapiccola B, Pignatti PF. First-trimester prenatal diagnosis of cystic fibrosis using the polymerase chain reaction: report of eight cases. Prenat Diagn 1989, **9**, 349–355.

18 Kogan SC, Doherty M, Gitschier J. An improved method for prenatal diagnosis of genetic diseases by analysis of amplified DNA sequences. Application to hemophilia A. N Engl J Med 1987, **317**, 985–990.

19 McIntosh I, Curtis A, Millan FA, Brock DJ. Prenatal exclusion testing for Huntington disease using the polymerase chain reaction. Am J Med Genet 1989, **32**, 274–276.

20 Newton CR, Kalsheker N, Graham A, et al. Diagnosis of alpha 1-antitrypsin deficiency by enzymatic amplification of human genomic DNA and direct sequencing of polymerase chain reaction products. Nucleic Acids Res 1988, **16**, 8233–8243.

21 Loche M, Mach B. Identification of HIV-infected seronegative individuals by a direct diagnostic test based on hybridisation to amplified viral DNA. Lancet 1988, **ii**, 418–421.

22 Rogers MF, Ou CY, Rayfield M et al. Use of the polymerase chain reaction for early detection of the proviral sequences of human immunodeficiency virus in infants born to seropositive mothers. New York City Collaborative Study of Maternal HIV Transmission and Montefiore Medical Center HIV Perinatal Transmission Study Group. N Engl J Med 1989, **320**, 1649–1654.

23 Nuovo GJ, Darfler MM, Impraim CC, Bromley SE. Occurrence of multiple types of human papillomavirus in genital tract lesions. Analysis by in situ hybridisation and the polymerase chain reaction. Am J Pathol 1991, **138**, 53–58.

24 Stetler-Stevenson M, Raffeld M, Cohen P, Cossman J. Detection of occult follicular lymphoma by specific DNA amplification. Blood 1988, **72**, 1822–1825.

25 Dobrovic A, Trainor KJ, Morley AA. Detection of the molecular abnormality in chronic myeloid leukaemia by use of the polymerase chain reaction. Blood 1988, **72**, 2063–2065.

26 Paabo S. Ancient DNA: extraction, characterization, molecular cloning and enzymatic amplification. Proc Natl Acad Sci USA 1989, **86**, 1939–1943.

27 Jeffreys AJ, Wilson V, Neumann R, Keyte J. Amplification of human minisatellites by the polymerase chain reaction: towards DNA fingerprinting of single cells. Nucleic Acids Res 1988, **16**, 10953–10971.

28 Shibata DK, Arnheim N, Martin WJ. Detection of human papilloma virus in paraffin-embedded tissue using the polymerase chain reaction. J Exp Med 1988, **167**, 225–230.

Techniques of molecular biology

Philippe J. Rocques

Section of Molecular Carcinogenesis, The Institute of Cancer Research,
15 Cotswold Road, Belmont, Sutton, Surrey, SM2 5NG, UK

Introduction

The term molecular biology is used to describe a set of techniques concerned with the study of genes and their functions, principally by manipulation of nucleic acid molecules (DNA and RNA). Only a broad overview of selected methods is presented here; some more exhaustive accounts are suggested under the heading References.

Nucleic acids can be modified in vitro using purified enzymes

Restriction endonucleases cut double-stranded DNA molecules at sites defined by specific base sequences

It is very useful to be able to cut DNA at sites of known base sequence and this is accomplished using restriction endonucleases. These are proteins purified or cloned from procaryotic species in which they protect the host cell from invasion by foreign DNA. The restriction endonucleases most commonly used in molecular biology are enzymes which recognise a palindromic sequence, called a restriction site, of four, six or eight bases and cleave the DNA by breaking phosphodiester bonds on both strands somewhere within the site.

Restriction sites four bases long statistically occur more frequently in a randomly chosen DNA than do longer sites, and it follows that four-cutter restriction fragments are much smaller than those which would be derived from the same initial DNA by digestion with an enzyme having a longer recognition sequence.

In general, six cutters tend to be most useful for Southern blotting and four cutters are used for making genomic libraries.

DNA polymerase synthesizes new DNA strands

DNA polymerase uses deoxynucleoside triphosphate precursors to elongate a short nucleic acid primer bound to a complementary template strand. All polymerases synthesize their products in the 5'–3' direction. Polymerases frequently occur as components of multi-enzyme complexes which also possess exo- and endonucleolytic activity. In the case of reverse transcriptase, the in vitro template can be RNA or DNA.

Polymerases are used in DNA sequencing, the polymerase chain reaction (PCR) and labelling of DNA by incorporation of radioactive nucleotides. The associated nuclease activities also have applications. Reverse transcriptase is used in the synthesis of cDNA from RNA.

DNA ligase joins double-stranded DNA molecules end to end

DNA ligase joins DNA molecules by catalysing the formation of phosphodiester bonds between double-stranded DNAs with 3' hydroxyl and 5' phosphate termini in the presence of ATP. DNA ligase is used in library construction and subcloning among other applications.

DNA kinase and phosphatase respectively add phosphate groups to and remove them from 5 termini of DNA molecules

DNA kinase and phosphatase are enzymes which respectively phosphorylate and dephosphorylate the 5' terminal position of a DNA, and can therefore be used to manipulate the competence of populations of DNAs to be ligated in vitro. DNA kinase is also used for end labelling of short DNAs (oligonucleotides) by transfer of a radiolabelled phosphate group from ATP.

Electrophoresis, blotting and hybridisation are techniques for the analysis of populations of nucleic acids

Electrophoresis allows separation of a population of DNA and RNA molecules according to size and conformation

Nucleic acids carry net negative charge, so they migrate through agarose or acrylamide gels in the direction of a positive electrode. The gel mobility of a double-stranded molecule decreases with increasing molecular weight (size). The situation is more complex for single-stranded molecules, since these can form secondary structure as a result of intrastrand base pairing. The gel mobility of a single-stranded molecule is therefore a function of both size and secondary structure conformation, the latter being base sequence dependent. A gel based on polyacrylamide provides greater resolution than is achieved with agarose. Such gels are used for sequencing (when it is necessary to separate large DNAs differing in size by a single nucleotide) and single-strand conformational polymorphism (SSCP) analysis. Sequencing gels and RNA gels are run under denaturing conditions (that is at high temperature and/or with formamide) to minimize secondary structure formation.

RNA or DNA in a gel can be visualised by staining with ethidium bromide and UV transillumination. Running nucleic acids of known sizes alongside the sample allows approximate determination of the sizes of bands, and it is possible to physically excise a portion of agarose containing a band of interest and extract DNA from it.

Single-stranded conformational polymorphism (SSCP) analysis is a method for detecting mutations in defined regions of DNA as electrophoretic mobility variants

SSCP analysis is carried out in two steps. In the first, a region of DNA in which mutations are sought is amplified by PCR. The PCR product DNA is then denatured (made single-stranded) by heating in formamide and electrophoresed on an acrylamide gel under non-denaturing conditions. This allows the formation of intrastrand hydrogen bonds producing secondary structure. PCR products amplified from genomic DNA containing a point mutation do not adopt the same conformation as wildtype products. The two products therefore have different gel mobilities, and the mutants can be identified by comparison with known normal controls following gel autoradiography.

SSCP detection works for PCR products up to about 500 bp, so that analysing a whole gene for mutations can require the use of multiple pairs of PCR primers in order to span the entire locus. It is not currently known exactly what proportion of all point mutations actually result in altered conformation, but it has been estimated that approximately 80% of mutations are detected by SSCP. This compares quite favourably with other far more labour-intensive methods of mutation detection.

Hybridisation is the formation of a double-stranded DNA between two single strands

Double-stranded DNA in vivo usually consists of two exactly complementary strands. If the base pairs in a few positions are not complementary then a duplex can still be formed, but one of reduced stability. Annealing together of two strands which may be exactly or nearly complementary is called hybridisation.

Southern and Northern blotting allow identification of particular components of a nucleic acid population following electrophoresis

Blotting preserves the distribution of nucleic acids resulting from electrophoresis in a permanent form accessible to further analysis. Following agarose electrophoresis, the separated DNAs or RNAs are transferred by capillary action onto a nylon membrane, to which they are covalently linked by UV irradiation, to make a Southern or Northern blot, respectively. In the case of Southern blotting, the DNAs to be separated are frequently fragments obtained by restriction enzyme digestion, whereas the aim in Northern blotting is to work with undegraded RNA. Blot hybridisation followed by autoradiography allows determination of the size of nucleic acid species complementary to a radiolabelled specific nucleotide sequence or probe.

A Southern blot of digested total genomic DNA can be used to detect gene amplification or deletion and chromosomal rearrangements, since these result in altered band sizes and/or intensities following probing. Northern blotting is used to analyse gene expression at the level of transcription, revealing for example the size and tissue distribution of mRNA hybridising to a probe.

Chromosomal in situ hybridisation is a powerful tool for physical mapping of DNA probes along a chromosome

A labelled probe can be hybridised to a spread of fixed metaphase or interphase chromosomes, allowing determination of the chromosomal localisation of sequences defined by the probe. This is technically easier with metaphase chromosomes, but interphase spreads give better resolution because the chromosomes are more elongated. This method can be used for example to map the positions of marker probes relative to the breakpoint of a chromosomal translocation.

Molecular cloning is the production in pure form of many identical copies of a chosen DNA molecule

Cloning vectors facilitate the insertion of foreign DNA into a cell and ensure its subsequent replication and sometimes expression

Most vectors have a number of features. They contain sequences ensuring their stable propagation in a population of dividing host cells. They provide some means of selecting for host cells containing vector, for example by conferring antibiotic resistance. They may contain elements called polylinkers with multiple restriction sites into which foreign DNA is ligated following restriction enzyme digestion. Polylinkers are frequently flanked by promoters allowing expression of cloned DNA.

Introduction of foreign DNA into bacterial cells requires temporary disruption of the cell membrane achieved either by heat shock or by electric shock

The most usual way of transforming *E. coli* cells is by calcium chloride treatment followed by heat shock. This temporarily makes the cells competent to take up foreign DNA. There is an alternative to this called electroporation, in which DNA enters cells following transient disruption of the cell membrane by electric shock. Electroporation can be far more efficient, but requires optimisation of conditions.

Plasmids are extrachromosomal DNA molecules which can be modified for use as cloning vectors

Plasmids are extrachromosomal DNA elements which occur naturally in a variety of organisms. Bacterial plasmids are closed circular DNAs far smaller than the bacterial chromosome, capable of independent replication within the bacterial cell and frequently conferring antibiotic resistance on the host. The plasmids used in molecular biology have been extensively altered by the addition of some or all of the features described above to make them more efficient for cloning purposes.

Bacterial viruses known as phage can also be exploited for cloning

A bacteriophage, or simply a phage, is a virus which infects and replicates in bacterial cells. Like plasmids, naturally occurring phage have been extensively modified for cloning purposes. The main difference between cloning in phage and plasmid lies in the method of introduction of the vector into the host cell. Transformation and electroporation are relatively inefficient processes, and do not work satisfactorily for DNA of size greater than about 20 kb. Far larger pieces of DNA can be inserted into bacteria using phage because use is made of the natural phage infection system.

Cosmid vectors are hybrids between phage and plasmids and have larger cloning capacities

Cosmids are vectors having properties of both phage and plasmids. They make use of the phage infection system, allowing insertion of very large pieces of DNA, and are able to propagate as plasmids once within the cell. The phage lambda packaging system will accept recombinant DNA ranging in size from about 20–50 kb. Because of this constraint, cosmids have had most of the phage genome removed to make room for larger inserts. Cosmids allow cloning of very large DNAs, ranging in size up to about 50 kb.

Yeast artificial chromosome vectors allow cloning of very large pieces of DNA

Yeast artificial chromosomes (YACs) allow cloning in yeast host cells of pieces of DNA comparable in size to normal yeast chromosomes. In practice, the process of preparation of intact high molecular weight DNA sets an upper bound of about 1 Mb, and most YAC clones range from 200 to 500 kb. YACs consist of cloned DNA flanked by two vector arms.

These must contain a centromere and yeast origin of replication (ARS) to ensure replication and segregation of the YAC, and each must have a telomere (a special sequence found at the ends of all linear chromosomes ensuring the integrity of terminal regions during replication). In addition, there is often some means for isolating short stretches of DNA from each end of the insert. These end clones allow determination of the orientation of a YAC relative to a chromosome by in situ hybridisation, and are also used for chromosome walking in YAC libraries. YACs are proving extremely useful in long range mapping of large regions of the human genome.

Libraries are random collections of fragments of genomic or cDNA from which selected DNAs can be cloned

A library is a collection of DNA fragments from which those of interest can be isolated or cloned. There are two main types of library, genomic and cDNA libraries.

Genomic libraries contain fragments of genomic DNA, generally obtained by restriction enzyme digestion. They contain many inserts which are not derived from protein coding sequence and which may be from regions of the genome very distant from the nearest gene. Complementary DNA, or cDNA, libraries contain DNA inserts obtained by reverse transcription of mRNA extracted from the cells or tissue of interest.

Identification of clones in libraries is carried out by hybridisation of labelled nucleic acid probes

This section describes the basic method of plaque purification of clones from phage libraries. The library is used to infect a bacterial culture, which is then plated on agar. Infected bacteria lyse, releasing phage particles which similarly kill surrounding cells. This results in a plaque containing many identical phage but no living bacteria. There are many such plaques on each agar plate, one corresponding to each bacterial cell initially infected. A replica of the plaque distribution is obtained by direct transfer of phage from the plates onto nylon membranes. The phage DNA is denatured and fixed to the membranes. The filters are then hybridised with a labelled DNA probe and autoradiographed, and the position of spots on the photographic film allows identification of the position of phage clones of interest. These are physically excised from the plate as plugs of agar containing phage, and subjected to one or two more rounds of plating and screening at lower density until a single clonal phage plaque can be picked.

Genomic DNA libraries are used principally for mapping work

Genomic libraries are mainly used for mapping projects, including attempts to clone genes from regions of the genome linked to known phenotypes. A probe might initially be shown by linkage analysis to map near a suspected disease gene. In order to move closer to the locus of interest, overlapping but distinct genomic clones are isolated from the library. Some of these new clones extend further towards the locus of interest. Repeated rounds of screening result in a contig (contiguous region) of clones spanning a portion of genome from the initial probe to the target locus. This approach is called chromosome walking.

cDNA libraries are used for studying the structure and function of protein coding regions

cDNA libraries are used for isolating coding sequence from a gene of interest. Reverse transcription of mRNA may be primed either using an assortment of random DNA hexamers or using oligo dT primers. Randomly primed libraries contain a collection of inserts not starting at any defined position within a given transcript. Oligo dT primed libraries have inserts which start at the 3' end of every message, since the primer anneals to the poly A tail.

Cloned cDNAs are used in conjunction with a variety of other techniques, including sequencing, mutagenesis and expression studies, all of which are described below.

DNA cloned in a plasmid can be subjected to a variety of experimental procedures such as mutagenesis and sequencing

Many cDNA and most genomic clones are too large for convenient manipulation. This problem is solved by inserting a fragment of the clone into a plasmid vector. The subcloned fragment can be analysed with relative ease.

Once a plasmid has been inserted into bacteria, a large culture can be grown starting from a single colony on a plate. This ensures that the culture is clonal. Large amounts of plasmid can be extracted from the culture.

A common method for investigating the function of a cloned cDNA or regulatory region is to introduce controlled mutations into the DNA. The functional effects of mutagenesis (deletions, insertions or base substitutions) can be assayed using for example gel retardation or expression systems.

DNA sequencing is carried out by the dideoxy sequencing reaction

This is a method for determining the sequence of bases in a DNA molecule. It is based on the fact that DNA polymerase can use 2'–3' dideoxynucleotides (ddNTPs) as substrates for incorporation into a nascent DNA strand, but cannot then elongate the strand further because of the lack of a 3'-OH terminus. DNA synthesis is carried out from a specific primer in the presence of a template (the DNA to be sequenced), all four deoxynucleotides (dNTPs), and a single ddNTP at far lower concentration. One or more of the dNTPs is radiolabelled and four such reactions are performed, one for each ddNTP. Strand elongation proceeds until stochastic incorporation of a ddNTP. Each reaction therefore eventually contains strand elongation products of different lengths, all starting at the primer and finishing at a dideoxynucleotide. The four sequencing reactions are run on a denaturing acrylamide gel in separate adjacent lanes. After autoradiography, the four lanes taken together contain a band for each base in the template. Because the ddNTPs incorporated are complementary bases in the template strand, the lane in which a band occurs indicates which base is present in a given position in the DNA being sequenced. The desired sequence can therefore be read directly from the autoradiograph. This is often achieved with a digitiser so that the sequence is immediately available in a computer file.

Expression of cloned genes is a direct method for studying gene function

Once a cDNA has been cloned and sequenced, it is often necessary to analyse its function further. Possible investigations range from in vitro work on the biochemistry of normal or mutant protein to determining the effects of the gene when expressed in various spatiotemporal patterns in transgenic mice.

Expression of a cloned cDNA in cultured cells can give valuable insights into the role the gene is playing in vivo, and allows determination of its interactions with other cellular components. Retroviral vectors are commonly used for such work.

Transgenic organisms allow the study of gene function or dysfunction in the context of the whole organism

The term transgenic is used to describe an organism which has had its genome artificially altered, either by insertion of a gene (a transgene) or by disruption of an endogenous gene. Transgenic animals provide opportunities for investigating how genes, including oncogenes, function in the context of whole organisms.

Embryonic stem cells can be genetically altered in culture and will colonise embryos into which they are injected

Embryonic stem (ES) cells are cultured from mouse inner cell mass (ICM) or delayed blastocyst. They proliferate when grown on layers of irradiated feeder cells or in the presence of a defined differentiation inhibitory activity (DIA). In the absence of these, they differentiate forming embryoid bodies. ES cells form tumours containing several differentiated tissue types when implanted into syngeneic mice. Most importantly, ES cells injected into blastocysts colonise the resulting embryo, giving a high frequency of chimaerism which can extend to the germ cells. These chimaeric mice can develop into fertile adults. The recently developed embryonic germ cells (EG) may provide a means of obtaining germline chimaerism in the majority of animals from which ES cells cannot be isolated.

Transgenic mice are constructed by either using embryonic stem cells or directly injecting DNA into a pronucleus

There are two principal methods for the construction of transgenic mice. One involves the manipulation in culture of the genome of an ES cell, followed by its injection into a blastocyst as above. Transgenes show simple Mendelian transmission, so breeding from germ cell chimaeric mice ultimately results in progeny homozygous for the novel genetic trait.

Another method of producing transgenics is to inject DNA into one of the pronucleii of a fertilised egg. The injected DNA typically integrates in many tandem copies at a random location.

Transgenic mice will be increasingly important in studying oncogene function

Two recent examples of the use of transgenic technology in cancer research are briefly mentioned. The Philadelphia chromosome associated with human CML and ALL results from a t(9;22)(q34;q11) chromosomal translocation. Sequences at the translocation breakpoint direct production of a bcr/abl fusion protein having abnormal tyrosine kinase activity. Injection into fertilised mouse eggs of a construct encoding this bcr/abl fusion under metallothionein promoter control results in mice which die of AML or ALL 10–58 days after birth. This clearly supports a causal relationship between the Philadelphia chromosome and human leukaemia.

Transgenic mice have been made which entirely lack any functional p53 gene, that is they are homozygous null mutants for p53. This was accomplished by first making a p53 heterozygous null ES cell. A targeting construct containing a disrupted segment of p53 was inserted into ES cells by electroporation. Various methods were used to select for homologous recombination between the construct and an endogenous p53 locus, and a suitable ES clone was used to generate chimaeric mice. These transmitted mutant p53 to their offspring at 50% frequency, and intercrossing the resultant heterozygotes produced 25% homozygous null mutants as expected. The p53–/p53– mice were apparently normal except for a strong predisposition to neoplastic disease. These findings suggest that while there is genetic redundancy in the normal function of p53, p53 homozygous mutation is a first step towards malignancy.

References

Old RW, Primrose SB. Principles of Gene Manipulation, 3rd edition. Blackwell Scientific, Oxford, 1985.

Watson JG et al. Molecular Biology of the Gene, 4th edition. Benjamin Cummings, Menlo Park, CA, 1987.

Sambrook J, Fritsch EF, Maniatis T., Molecular Cloning, 2nd edition. Cold Spring Harbour Laboratory Press, Cold Spring Harbor, NY, 1989.

Glossary

Collated by Philippe Rocques, Philip Mitchell and John Yarnold

Active genes   Genes transcribed in the cell type in question. Some genes are active in all cells (housekeeping genes), whilst others are specific to one cell type (e.g. haemoglobin gene in erythrocytes). Associated conformational features include looser duplex winding around nucleosomes*, less histone* H1 binding and situation in euchromatin.

Adduct   *See* Drug adducts.

Alkylation   Process of covalent binding of reactive alkyl group of drug (e.g. CH_2Cl) to biological molecules (DNA bases or proteins) which have an excess of electrons.

Alleles   Alternative DNA sequences at a locus, can be coding or non-coding.

Amplification   Increase in copy number of a chromosomal region, typically by tandem* duplication.

Anchorage dependence   The dependence of normal cells on an appropriate surface/substrate on which to grow in culture.

Annealing   Complementary base pairing of homologous single strands of nucleic acids, e.g. attachment of primers* to denatured target DNA in PCR.

Antisense   A nucleotide sequence (RNA or DNA) complementary* to the coding sequence (sense strand*).

---

*Terms marked with an asterisk are themselves entries in the Glossary.

317

Apoptosis   An active mechanism of cell death in which DNA degradation and nuclear destruction precede loss of plasma membrane integrity and cell necrosis.

Autocrine   A mechanism of growth stimulation involving the binding of a growth factor* secreted by a cell to its own plasma membrane receptors, cf. Paracrine* .

Bacterial transformation   The uptake of foreign DNA by a bacteria which may result in e.g. antibiotic resistance or some other phenotype.

BSO   Buthionine sulphoximine, an inhibitor of γ-glutamyl cysteine synthetase, leads to depletion of GSH*.

Carcinogen   A physical or chemical agent which has a causal role in the development of cancer (these are often, but not always mutagens*).

CDC genes   Cell Division Cycle proteins involved directly in the control of the cell cycle, e.g. CDC 2 (p34) is a kinase* enzyme.

cDNA library   A collection of DNAs produced by reverse transcription* from the mRNA* of a cell population of interest and inserted into a suitable vector*.

cDNA   DNA complementary* to RNA and synthesised from it by reverse transcription*.

Cell line   A cell culture composed of a single immortalised* stem cell population.

Cell strain   A cell culture composed of a mixture of clonal populations.

Cellular oncogene   Proto-oncogene* altered by mutation* which leads it to acquire an altered cellular function that contributes to carcinogenesis.

Centimorgan   Unit representing a recombination* frequency of 1%, i.e. approximately 1 million base pairs (Morgan "coined" the term).

Centromere   Specific DNA sequences that attach the chromosome to the mitotic spindle during M phase.

Chromatin   Chromosomal DNA together with a variety of associated proteins, the most abundant of which are histones*. *See* DNA chromatin.

Cleavable complex   DNA–drug adduct* or intercalation which stabilises topo-isomerase* II binding to DNA and which, on deproteination, reveals strand breakage.

Cloning   The generation of multiple identical copies of a DNA sequence by replication* in a suitable vector*, e.g. phage* or plasmid*.

Codon   A triplet of nucleotides coding for one amino acid.

Complementary bases   Bases that are hydrogen bonded specifically in DNA duplex or in DNA/RNA heteroduplex.

Complementation   Is said to occur when a cell deficient in a particular function, e.g. repair of UV DNA damage, is restored to normal function by addition of foreign DNA, by a technique such as cell fusion. This process has been used to show that a number of different gene defects can be involved in single human DNA repair disorders, e.g. xeroderma pigmentosum.

Constitutional deletion   Deletion inherited in the germline* from one or other parent and present in every cell of the body (usually examined in lymphocytes), cf. Somatic deletion*.

Contact inhibition   Inhibition of cell division in cell culture by cell–cell contact.

Copy number   The number of copies of a gene present in genomic* DNA.

CpG islands   Sequences "rich" in the dinucleotide CpG; often in the 5' region of genes and possibly involved in transcriptional regulation.

Cyclic AMP   A ubiquitous intracellular messenger* synthesised from ATP by plasma membrane bound adenylate cyclase (activated by a G protein*).

Cytokines   Proteins that act as intercellular signals to coordinate the immune response.

Deletion   Loss of a segment of a chromosome (*see* Constitutional deletion and Somatic deletion).

Denaturation of DNA   Melting (separation) of the complementary* strands caused by high temperature or chemical conditions, usually reversible.

Denaturation of protein   Loss of higher order structures caused by high temperature or chemical conditions, usually irreversible.

Determination   A commitment to follow a given developmental lineage (pathway).

Differentiation   An increase in specialisation towards a specific function.

DNA chromatin   Structural packing of DNA double helix into nucleus. Hierarchy of helix coiled around histones (to form a nucleosome*); twisting of these into a 30 nm wide fibre; supercoiling of this fibre into loops attached to the nuclear matrix and coiling of loops/matrix into chromosome bands.

DNA cloning   A technique involving the integration* of specific DNA sequences into a self-replicating element (plasmid* or virus) that reproduces itself in bacteria to generate huge numbers of identical copies.

DNA library   Collection of different cDNAs* or fragments of genomic* DNA propagated in a cloning vector* (phage or plasmid*) from which specific sequences can be isolated (cloned*).

Dosage effect   The effect on a cell's morphology/behaviour from having more or less than the usual diploid number of normal genes.

Double minutes   Chromosomal fragments visualised on metaphase spreads, associated with amplification*, e.g. multidrug resistance (MDR), *myc* oncogene.

Down-regulation   Process by which a cell loses its sensitivity to growth factor* stimulation (often by endocytosis of growth factor receptors*).

Downstream   Beyond the 3' end of a gene sequence, cf. Upstream*.

Drug adducts   DNA modified by covalently bound drug or reactive sidegroup of drug.

EcoRI   A restriction endonuclease originally isolated from a strain of *E. coli*.

Endocrine   A mechanism of growth stimulation involving the secretion of a growth factor* which binds to a specific cell receptor* after diffusing through the circulation.

Epitope   Antigenic determinant (there may be several per molecule).

Eucaryotic cell   Cells which have a nucleus (yeast, mammalian cells), cf. Procaryotic cells* (bacteria).

Euchromatin   The decondensed form of chromatin* typical of the interphase nucleus.

Exons   Transcribed sequence not spliced out of mature RNAs (cf. Introns).

G protein   A GTP binding protein involved in transmitting a signal from a cell surface receptor to an intracellular effector.

Gel electrophoresis   Technique for separating molecules of DNA, RNA or protein according to relative size and charge by passing them through a porous gel matrix under the influence of an electric field.

Gel shift or "retardation" analysis   Technique used to detect specific binding of a protein to a particular sequence of DNA. The resulting DNA–protein complex migrates more slowly in gel electrophoresis than "unbound" DNA, i.e. it is retarded.

Gene   A region of genomic DNA specifying the coding and controlling sequences for the expression of a protein or RNA product.

Genetic engineering   The processes by which genes can be isolated from cells and manipulated (e.g. mutated*) in vitro before re-introduction into the same or different species.

Genome   The genetic complement of a cell organelle, species etc., e.g. nuclear genome, mitochondrial genome, *Drosophila* genome.

Genotype   The hereditary information encoded in nucleic acid, cf. Phenotype*.

Germline deletion   *See* Constitutional deletion.

Germline mutation   Mutation inherited from one or other parent and present in every cell of the body.

Glutathione-*S*-transferases   Family of enzymes that are responsible for conjugation reactions involving GSH*.

Growth factor Proteins (first messengers*) that bind to specific cell surface growth factor receptors* and modify cell growth, e.g. EGF, PDGF.

Growth factor receptor Proteins that span the plasma membrane with an extracellular growth factor* binding domain and an intracellular signalling domain that is activated by growth factor binding.

GSH Reduced glutathione; conjugation to GSH* may be an important step in detoxification for a number of drugs.

Heterochromatin Regions of highly condensed chromatin* in the interphase nucleus visible with a light microscope, cf. Euchromatin*.

Histones A class of nuclear proteins involved in maintaining the higher order structure (folding) and function of genomic* DNA.

hnRNA Heterogeneous nuclear RNA, the immediate product of transcription*, i.e. before the splicing* out of introns* to produce messenger RNA* (mRNA).

Homeobox genes A family of transcription factors which all possess a highly conserved 180 bp sequence coding for the "homeobox" DNA binding domain.

Homogenously staining regions Regions of chromosome visualised on metaphase spreads corresponding to hugely amplified* ene sequences (*see also* Double minutes).

Homozygous deletion Deletion of both alleles at a locus.

Housekeeping genes Genes that are expressed in most cell types, cf. tissue-specific genes which are expressed only in selected cell types.

Hybridisation The base pairing of complementary* single strands of nucleic acid that leads to the double-stranded molecule.

Hydrogen bonding *See* Weak bonds.

Immortalised cells Cells not restricted by a specific number of cell divisions.

Immunotherapy Therapies designed to enhance the immune response to infective disease or tumours by vaccination, cytokine administration, adoptive transfer of immune cells or antibody administration.

In situ hybridisation The use of labelled single-strand RNA or DNA probes* to detect the presence of complementary* sequences in tissue sections.

Initial induced damage Ionising radiation-induced damage after chemical modification (indirect ionisation and radical scavenging by thiols or DNA proteins) but before enzymatic repair (i.e. within microseconds of a photon/nucleus interaction).

Integration Incorporation of foreign DNA (e.g. viral) into host cell DNA.

Interleukin 2 A protein secreted by activated T lymphocytes which stimulates proliferation of lymphocytes and activates cytotoxic functions of macrophages and lymphocytes.

Introns Transcribed* sequences spliced* out of mature mRNAs*.

Jumping A modification of the walking* technique. Clones* with large interstitial deletions* are used so that areas of the genome some distance apart can be covered.

Karyotype The chromosomal composition of a cell.

kb Kilobases.

Kinase An enzyme that catalyses the addition of a phosphate group onto a specific residue of a protein or nucleic acid.

Ligand A molecule (e.g. growth factor*) that specifically binds to a receptor.

Lineage The developmental ancestry of a particular cell type.

Linkage (genetic or locus linkage) Co-segregation of the two genetic loci*, e.g. coding or non-coding on the same chromosome, at frequencies greater than would be expected at random.

Locus Position on a chromosome.

Loss of heterozygosity Loss of one allele* at a locus followed by duplication of information from the remaining locus leading to homozygosity.

Matrix   *See* Nuclear matrix.

Messengers   The term applied to molecules (generally small) that can transmit signals between and within cells.

Methylation   Modification of a base by addition of a methyl group. Conversion of cytosine to *S*-methylcytosine is thought to be associated with transcriptional inactivity of a gene in or near which it occurs.

mRNA   Messenger RNA, the product of DNA transcription* and hnRNA* splicing* that serves as a template for protein translation*.

Multipotent   A cell with the capacity to differentiate along one of several cell lineage pathways (cf. unipotent, a cell with the capacity to differentiate along only one pathway).

Mutagen   A physical or chemical agent which introduces a non-lethal change in the cell's DNA sequence.

Mutation   Any alteration in DNA sequence, as a result of point mutation*, deletion*, translocation*, etc.

N-CAM   Neural cell adhesion molecule, a plasma membrane glycoprotein expressed on the surface of nerve and glial cells involved in cell–cell adhesion (a member of the immunoglobulin family).

Nested primer   A primer used to increase specificity of sequencing or re-amplification of a PCR product, located between the two original primers.

Non-coding DNA   The 95%+ of the cell's DNA which does not code for amino acid sequences or structural RNAs.

Northern blot   Standard technique for identifying specific RNA molecules (*see* Southern blot, named after the man who first invented this method for DNA analysis).

Nuclear matrix   Scaffold postulated as structural support for chromatin loops and many nuclear enzymes, e.g. topoisomerases.

Nucleoid   Histone*-free DNA prepared from a cell lysed in non-ionic detergent and high salt (2 M NaCl). Nuclear matrix* attachments are preserved.

Nucleosomes   Histone* complexes (octamers) around which the DNA helix is wound as the first level of DNA packing*.

O⁶-alkylguanine-DNA alkyltransferase   An enzyme that removes alkyl groups from the O⁶ position on guanine following alkylation. It is inactivated by this transfer, i.e. acts as a suicide inhibitor.

Oligonucleotide   A sequence of several nucleotides.

Open reading frame (ORF)   A sequence of translatable codons* not interrupted by stop codons*, which could therefore code for a polypeptide.

p, q   The short arm (p) and long arm (q) of a chromosome.

P-glycoprotein   Membrane glycoprotein coded by MDR which functions as an active drug efflux pump.

Packing   The process by which 50 cm of chromosomal DNA is folded and compressed into a specific higher order structure in order to fit into the volume of a cell nucleus.

Paracrine   A mechanism of growth stimulation involving the secretion of a growth factor* that interacts with a specific plasma membrane receptor* on a neighbouring cell, cf. Endocrine*.

Phage (bacteriophage)   A bacterial virus.

Phenotype   The biological expression of a cell or organism's genotype*, e.g. cell morphology, surface receptors expressed.

Phosphorylation/dephosphorylation   The addition/removal of a high energy phosphate group to/from a specific residue in a protein, resulting in the modulation of some specialised function, e.g. signal pathways*.

Physical map   A map based on actual distances in base pairs between loci*, as opposed to a genetic map which allocates distances between loci* on the basis of recombination* frequency, or equivalently, genetic linkage* data.

Plasmid   A double-stranded circle of DNA capable of being autonomously replicated in bacteria; useful for DNA cloning*.

Point mutation   Substitution of one base by another.

Polyglutamation   Metabolism of folates or antifolates involving the addition of glutamic acid residues. Leads to retention in the cell and may alter enzyme kinetics.

Polymerase chain reaction (PCR)   An in vitro method that uses enzyme synthesis to exponentially amplify specific DNA sequences.

Polymerases   Enzymes that catalyse the addition of deoxyribonucleotides (G, C, A or T DNA) or ribonucleotides (G, C, A or U in RNA) to the 3' end of a nucleotide chain as part of DNA replication* or RNA transcription*.

Polymorphism   Multiple alternative forms of a protein, e.g. G6PD polymorphism, an RNA or DNA sequence occurring naturally in a population. Used in linkage* analysis.

Post-mitotic   Describing a mature cell that can no longer undergo cell division, e.g. neuron.

Primer   Short oligonucleotide that binds to specific single stranded target nucleic acid sequence enabling polymerase to initiate strand synthesis.

Probe   A short, specific DNA sequence labelled with ^{32}P or biotin that can be used to detect complementary* sequences in "test" DNA.

Procaryotic cells   Cells lacking a cell nucleus, e.g. bacteria.

Promoter/enhancer sequences   DNA sequences that are control points for gene transcription*. Promoter sequences are usually in the vicinity of upstream exon or in the 1st exon*; enhancer sequences may be many kbp* upstream or downstream* of the gene.

Proofreading ability   The ability of an enzyme to detect and correct mistakes that it has made in DNA synthesis.

Protein kinase   Family of enzymes catalysing the transfer of a high energy phosphate from ATP to specific residues on proteins; one of the major mechanisms for regulation of protein function.

Proto-oncogene   A cellular gene whose alteration has been shown to be involved in malignant transformation.

Provirus   A viral genome integrated into a host cell genome.

Ras   A GTP-binding regulatory protein involved in growth factor* stimulation (*see* G protein).

Receptor   Cell surface protein that binds a specific molecule (ligand) and transmits subsequent intracellular signals.

Recombination   "crossing over", a mechanism of exchange of genetic material between a pair of homologous chromosomes.

Renaturation of DNA   The ability of denatured* DNA to resume its normal double-strand conformation (*see* Denaturation*).

Repetitive sequences   Approximately 30% of genomic* DNA consists of repeated, non-coding* nucleotide sequences, i.e. tandem repeats* or satellite DNA, with no known functions.

Replication   Duplication of genomic* DNA during S phase.

Restriction endonuclease   A group of endonucleases each of which cleaves double-stranded DNA at a specific "recognition site" ("restriction site") determined by the exact DNA sequence. Names indicate the bacterium of origin, e.g. the endonuclease *Eco*RI originates from *E. coli.*

Restriction fragment-length polymorphism (RFLP)   A polymorphism* in the size of restriction fragments due to a sequence difference between alleles*, usually in non-coding* regions.

Restriction fragments   The products of digesting DNA with restriction endonucleases*.

Restriction map   Schema showing the positions of cutting sites of specific restriction enzymes in DNA: often used as a way of characterising specific genomic* sequences, many kb* in length.

Retroviral transduction Incorporation of part of a host genome*, e.g. a proto-oncogene*, into the genome of a newly formed retroviral particle, the presumed origin of the retroviral oncogenes.

Retroviruses RNA viruses which transcribe* their RNA into DNA using reverse transcriptase* as part of their intracellular life cycle.

Reverse transcriptase An important enzyme used by retroviruses*. The purified enzyme is useful in synthesis of cDNA* in vitro.

RNA Ribonucleic acid.

Second messenger Cytoplasmic molecules that transmit chemical signals across the cell following growth factor* (first messenger) binding/activation to a cell surface receptor*.

Senescence Programmed cell ageing.

Sense strand A sequence of nucleotides coding for a sequence of amino acids, cf. Antisense*.

Signal pathway Sequence of chemical interactions that selectively amplify and transmit messages, e.g. growth stimulus, across the cell.

Somatic deletion Deletion* arising de novo in a somatic cell, cf. Germline deletion.

Somatic mutation Mutation* arising de novo in a somatic cell, cf. Germline mutation*.

Southern blot Standard technique for identifying specific DNA sequences; typically chromosomal DNA is digested with a restriction enzyme and the DNA fragments are separated by gel electrophoresis*. The separated fragments are denatured (converted to single stranded form) and transferred (blotted) onto special membrane (nitrocellulose filter). A radioactively labelled probe* will hybridise* to a complementary* sequence on the filter which is detected by autoradiography. In Northern blotting, RNA molecules are separated by electrophoresis without prior digestion.

Splicing The process whereby introns* are removed from freshly transcribed RNA (heterogeneous nuclear or hnRNA*) to produce messenger RNA* (mRNA).

Start/stop codons   The AUG sequence in mRNA* coding for the amino acid methionine marks the starting point of translation*. UAA, UAG and UGA are stop codons in mRNA which signal the end of the protein.

Stem cell   A cell that undergoes unequal division to produce dissimilar daughters; also used to mean any cell capable of indefinite multiplication.

Superfamily   Structurally related genes arising by duplication and divergence of ancestral genes.

Supercoiling   Increase or decrease in the number of turns of one strand of DNA about the other in the double helix due to twisting of the DNA about its own axis. DNA packed on a nucleosome is supercoiled and the degree of supercoiling is altered in vivo by topoisomerases.

Tandem duplication   *See* Tandem repeat.

Tandem repeat   DNA sequences repeated head to tail in genomic DNA. Responsible for homogeneously staining regions (HSRs)

Taq polymerase   The heat stable enzyme used in PCR*, isolated from algae that live in hot springs (*see* Polymerases).

Telomere   Specific DNA sequences located at both ends of a chromosome.

Topoisomerase   Enzymes that relax DNA packing* and open up the helix during replication and transcription*. Type I: DNA enzyme that uses a reversible single strand break to catalyse relaxation of negative supercoiling. Type II: this enzyme uses the energy from ATP to produce negative supercoiling via a reversible double-stranded break.

Transcription   Copying of genomic* DNA sequences into complementary* RNA.

Transcription factors   Specific regulatory proteins that control gene expression (transcription*) by recognising and binding to specific DNA promoter–enhancer* sequences nearby.

Transduction   The transfer of a chemical signal across the cell, e.g. conversion of a growth factor* (first messenger) signal outside the cell into a cytoplasmic signal carried by second messenger molecules.

Transfection   Uptake of foreign DNA by a cell (in vitro).

Transfer   *See* Southern blot.

Translation   Protein synthesis based on an mRNA* template.

Translocation   Exchange of chromatin* between chromosomes.

Tumour suppressor gene   Gene whose normal function is involved in the suppression of cell proliferation.

Tyrosine kinase   *See* Protein kinase.

Upstream   Beyond the 5' end of a gene sequence, i.e. preceding the start of coding sequence, cf. Downstream*.

Vector   An independently replicated DNA molecule, e.g. a phage or plasmid*, into which a specific DNA sequence can be integrated* and replicated*, e.g. in a bacterial host.

Viral oncogene   Gene in a virus responsible from its oncogenic effects in an animal host.

Walking   A technique of mapping segments of DNA (up to several hundred kb*) through the identification of overlapping DNA fragments in a genomic DNA library (*see* DNA library).

Weak bonds   Non-covalent interactions (ionic bonds, hydrogen bonds and van der Waal's attractions) that are responsible for the three-dimensional shape of protein molecules and for protein–protein interactions, e.g. antigen–antibody binding.

Western blot   A technique for identifying specific protein species, analogous to Southern and Northern blotting.

Index

- human papillomavirus and, 139–140
PI-3 kinase, 73
*pim*1 gene, 28
Pituitary tumours, 36, 51, 90
Plasmid(s), 311, 313–314
- reactivation assay, 228
Platelet-derived growth factor (PDGF), 215, 218–219
Platinum coordination compounds, 223, 225–226, 231, 235
- resistance to, 227–228, 231, 233
Point mutations, 14
- in dominant oncogenes, 16
- in G proteins, 90–91
- in ras proteins, 11, 16, 94
- in tumour suppressor genes, 20–21, 33, 100–101
Polymerase chain reaction (PCR), 299–304, 309
- BCL-2 gene, 59–61
- monoclonal antibodies and, 253–255, 257
- principles of, 299–301
Polymorphisms, DNA, 283–284
- polymerase chain reaction and, 302, 304
Polyposis coli, familial, 20–21, 35, 113–115, 126, 128–129
Post-mitotic cells, 156, 159–160
Potentially lethal damage repair, 215
pRb *see* Retinoblastoma gene
Predisposition to cancer *see* Inherited cancer predisposition
Prenatal diagnosis
- familial polyposis, 114–115
- with polymerase chain reaction, 303
Primary (early) response genes, 216–217, 219
Prodrug therapies
- ADEPT, 259
- VDEPT, 269–270
Proliferating cell nuclear antigen, 230
Promoters, 280
Protein kinase C, 71–72, 218

Proto-oncogenes, 3–6
- *see also individual genes*
Proviral integration, 27–28, 265
Pseudomonas exotoxin, 25

rab proteins, 91, 94
Radiation/radiotherapy
- cytokine, 219
- DNA damage induced by, 169–175, 177–178, 180–182, 193
  - cytokines/growth factors and, 213–219
  - in vitro studies of, 195
  - repair of, 172–174, 179, 183–184, 193, 215
- retinoblastoma, 117
- *see also* Ataxia telangiectasia; Ultraviolet radiation
Radioimmunoconjugates, 258
raf gene *see* c-*raf* gene
ral gene, 36
- product, 91
rap protein, 91
ras genes
- characterization of, 30–31
- gene cooperation, 129–130
- gene therapy, 96, 264, 271
- products *see* ras proteins
- radio-resistance and, 174
- transgenic modelling of, 41–45, 130
- *see also individual genes*
ras proteins, 5, 10–11, 91–95
- GTPase activating proteins and, 73, 92, 94–95
- neurofibromatosis and, 11, 92, 95
- point mutations in, 11, 16, 94
- therapeutic strategies and, 96
*Rec*A gene, 192
Recessive oncogenes *see* Tumour suppressor genes
Recombination, DNA repair and, 192–193
Renal cancer, 79, 243
Repair patches, 184
Restriction enzymes, 195, 284, 307–308